Wood Deterioration and Preservation

ACS SYMPOSIUM SERIES **845**

Wood Deterioration and Preservation

Advances in Our Changing World

Barry Goodell, Editor
University of Maine

Darrel D. Nicholas, Editor
Mississippi State University

Tor P. Schultz, Editor
Mississippi State University

American Chemical Society, Washington, DC

Library of Congress Cataloging-in-Publication Data

Wood deterioration and preservation : advances in our changing world / Barry Goodell, editor, Darrel D. Nicholas, editor, Tor P. Schultz, editor.

p. cm.—(ACS symposium series ; 845)

Developed from a symposium sponsored by the Cellulose, Paper, and Textile Division at the 221st National Meeting of the American Chemical Society, San Diego, California, April 1–5, 2001.

Includes bibliographical references and index.

ISBN 0–8412–3797–2

1. Wood—Deterioration—Congresses. 2. Wood—Preservation—Congresses.

I. Goodell, Barry. II. Nicholas, Darrel D. III. Schultz, Tor P. 1953- IV. American Chemical Society. Meeting (221st : 2001 : San Diego, Calif.) V.Series.

TA422 .W68 2003
674′.386—dc21 2002028331

The paper used in this publication meets the minimum requirements of American National Standard for Information Sciences—Permanence of Paper for Printed Library Materials, ANSI Z39.48–1984.

Distributed by Oxford University Press

PRINTED IN THE UNITED STATES OF AMERICA

Foreword

The ACS Symposium Series was first published in 1974 to provide a mechanism for publishing symposia quickly in book form. The purpose of the series is to publish timely, comprehensive books developed from ACS sponsored symposia based on current scientific research. Occasionally, books are developed from symposia sponsored by other organizations when the topic is of keen interest to the chemistry audience.

Before agreeing to publish a book, the proposed table of contents is reviewed for appropriate and comprehensive coverage and for interest to the audience. Some papers may be excluded to better focus the book; others may be added to provide comprehensiveness. When appropriate, overview or introductory chapters are added. Drafts of chapters are peer-reviewed prior to final acceptance or rejection, and manuscripts are prepared in camera-ready format.

As a rule, only original research papers and original review papers are included in the volumes. Verbatim reproductions of previously published papers are not accepted.

ACS Books Department

Contents

Detection and Test Methodology

New Preservation–Protection Systems

Indexes

Preface

A new understanding of how wood degrading agents attack wood, combined with escalating restrictions on traditional wood preservative chemicals, are the two primary issues that spurred the three co-editors' interests in producing this text and the associated American Chemical Society (ACS) Cellulose, Paper, and Textile Division Symposium *Current Knowledge of Wood Deterioration Mechanisms and Its Impact on Biotechnology and Wood Preservation.* This symposium was held April 2001 in San Diego, California, at the ACS National Meeting. The success of the symposium led to the approval by the ACS to develop this book.

The objective of this book is to provide an overall view of our current understanding of wood degradation processes and new developments in the rapidly changing field of wood protection. The later is particularly important in light of the increasing restrictions on the traditional wood preservatives, creosote, pentachlorophenol, and chromated copper arsenate, which currently account for about 95% of the wood preservatives used in North America. However, within two years of the publication of this text use of these chemicals will likely account for less than 50% of the preservative market globally and, thus, the later part of this text explores potential alternative biocides and protection methodology.

In addition to chapters written by the symposium speakers, chapters were also contributed by those who did not speak at the symposium to provide an encompassing view of this rapidly changing field. Their chapters include information ranging from wood preservative chemistry to marine borer biology.

The chapters are arranged first to provide the reader with an understanding of the organisms that cause deterioration and the mechanisms they employ in breaking down wood. This is followed by methods used to detect deterioration or to identify the specific organism involved, and the final section discusses the protection of wood products.

It was an enjoyable and truly cooperative effort by all three editors to pull these chapters together into a comprehensive text. With authors contributing from around the world, the three editors feel that the text should provide the readers with an up-to-date review of our rapidly changing field.

We acknowledge the international team of reviewers, listed separately after the Preface, who made this text possible by their thorough and professional reviews.

We are also indebted to our sponsors without whose help the symposium and book could not have been realized:

American Chemical Society Cellulose, Paper, and Textiles Division; Chemical Specialities, Inc.; Hickson Corporation (now Arch Wood Protection, Inc.); Janssen Pharmacenica, Inc.; Lonza Group, Ltd.; Mississippi State University, Forest Products Laboratory; Osmose, Inc.; Rohm and Haas Company; University of Maine, Wood Utilization Research Program; and Weyerhaeuser Company, Inc.

We hope you enjoy the book.

Barry Goodell
Professor, Wood Science and Technology
Cooperating Professor, Chemical Engineering and the
Advanced Engineered Wood Composite Center
The University of Maine
Orono, ME 04469–5755

Darrel D. Nicholas
Tor P. Schultz
Forest Products Laboratory
Mississippi State University
Mississippi State, MS 39762–9820

International Reviewers

Terry Amburgey
Aria Amirbahman
Susan Anagnost
Dimitris Argyropoulos
H. Michael Barnes
Robert Blanchette
Carol Clausen
David Clifford
Laurie Cookson
Simon Cragg
Susan Diehl
Dan Distel
Judith Eglin
Phillip Evans
Mike Freeman
John French
Irving S. Goldstein
Michael Hale
Richard Hemingway
William Henry
Terry Highley
Claudia Husseneder
D. Pascal Kamdem
Barbara Illman
Shunji Itakura
Jöran Jermer
Lehong Jin
Peter Laks
Stan Lebow
Laichang Li
W. "Pete" Ma
Shawn Mansfield
Craig McIntyre
Paul Merrick
Jeffery J. Morrell
Richard Murphy
Paul Morris
Michael Paice
Ralph Paisley
Andrzej Paszczynski
John Palfreyman
Jenine Powell
T. Powell
Kenneth G. Ross
Elmer L. Schmidt
Olaf Schmidt
Herbert Schroeder
Terry Sellers, Jr.
James Shaw
Thomas Shelton
W. Ramsay Smith
David Stanbury
Ming Tien
Alex R. Valcke
Terrence Wagner
Andrew Zahora

Introduction

Chapter 1

Introduction to Wood Deterioration and Preservation

Barry Goodell[1], Darrel D. Nicholas[2], and Tor P. Schultz[3]

[1]Wood Science and Technology Department, University of Maine, Orono, ME 04469–5755
[2]Department of Forest Products and [3]Forest Products Laboratory, Mississippi State University, Mississippi State, MS 39762–9820

Wood-degrading fungi, insects, bacteria and marine borers cause damage resulting in billions of dollars being spent on repair and replacement of wooden structures every year. The equivalent of one-tenth of the forest products produced every year is estimated to be destroyed by these agents. Although wood degradation results in an enormous waste of resources, without wood degrading organisms our world would be buried under cellulose and lignin debris, as these organisms are among the few that efficiently recycle lignocellulosic carbon. Further, in recent years some of the mechanisms employed by microorganisms to degrade wood have been used in bioindustrial processes to benefit humans. For example, fungal-based oxidative reagents are being examined in biopulping and biobleaching processes, and microbial enzymes from wood degrading organisms have been used in systems ranging from wastewater cleanup to the production of fuels from biomass. Wood degrading agents therefore cause many problems, but also greatly benefit mankind. With greater knowledge of their capabilities and potential we will be able to find better controls for their unwanted actions and direct their biochemical mechanisms to desirable applications. The chapters in this text discuss not only the mechanisms that fungi, bacteria, termites and marine borers use to degrade wood but also ways to harness degradative processes for humankind's benefit. Additional chapters of the text are devoted to methods for detecting these organisms and developing environmentally benign methods to protect wood against the many organisms that can attack and degrade wood.

This book is important because the field of wood protection is rapidly and dramatically changing. This text, and the associated Symposium from which most of the chapters come, attempts to provide some perspective on these changes. Historically, humans have battled wood deterioration agents through the ages. The story of the 'little Dutch boy' plugging the dike in Holland relates back to the concerns over failures of wooden dikes that were often riddled with marine borers. The Spanish Armada was defeated off the coast of England in the 1580's in large part because their wooden ships were so badly deteriorated that they broke up in storms. We have relied on chemical treatments to protect wood from deterioration since the time of the earliest recorded history. Heating wood in an anoxic environment, and the application of cedar oil or copper salts to prevent decay and insect attack, have long been known.

For the past 150 years we have used pressure systems to force preservative chemicals such as creosote or water-borne zinc chloride into wood. However, environmental concerns have recently dramatically changed the way we view wood preservative chemicals. This is particularly true because the market has changed in the past 30 years from mainly producing products for industry such as utility poles and railroad ties, to principally manufacturing products for residential construction (70% of the treated wood market) such as decking. However, arsenical-based preservatives were banned in Germany some 20 years ago, setting in motion a world-wide effort to ban or restrict the use of the three most commonly used, traditional preservatives, the arsenicals, pentachlorphenol, and creosote. Planned restrictions on the use of these chemicals, primarily for residential applications in Asia, Europe, and North America, has already caused the wood preservation industry to rush to find more environmentally acceptable methods to protect wood. In the short-term several copper:organic mixtures are poised to fill the void in water-borne treatments. These chemicals, however, are also now being scrutinized by the public because of the potential leaching of copper, which in some cases can be 10-times that which occurs with arsenical-based preservatives such as chromated copper arsenate (CCA). In addition, because metals cannot be broken down in the environment, the disposal of any wood treated with a metal-based preservative will be more expensive and difficult in the future. Thus, an active search is underway to develop environmentally acceptable organic preservatives that contain no metals.

Currently, the driving forces in wood preservation include both regulatory-based and market/cost-based issues. In the short-term, regulatory pressure may drive the industry to use more environmentally acceptable chemicals that may not protect wood as well as traditional preservative chemicals. As knowledge of wood protection methods develops, and methods to improve treatment longevity using benign chemicals become better understood, both environmental acceptance and long-term effectiveness will improve. In the short-term, however, wood may be protected with less effective biocides. This may lead to failures with subsequent liability problems for specific manufacturers, and market loss for forest products companies in general. The wood products industry has already seen an influx in the use of metal studs for residential construction and plastic/wood as a substitute for wooden decking.

Both water- and oil-borne wood preservatives are currently used for the treatment of wood. Water-borne chemicals have an advantage in that the water carrier is low-cost, safe, and the wood is normally left with a clean surface. Unfortunately, water swells wood and water-borne preservatives are often used for some wood products like large timbers, where dimensional change during treatment cannot be tolerated. For many industrial applications such as bridge timbers, utility poles and marine applications, the oil carriers used with some preservatives are preferable because of the properties imparted to the wood by the oil treatment, including water repellency and greater efficacy. With the restricted use of water-borne CCA, and increasing restrictions on traditional oil-borne chemicals such as pentachlorophenol and creosote, new preservative and wood protection systems that can replace these older chemicals must be developed. In some cases, the addition of water repellents may allow water-borne chemicals to partially substitute for some of the traditional oil-borne treatments. The use of water repellants has been shown to improve the performance of preservatives, particularly in above-ground applications such as decking.

Also, environmental concerns, disposal issues and public perceptions – even perceptions that may be erroneous - are directing researchers and the industry to develop environmentally acceptable methods to protect wood. The problem is to better understand how wood deteriorating organisms attack wood so that they may be safely controlled, yet allow the wood product to be safely disposed of or recycled at the end of the product's life.

One of the greatest challenges to protecting wood from the variety of degradative agents found in the environment is the length of time that wood must be protected. Unlike the protection of crops or other commodities where protection against pathogens is normally needed only for a relatively short-term period of weeks to months, wood is expected to last for decades without supplemental protection. The use of long-lasting biocides as wood preservatives is often at odds with the development of environmentally friendly protection strategies. Biocides and other chemicals that persist in the environment without degrading proved excellent for protecting wood, but were later found to diffuse or move to non-target sites where they became long-term pollutants. The use of biocides that degrade rapidly, however, is not consistent with the need to protect wood for the long times expected by consumers. What is needed therefore, are targeted biocides that are resistant to degradation and that are specific only to the biochemical mechanisms used by wood degradation agents. The development of these 'site-specific' biocides, and/or wood-modification processes that are permanently fixed in the wood to prevent leaching of chemicals into the environment, is also critical.

Development of site specific systems that target biochemical mechanisms specific to wood degradation agents is also consistent with consumers' apparent need for frequent changes and remodeling of structures. Today, the average life of an outdoor deck in the USA is only about 13 years. This relatively short life span is not the result of decay or other deterioration, but stems from the desire of new or existing owners to redesign homes and other structures including the replacement of decks. This has resulted in the disposal of large volumes of treated wood products in landfills, and a search for ways to degrade and detoxify these products that were treated specifically to resist this type of environmental breakdown. Alternately, methods to recycle these products are being explored. The development of targeted preservatives or systems that can undergo degradation or be easily recycled would benefit this effort. The public will also require any preservative system to be low cost and impregnated into wood using a non-polluting carrier. Furthermore, to be competitive with other construction materials, any new wood protection system must be economical. However, new organic-based systems will likely consist of biocides that are relatively expensive and, because of solubility problems, may only be compatible with hydrocarbon solvents. Development of co-solvent systems providing improved solubility with aqueous solutions may be one remedy to this problem.

While the field of wood preservation has rapidly changed in the past decade, so too has our understanding of wood decay and deterioration processes. Recently uncovered biochemical mechanisms employed by wood decay microorganisms are reviewed in this text. These mechanisms help to explain the unique ability of fungi and bacteria to break down the very complex biopolymer that we know of as wood. New insights into marine-boring organisms and termites are also changing the way we think about their methods for attacking wood. In addition, new methods to detect and monitor decay and insect attack are being developed. These methods will help us to better understand how the various organisms attack wood and, to detect it in-situ to better utilize protection and remediation strategies.

Our evolving understanding of wood degradation by bacteria, fungi, insects and marine organisms has allowed us to think about new methods to protect wood that are less harmful to the environment. For example, the use of relatively benign and low cost antioxidants for combating the free radicals produced by wood-degrading fungi has shown promise to economically enhance the efficacy of relatively-expensive organic biocides (see Chapter by Green and Schultz). In addition, synergistic systems for the protection of wood, where two or more chemicals may be used in relatively low dosages to protect wood because of the increased effectiveness of both when used together, has the potential to reduce preservative levels used in treated wood. This provides both economic and environmental benefits. Other methods to protect wood continue to be explored. The chemical modification of wood to circumvent the biochemical mechanisms that microorganisms, insects and marine borers have developed specifically to attack the holocellulose and lignin components has proven to be effective if the component chemicals of wood are adequately disguised in this process. (See the chapter on Wood Composite Protection for a discussion of chemical modification of wood). The recent resurrection of thermal treatments to protect wood in Europe also shows potential for limited applications. Although thermal treatments have shown some success in protecting wood in above-ground, low decay and insect hazard environments, the protection of wood in ground contact or moist conditions still requires chemical protectants.

In our effort to define the contents of the Symposium as well as this text, we drew on the expertise of some of the leading scientists on wood deterioration and protection from around the world. To protect wood we must understand how organisms break wood down. For this reason we also looked beyond the borders of our own field. In this regard, this text begins with a description of free radical mechanisms, including a general overview of free radicals in the environment by Barry Halliwell followed by one specific to white rot degradation of wood from N. Scott Reading, Kevin Welsh and the laboratory of Steven Aust. As discussed above, given our current state of knowledge free radical reactions clearly appear to be the key to understanding degradation mechanisms, and methods to control free radical reaction chemistry may well guide the development of future wood protection systems.

An excellent overview by Geoffrey Daniel on bacteria and fungal degradation patterns, visualized through microscopy and electron microscopy, opens the section of the text of Wood Deterioration Processes specifically related to microbial degradation. This is followed by chapters on White Rot and Brown Rot decay by Kurt Messner and co-authors, and Barry Goodell, respectively. This section continues with works by Timothy Filley on lignin degradation by decay fungi, and a chapter by Akio Enoki, Hiromi Tanaka, Shunji Itakura on a unique hypothesis relating fungal degradation by white rot, brown rot and soft rot fungi. Jaime Rodriguez, André Ferraz and Maricilda de Mello's chapter focuses on metal based reactions that occur in the fungi, whereas William Henry looks at the chemistry of metals from the perspective of an organometallic chemist. The final chapters of this section by Kaichang Li, and William Kenealy and Thomas Jeffries, focus on fungi and fungal mechanisms that have been employed in biotechnological applications including the use of mediators, and fungal enzymes for bio-pulping and related processes.

Three chapters on termites and marine borers complete the section on Wood Deterioration Processes. Shelton and Grace give a broad overview of the termites in their chapter. Daniel Distel and Simon Cragg provide chapters on the marine borers reviewing marine boring bivalves and arthropods, respectively, to provide a badly needed update on the modes of action of the marine boring organisms and their distribution.

The third section of the book concerns methods for the detection and monitoring of the agents of wood deterioration. Darrel Nicholas' and Douglas Crawford's chapter starts the section with an overview on the early stage progress of wood decay, and the development of an accelerated method for decay detection based on mechanical properties. Following this are two chapters by Susan Diehl and Carol Clausen that focus on chemical and immunological methods for decay detection. Barbara Illman follows with an update on the use of synchrotron technology as a research tool to detect both decay in wood as well the valence state of metals used in wood preservatives. Jody Jellison and Claudia Hussenender and their co-authors round out this section with chapters on DNA methods for the detection and monitoring of wood degrading fungi and termites, respectively.

The final section of the text concerns the important topic of wood protection and wood preservation systems. The first chapter by Alan Preston provides a perspective and introduction to current problems and the long-term approach for wood preservative development from an industrial perspective. Rick Green and Tor Schultz then introduce information on the relatively new development of adding non-biocidal chelators and anti-oxidants to wood preservatives to synergistically enhance biocidal activity. This has the benefit of reducing the amount of the relatively expensive bioactive chemicals needed to protect wood. Liam Leightley then adds to this perspective, from an industrial standpoint, in looking at synergistic wood preservative combinations with near-term commercialization potential. Douglas Gardner, Cihat Tascioglu, and Magnus Wålinder follow this with a much-needed overview on the state of wood protection in the wood composites field. The final chapter by Tor Schultz and Darrel Nicholas provides an encompassing summary of the pros and cons for many of the newer non-arsenical wood preservatives that have recently been developed, or that show potential. This final chapter provides guidance for the future in reviewing wood preservatives that are under development now. Some of these chemicals will become the next generation of wood preservatives. Others will fall by the wayside, perhaps to be looked at again when problems develop with existing protection methods.

It is impossible to predict the future, but this book hopefully sheds light on recent developments in our understanding of wood deterioration mechanisms, new ways to detect the agents of deterioration, and trends in methods for wood protection.

The co-editors look forward to your comments.

Wood Deterioration Processes

Fungal Free Radical Mechanisms

Chapter 2

Free Radical Chemistry as Related to Degradative Mechanisms

Barry Halliwell

Department of Biochemistry, National University of Singapore, Medical Drive, MD7 #03-15, Singapore 119260

Fungal systems have harnessed the power of reactive oxygen species to achieve breakdown of one of the world's toughest substances, wood. To attain this goal, powerful oxidizing agents must be used. However, such agents present a risk to the fungal cells generating them. If they are highly reactive, there is a further risk that they will be dissipated in non-productive reactions with environmental biomolecules unless they are selectively delivered to their sites of action. Let us examine the basic chemistry of some reactive species to see what might be possible.

Oxidations by oxygen

Oxygen itself is a biradical, containing two unpaired electrons. The parallel spin of these two electrons makes it difficult for oxygen to react directly with non-radicals. As a result, direct oxidation of most biomolecules with O_2 is slow. Wood is the perfect example; it resists direct oxidation by O_2 for centuries. However oxygen does react fast with other free radicals, such as carbon-centred radicals (*1*)

$$— C^{\bullet} + O_2 \rightarrow — CO_2^{\bullet}$$

peroxyl radical

or sulphur-centred radicals

$$— S^{\bullet} + O_2 \rightarrow — SO_2^{\bullet}$$

Thus once a free radical chain reaction has started, O_2 usually becomes a ready participant. The classic example of this is lipid peroxidation (1), in which a reactive radical ($R^\bullet$) abstracts a hydrogen atom from a polyunsaturated fatty acid residue (LH), leaving a carbon-centered radical ($L^\bullet$)

$$LH + R^\bullet \rightarrow L^\bullet + RH$$

followed by reaction of this carbon radical with O_2

$$L^\bullet + O_2 \rightarrow LO_2^\bullet$$

to generate a peroxyl radical that propagates the reaction by abstracting hydrogen from previously-unmolested fatty acid residues

$$LO_2^\bullet + LH \rightarrow \underset{\text{lipid peroxide}}{LO_2H} + L^\bullet$$

Oxygen metabolism

Oxygen taken up by aerobic organisms is reduced to water, a 4-electron process

$$O_2 + 4H^+ + 4e^- \rightarrow 2H_2O$$

In eukaryotes, this is carried out in mitochondria. Chemically, it is impossible to add four electrons to O_2 at once; it must be done stepwise (*1*)

$$O_2 + 1e^- \rightarrow O_2^{\bullet-} \text{ (superoxide radical)}$$

$$O_2^{\bullet-} + 1e^- + 2H^+ \rightarrow H_2O_2 \text{ (hydrogen peroxide)}$$

$$H_2O_2 + 1e^- \rightarrow OH^- \text{ (hydroxyl ion)} + OH^\bullet \text{ (hydroxyl radical)}$$

$$OH^\bullet + 1e^- + H^+ \rightarrow H_2O$$

$$OH^- + H^+ \rightarrow H_2O$$

This series of reactions (except the last) is catalysed by mitochondrial complex IV, cytochrome oxidase. This multi-enzyme complex also has the unenviable task of sequestering reactive partially-reduced intermediates on its active sites until reduction to H_2O is completed, preventing their release to cause havoc in the rest of the cell (1).

Superoxide

The first reduction step, oxygen to superoxide, is fairly easy and many systems with a reduction potential of –0.16V or less can achieve it. Thus some $O_2^{\bullet -}$ is made free in mitochondria by escape of electrons onto O_2 from electron carriers (*1*). This superoxide is largely or entirely removed by mitochondrial superoxide dismutase enzymes (Mn SOD). These are essential to survival, as revealed by the observation that transgenic mice lacking them usually die soon after birth (*2*). Several biomolecules autoxidize to make $O_2^{\bullet -}$, including dopamine, adrenalin, tetrahydropteridines and thiols (*1*).

Superoxide, as a pure chemical, is not very reactive: it does not attack DNA, lipids, or most proteins, for example (*1*). So why is its removal, particularly in mitochondria (*2*), so essential? Superoxide can directly attack certain iron-sulphur proteins, including the Krebs cycle enzyme aconitase, inactivating them and releasing iron (*3*). Superoxide also mobilizes iron from the storage protein ferritin (*4*).

$$\text{Ferritin }(Fe^{3+})_n + O_2^{\bullet -} \rightarrow \text{Ferritin }(Fe^{3+})_{n-1} + Fe^{2+}_{(aq)} + O_2$$

Superoxide can protonate (*1*)

$$O_2^{\bullet -} + H^+ \rightleftharpoons \underset{\text{hydroperoxyl radical}}{HO_2^{\bullet}}$$

Hydroperoxyl radical, the simplest peroxyl radical ($RO_2^{\bullet}$, where R=H), is much more reactive than $O_2^{\bullet -}$ (*1*). The pKa for $HO_2^{\bullet}$ is 4.8, which means that little $HO_2^{\bullet}$ is present in $O_2^{\bullet -}$ - generating systems at pH 7.4. Nevertheless, more will be present at low pH. Superoxide cannot cross most biological membranes, whereas $HO_2^{\bullet}$ should be able to. Like other peroxyl radicals, it can attack unsaturated fatty acid residues (*1*).

$$LH + HO_2^{\bullet} \rightarrow L^{\bullet} + H_2O_2$$

Nitric oxide and peroxynitrite

Superoxide can react very fast with nitric oxide ($NO^{\bullet}$). Like $O_2^{\bullet -}$, $NO^{\bullet}$ is a selectively-reactive radical with multiple biological roles, although cytotoxic if generated in large excess (*5*). It is generated in living organisms both enzymically from L-arginine by the activity of nitric oxide synthase proteins, and non-enzymically by the reaction of nitrite with acid (*5,6*).

$$NO_2^- + H^+ \rightarrow HNO_2 \rightarrow \text{oxides of nitrogen}$$

Reaction of $O_2^{\bullet -}$ with $NO^{\bullet}$ produces peroxynitrite, $ONOO^-$

$$O_2^{\bullet -} + NO^{\bullet} \rightarrow ONOO^-$$

Its protonated form, peroxynitrous acid (ONOOH) is a powerful oxidising, hydroxylating and nitrating species that reacts with many biomolecules including DNA, proteins and lipids *(6,7)* to cause damage.

Hydrogen peroxide

Non-enzymic or superoxide dismutase-catalysed dismutation of $O_2^{\bullet -}$ generates H_2O_2

$$2O_2^{\bullet -} + 2H^+ \rightarrow H_2O_2 + O_2$$

H_2O_2 is also produced directly by several enzymes, such as urate-, glycollate-, cellobiose - and monoamine oxidases (*1,8*). Like $O_2^{\bullet -}$ and $NO^{\bullet}$, H_2O_2 is not highly reactive. Its levels are kept down by catalase and peroxidase enzymes, but it is not eliminated entirely *in vivo* because it plays an important role as an intracellular and extracellular signal transduction molecule (*1,9*). Facilitating this role are both its poor reactivity, which allows H_2O_2 to survive to reach its cellular target (unless intercepted by catalases or peroxidases on the way), and its ability to pass unhindered through membranes, probably using the aquaporin channels (*1,10*). The reactivity of H_2O_2 can be increased in three ways. First, some peroxidase enzymes can use H_2O_2 to generate reactive products (1). Thus plant peroxidases oxidize a wide range of substrates, including phenols to phenoxyl (aromatic alkoxyl) radicals (*1*). In neutrophils, myeloperoxidase oxidizes chloride ions to the reactive oxidizing and chlorinating agent hypochlorous acid, aiding the killing of micro-organisms (*1*).

$$H^+ + Cl^- + H_2O_2 \xrightarrow{\text{MPO}} H_2O + HOCl$$

Second, UV light can split H_2O_2 into hydroxyl radical by homolytic fission of the O-O bond (*1*).

$$H_2O_2 \xrightarrow{\text{UV}} 2OH^{\bullet}$$

Transition metal ions

Third, H_2O_2 can react with several metal ions to generate $OH^{\bullet}$. The best studied are iron and copper ions, but some chromium, vanadium and nickel species are also effective (1). Interestingly, Mn^{2+} is not effective in forming $OH^{\bullet}$ from H_2O_2 (1,11).

$$\underset{(Cu^{+})}{Fe^{2+}} + H_2O_2 \rightarrow \left[\begin{array}{c}\text{Intermediate} \\ \text{Oxo-iron} \\ \text{or oxo-copper} \\ \text{complex}\end{array}\right] \longrightarrow \underset{(Cu^{2+})}{Fe^{3+}} + OH^{\bullet} + OH^{-}$$

The ligands to the metal ions can influence the rate of the reaction with H_2O_2 and the amount of $OH^{\bullet}$ generated (*1*).

Hydroxyl radical is indiscriminately-reactive: it can attack whatever biomolecules are next to it whenever $OH^{\bullet}$ is formed. Its diffusion distance is essentially zero (*1*). By contrast, H_2O_2 can pass through intercellular and plasma membranes, causing little if any damage(*1,8,10*). If H_2O_2 meets a strategically-placed peroxidase enzyme (see above) or transition metal ion, it can cause site-specific damage (*1,8*). For example, H_2O_2 does not react with DNA. However, cells treated with H_2O_2 show oxidative DNA damage. This is because the H_2O_2 penetrates to the nucleus and reacts with iron (or possibly copper) ions bound to DNA, and the resulting $OH^{\bullet}$ causes instant oxidative DNA damage (*1,12*).

Transition metal ions also promote other free radical reactions: indeed metalloenzymes such as peroxidases, hydroxylases, and oxygenases are harnassing and expanding this power. Thus ions of iron, copper, manganese etc can accelerate autoxidation reactions, by catalysing single electron transfers between the substrate and O_2 (*1*). Iron and copper ions can decompose not only H_2O_2, but also lipid peroxides (*1*)

$$LOOH + Fe^{2+} (Cu^{+}) \rightarrow \underset{\text{alkoxyl radical}}{LO^{\bullet}} + Fe^{3+} (Cu^{2+}) + OH^{-}$$

$$LOOH + Fe^{3+} (Cu^{2+}) \rightarrow \underset{\text{peroxyl radical}}{LOO^{\bullet}} + Fe^{2+} (Cu^{+}) + H^{+}$$

Thus controlling metal ion availability can have a powerful effect on oxidative damage rates (*1*).

Summary

A wide range of free radicals and other reactive oxygen/ nitrogen/chlorine species can be formed in living organisms. Their reactivity varies from highly-selective ($O_2^{\bullet-}$, $NO^\bullet$, H_2O_2) through intermediate ($RO_2^\bullet$) to highly-reactive (ONOOH, $RO^\bullet$, HOCl) to indiscriminately reactive ($OH^\bullet$). Their diffusibility varies from almost zero ($OH^\bullet$) through limited by reactivity and/or charge ($RO_2^\bullet$, $RO^\bullet$, HOCl, ONOOH, $O_2^{\bullet-}$) to highly diffusible (H_2O_2, O_2, $NO^\bullet$). Organisms have often evolved to use some of these less-reactive species as signalling molecules ($O_2^{\bullet-}$, H_2O_2, $NO^\bullet$) whereas species such as $OH^\bullet$ can only be employed for useful purposes if they can be transiently formed or held in specific microenvironments. In the case of $OH^\bullet$ and ONOOH this can be achieved by selective location of peroxidases or of metal ions catalytic for Fenton chemistry, or by co-generation of $O_2^{\bullet-}$ and $NO^\bullet$ respectively.

References

1. Halliwell, B; Gutteridge, J.M.C. *Free Radicals in Biology and Medicine* Oxford University Press, Oxford, UK, 1999, chapters 1, 2, 3 and 4.
2. Li, Y; Huang, T.T; Carlson E.J. *et al. Nature Genet.* **1995,** *11*, 376-381
3. Liochev, SI. *Free Rad. Res* **1996**, *25*, 369-384.
4. Bolann, B.J.; Ulvik R.J.; *Eur. J. Biochem.* **1990**, *193*, 899-904.
5. Bredt, D.S. *Free Rad. Res.* **1999,** *31*, 577-596
6. Halliwell, B; Zhao, K.; Whiteman, M. *Free Rad. Res.* **1999,** *31*, 651-669.
7. Beckman, J.S.; Koppenol, W.H. *Am. J. Physiol.* **1996**, *271,* C1424-C1437.
8. Halliwell, B; Clement, M.V.; Long, L.H. *FEBS Let.* **2000**, *486*, 10-13.
9. Sen, C.K.; Sies, H.; Baeuerle, P.A.; eds. *Antioxidant and Redox Regulation of Genes* Academic Press, San Diego, CA, 2000.
10. Henzler, T.; Steudle, E.J. *Exp. Bot.* **2000**, *51,* 2053-2066.
11. Gutteridge, J.M.; Bannister, J.V. *Biochem J.* **1986**, *15*, 225-228.
12. Spencer, J.P.E; Jenner, A.; Chimel, K.; Aruoma, O.I.; Cross C.E.; Wu, R; Halliwell, B. *FEBS Lett.* **1995**, *374,* 233-236.

Chapter 3

Free Radical Reactions of Wood-Degrading Fungi

N. Scott Reading, Kevin D. Welch, and Steven D. Aust*

Chemistry and Biochemistry Department, Biotechnology Center, Utah State University, Logan, UT 84322–4705
***Corresponding author: email: sdaust@cc.usu.edu**

Wood-rotting fungi catalyze, through a unique system of enzymes and chemicals, the degradation and mineralization of the structural polymers in wood. The most complex of these polymers is lignin, which is a heterogeneous, stereo-irregular, three-dimensional polymer that is very resistant to enzymatic degradation due to the lack of repeating and readily hydrolyzable bonds. The biodegradation systems of the wood-rotting fungi are based largely on free-radical reactions catalyzed by a variety of extracellular enzymes. The wood-degrading systems employed by these fungi can catalyze both direct and indirect (or mediated) oxidative and reductive reactions. Fungi are able to catalyze other reactions that support the free-radical nature of these biodegradative systems and together favor depolymerization and degradation rather than synthetic or polymerization reactions.

Of the more than a million species of fungi, only a relative few, specialized and ubiquitous fungi, belonging to the phyla *Basidiomycota* and *Ascomycota* have the ability to degrade wood (*1*). These wood-decaying fungi are frequently grouped according to the characteristics of the wood during degradation, which reflects fundamental differences in their enzymatic and non-enzymatic activities. These are the white-rot, brown-rot, and soft-rot fungi. The white-rot fungi comprise the largest group of wood-rot fungi, having the capability to degrade and mineralize all the major components of wood (cellulose, hemicellulose, and lignin), resulting in wood that becomes progressively more fragile as it decays. The white color associated with these fungi is a result of the rapid degradation of lignin, which exposes the more slowly degraded cellulose that is white in appearance. Brown-rot fungi comprise approximately 10% of all wood decay fungi and primarily attack conifers. Most of these fungi are found in the class *Polyporaceae*. Brown-rot fungi primarily degrade cellulose and hemicellulose, leaving partially oxidized, brown lignin, which results in weakened wood structure such that the wood breaks up into cubic particles that comprise a major component of humus. Soft-rot fungi are the least specialized of the wood-rot fungi, often requiring significant levels of nitrogen from the surrounding environment to effectively decay wood. Soft-rot fungi typically occur in wood with high water content and degrade primarily cellulose and hemicellulose. These fungi primarily belong to the phyla *Ascomycota* and are common decomposers of cellulose in soil.

WOOD STRUCTURE AND FREE-RADICALS

Wood-rotting fungi need to be specially adapted to overcome three major defense strategies in order to degrade wood. Two of these defense strategies have a biological basis, while the third is chemically based. The biological defenses against wood degradation are nutrient availability and the presence of compounds toxic to fungi. Wood typically has a very low content of nitrogen and phosphorus (*2*), two elements that are important for microbial growth. The average nitrogen content for hardwoods and softwoods is 0.09% of the dry weight of wood with an average carbon to nitrogen ratio of 600:1 (*2*). Wood-rotting fungi have adapted to this constraint by using nutrient limitation as a key factor involved in expression of their wood-degrading systems (*3*). The presence of potentially toxic chemicals within non-living heartwood, such as tannins in deciduous trees and a variety of phenolic compounds in coniferous trees, prevent or limit wood-rot fungi from colonizing living trees (*1*).

The third defense strategy used by trees to decrease their susceptibility to degradation is the formation of complex organic compounds that limit the availability of easily usable substrates, such as simple sugars and starches. The principle components of wood are cellulose, hemicellulose and lignin. Cellulose comprises 40-50% (dry weight) of wood and is composed of long, linear chains of β-1,4-linked glucose (*1*). Cellulose polymers hydrogen bond to form fibrils. Hemicellulose comprises 25-40% (dry weight) of wood and is a complex combination of relatively short polymers of xylose, arabinose, galactose, mannose, and glucose (*1*) with acetyl and uronic extensions. Hemicellulose hydrogen bonds to cellulose fibrils and also covalently links with lignin, creating a complex web of bonds that provide structural strength to wood. Lignin comprises 20-35% (dry weight) of wood and is a complex, stereoirregular, three-dimensional polymer composed of *p*-hydroxycinnamyl alcohols, most notably *p*-coumaryl, coniferyl, and synapyl alcohols, which are cross-linked to each other with a variety of chemical bonds (*1*). Lignin forms an amorphic complex with hemicellulose to encapsulate cellulose, thereby reducing the bioavailability of these two cell wall constituents. The lack of readily hydrolyzable, repeating linkages makes lignin particularly difficult to biodegrade and thus acts as an effective barrier against microbial attack.

The biosynthesis of lignin in plants is believed to occur by a peroxidase-catalyzed, free-radical polymerization of the *p*-hydroxycinnamyl alcohols. The relative concentration of each phenylpropanoid precursor, which differs with each plant species and particular cell types, determines the final structure of lignin. Peroxidases oxidize the phenylpropanoid compounds to generate radicals that combine randomly with other lignin precursors and lignin substructures to form a chemically unique, three-dimensional polymer linked through a variety of carbon-carbon and carbon-oxygen bonds. The number and variation of chemically unique structures that comprise lignin make its degradation a complicated process that requires a nonhydrolytic system that is extracellular, and nonspecific. The lignin degradation system developed by wood-degrading fungi (in particular white-rot fungi) is based largely on free-radical reactions catalyzed by a relatively small number of enzymes. Thus, the fungal, free-radical based, enzymatic system, in conjunction with other extracellular fungal activities, is able to degrade the large number of diverse chemical structures that are found in wood.

The process of lignin synthesis results in a highly oxidized polymer. Therefore, to effectively degrade and mineralize lignin reductive as well as oxidative reactions are required, both of which must occur aerobically. These reactions must be balanced or otherwise controlled to prevent redox cycling and free-radical based polymerization of the degradation products. Additionally, the oxidizing and reducing equivalents need to be unique and continuously produced since extracellular regeneration would be improbable. This precludes the use of common biological compounds for reducing or oxidizing equivalents, such as NADPH, which would be difficult to regenerate once released extracellularly. An advantage of the extracellular formation of the free-radical species is their ability to diffuse away from their site of origin and mediate indirect reactions with the insoluble lignin polymer. Thus, the smaller, diffusible radicals achieve a greater area of reactivity than could be achieved by reactions catalyzed by enzymes or the fungi directly. This also causes free-radical reactions to occur away from the fungal mycellium preventing self-inflicted damage to the fungus.

INITIATION OF FREE-RADICAL REACTIONS

The extracellular enzymes of wood-degrading fungi involved in the production of free-radicals include lignin peroxidases (LiP), manganese-dependent peroxidases (MnP), cellobiose dehydrogenase (CDH) and laccases. Other factors related to free-radical generation by these enzymes are hydrogen peroxide (H_2O_2), oxalate, small molecule mediators, methyl transferases, and the plasma membrane redox potential. The free-radical reactions generated by these systems lead to both direct and indirect (mediated) oxidations and reductions. The ligninolytic peroxidases have been isolated from white-rot fungi, laccases have been isolated from the white-rot and brown-rot fungi, and CDH has been isolated from species from each group of wood-rotting fungi (4). The unique combination of the ligninolytic peroxidases and laccases to the white-rot fungi may be significant in that these fungi are the most efficient among the wood-rotting fungi at lignin degradation.

Peroxidases

The mechanism of oxidation applicable to all peroxidases, shown in Figure 1A, has been reviewed in detail for the fungal peroxidases (*5*). The ferric form of a peroxidase is oxidized by H_2O_2 via two electron transfer. The oxidized enzyme can then perform two sequential, one-electron, direct oxidations of reductants that interact with the peroxidase. The fungal peroxidases are unique in that they are strong oxidants with reduction potentials of 1.4 volts (V) (*6*), which are higher than other peroxidases (i.e., 0.8 V for horseradish peroxidase) (*7*). As a result LiP is able to directly oxidize a greater range of chemicals, with higher reduction potentials, such as veratryl alcohol (VA) (1.2 V) and a variety of environmental pollutants (*5*). Additionally, LiP has been shown to oxidize a variety of large molecular weight, soluble or insoluble polymers, including lignin (*3*) and synthetic polymers (*8*), that would not have access to the oxidized heme center. This is attributed to LiP having a surface accessible reaction site for the oxidation of chemicals (*9*). Unlike LiP, MnP does not directly oxidize lignin or other large molecules and is considered to be entirely dependent on manganese to complete its catalytic cycle. Thus, it is generally considered that MnP accomplishes oxidations through mediation reactions using the powerful oxidant Mn(III) (*10*). However, MnP has been shown to react with some easy to oxidize phenolic and aromatic compounds in the absence of manganese (*11*).

Laccases

Oxidations by laccases involve the reduction of molecular oxygen (O_2) to water with the concomitant one-electron oxidation of aromatic substrates (Figure 1B),

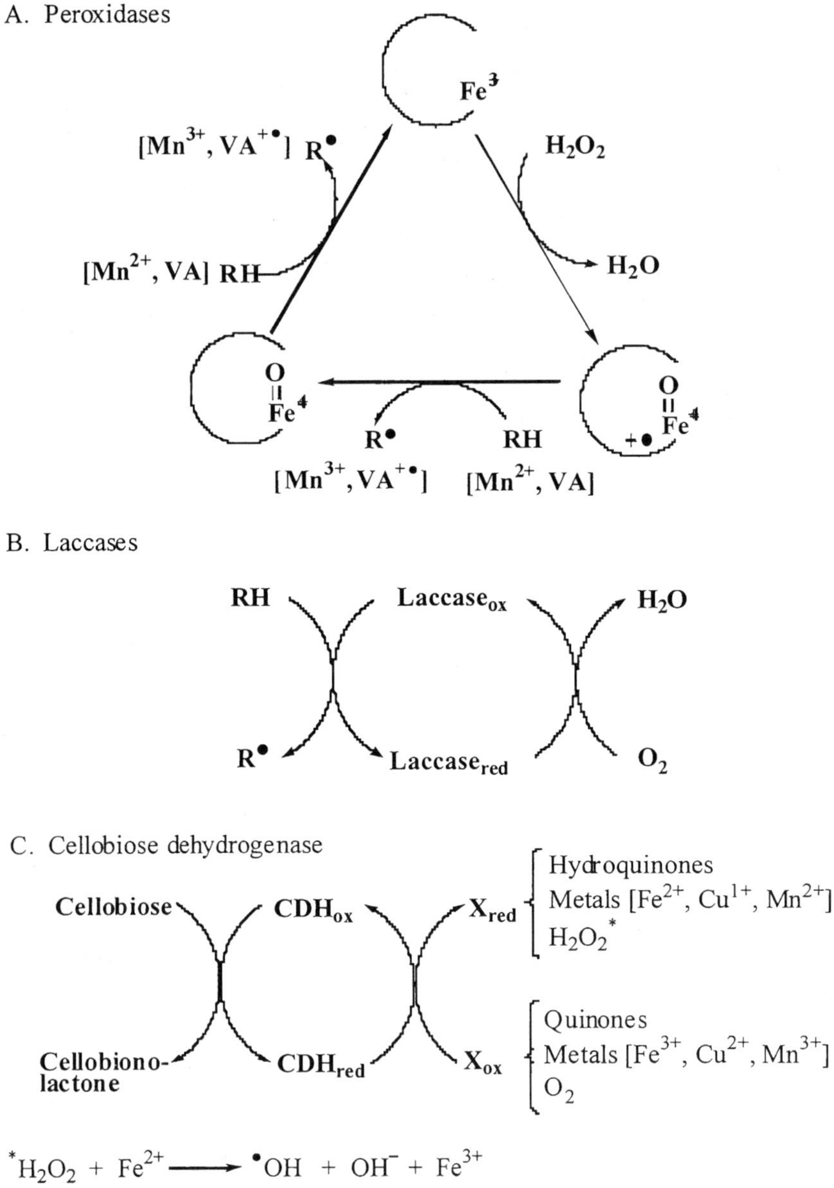

*Figure 1. Direct oxidative and reductive reactions catalyzed by enzymes from the wood-rot fungus Phanerochaete chrysosporium. *Fenton reagent, formed by CDH, results in the hydroxyl radical.*

which include polyphenols, methoxy-substituted monophenols, aromatic amines, and other relatively easily oxidized aromatic compounds (*12*). The products of these oxidations are oxygen-centered radicals and cation radicals that usually react further nonenzymatically. Reactive oxygen species are not released during the reduction of O_2 by laccases. The reduction potential of laccase is estimated to be 0.8 V (*13*), which is generally thought to be too low to oxidize the nonphenolic subunits of lignin. Laccases have been shown to oxidize lignin and lignin-model compounds mainly through the oxidation of phenolic subunits leading to Cα oxidations, Cα-Cβ cleavage, and alkyl-aryl cleavage (*14*). Although the mechanisms governing laccase degradation of lignin are not well understood, mediation or indirect oxidations are believed to occur (*15*, *16*).

Cellobiose Dehydrogenase

Cellobiose dehydrogenase preferentially oxidizes cellobiose, the product of cellulose hydrolysis by cellulases, and uses a variety of substrates as electron acceptors (Figure 1C). The substrates used as electron acceptors include quinones, phenoxyl radicals, cation radicals, transition metals (Fe, Mn, Cu), and O_2 (*4*). Some of these are oxidation products of lignin degradation produced by the peroxidases and laccase. This may suggest the possibility of redox cycling of compounds between these enzymes, setting up a futile cycle where a product of oxidation is simply reduced by CDH to regain the original substrate, a condition that, as mentioned earlier, needs to be avoided. Indeed, we suggest that the reductive reactions catalyzed by CDH are controlled reactions that facilitate degradation both of cellulose and lignin. The oxidation of cellobiose to cellobionolactone is thought to relieve product inhibition of cellulases, thereby promoting cellulose degradation (*17*). Cellobiose dehydrogenase is thought to enhance lignin degradation by minimizing repolymerization of phenoxyl radicals produced during the oxidation of lignin by the peroxidases or laccases (*18*). It has been suggested that proteolytic cleavage of CDH is a control measure to prevent the formation of a futile redox cycle between the peroxidases and CDH. Inhibition of fungal peroxidase activity by CDH was attributed to the direct reduction of either the oxidized peroxidase intermediates or the oxidized products of catalysis (*4*). During peroxidase production in ligninolytic cultures of *Phanerochaete chrysosporium* CDH is proteolytically cleaved into two domains, one containing a heme and the other a flavin prosthetic group (*19*). The flavin domain remained active in oxidizing cellobiose and reducing quinones but did not inhibit peroxidase activity (*4*).

PROPOGATION OF FREE-RADICAL REACTIONS

Many of the primary substrates of the wood-degrading enzymes are not wood structures (cellulose, hemicellulose, or lignin) which are large, bulky and generally insoluble, but small molecules that are either produced by the fungi (H_2O_2, VA, oxalate), available in wood (metals [Mn, Fe, Cu], O_2), or released during wood degradation (phenolic, other aromatic compounds, quinones). These small molecules can act as mediators to promote redox chemistry and, depending on the enzyme system that initiates the reaction, these chemicals can be the primary reactants from which secondary and tertiary redox reactions are derived. Thus, from either a primary oxidative or reductive enzymatic system a number of oxidative and reductive equivalents can be formed. Therein lies the value of mediation to the wood-rotting fungi, which have to degrade a variety of chemicals with diverse structures and oxidation states.

Veratryl Alcohol Cation Radical and Manganese (III)

Veratryl alcohol and manganese are considered to be the primary physiological substrates for the fungal peroxidases. They act as the reductants to the oxidized intermediates of LiP and MnP, respectively, as illustrated in Figure 1A. The result of this interaction would be the formation of two powerful oxidants, the veratryl alcohol cation radical ($VA^{\bullet+}$) and Mn(III), which could mediate the oxidation of other chemicals such as lignin, organic acids, hydroquinones, and other compounds (Figure 2A). One of the stumbling blocks for VA involvement in lignin degradation was proof that the $VA^{\bullet+}$ was produced during catalysis. The $VA^{\bullet+}$ intermediate was detected during VA oxidation by LiP (*20*) and it was demonstrated that through association of the radical with the peroxidase $VA^{\bullet+}$ was stabilized (*6*) so that it could act as a mediator of oxidation (*21*, *22*).

An excellent example of mediation is provided by MnP. The primary substrate for MnP is Mn(II), which is oxidized to Mn(III) at a significant rate (k_{cat} = 200-300 s^{-1}) (*10*). Since Mn(III) is a strong oxidant a chelator such as oxalate (*23*), is required to separate it from the enzyme to act as a diffusible mediator for the oxidation of other chemicals (*24*, *25*). However, Mn(III) has reduced redox potential because of its chelation by oxalate. Because chelated Mn(III) has a reduced oxidation capacity, MnP would not be able to oxidize some chemicals that might be oxidized by LiP, either directly or mediated by the $VA^{\bullet+}$. This follows the hierarchy of free-radical reactions, that mediators can only react with chemicals that have lower reduction potentials. Therefore, it must be realized that mediation cannot be proposed for the oxidation of chemicals with higher reduction potentials than the mediator.

Carboxylate Anion Radical

The primary oxidations catalyzed by the peroxidases can also give rise to reductive radicals through secondary reactions. The importance of reductive equivalents to the degradative ability of the wood-rotting fungi has been

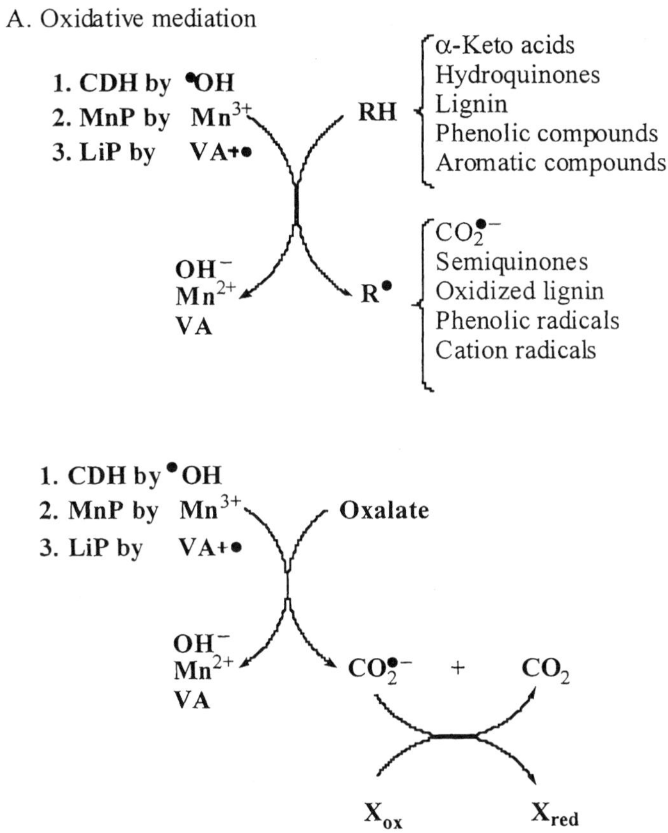

Figure 2. (A) The oxidation products of catalysis by the enzymes from the white-rot fungus Phanerochaete chrysosporium can mediate additional oxidative reactions. (B) The oxidation of an organic acid, such as oxalate, results in the generation of a strong reductant, which can mediate additional reductive reaction (e.g., see reference 26).

demonstrated for many environmental pollutants (*26*). Of physiological significance is the carboxylate anion radical ($CO_2^{\bullet -}$), which is derived from the oxidation of oxalate (*27*). Readily available in wood and produced by fungi (*28*), oxalate is very susceptible to oxidation (as a di-α-keto acid it is an excellent reductant). As shown in Figure 2B, oxalate can be oxidized by $VA^{\bullet +}$, Mn(III) or the hydroxyl radical ($^{\bullet}OH$) and the charge distribution on the radical results in homolytic cleavage of the C-C bond with the concomitant formation of $CO_2^{\bullet -}$ and CO_2. The $CO_2^{\bullet -}$ has a reduction potential of -1.9 V (*29*) and can reduce O_2 and other chemicals (*26*). The reaction equilibrium is entropically "pulled" towards oxidation of oxalate and is irreversible since one product escapes as gaseous CO_2. Other α-keto acids such as EDTA, can be similarly oxidized resulting in the formation of powerful reductants (*26*, *27*).

Semiquinone Radicals

Quinones are formed during the degradation of lignin by white-rot fungi (*30*) and are able to polymerize, thereby potentially hindering the degradation process. The white-rot fungi cope with this potential problem first by reducing quinones (31). One extracellular quinone reductase is CDH, which reduces quinones to hydroquinones. Hydroquinones are readily oxidized by Mn(III) and $VA^{\bullet +}$ to semiquinones which can be used for reduction reactions (*32*), such as reduction of phenoxyl radicals produced during lignin degradation to prevent their repolymerization. Additionally, semiquinone radicals can reduce Fe(III), O_2, or cause the reductive homolytic cleavage of H_2O_2 leading to the formation of the $^{\bullet}OH$ (*25*). Quinones can also be reduced to hydroquinones by the membrane redox potential, which then are susceptible to methylation by methyl transferases leading to degradation, as described below.

Superoxide Anion Radical

Molecular oxygen can act as the electron acceptor in reactions with CDH (*5*), semiquinone radicals, and $CO_2^{\bullet -}$ (*33*). The enzymatic reduction of O_2 by CDH results in a two-electron reduction to form H_2O_2 (Figure 1C), while non-enzymatic reduction occurs at diffusion limited rates resulting in the formation of superoxide anion radical ($O_2^{\bullet -}$). Superoxide can act as a reductant or an oxidant depending on the pH and has been implicated in the degradation of lignin and environmental pollutants (*34*, *35*). Superoxide is able to dismutate to form H_2O_2 and can also reduce Fe(III), providing the conditions for Fenton chemistry.

Fenton Chemistry

The reaction of H_2O_2 and Fe(II) results in the formation of $^{\bullet}OH$ (Figure 1C), which is one of the most potent oxidizing agents with a reduction potential of 2.3 V. Prior to the discovery of the enzymatic mechanisms for lignin degradation in white-rot fungi the $^{\bullet}OH$ was assumed to be largely responsible for the biodegradative ability of the wood-degrading fungi. Wood degrading fungi have a diverse number of mechanisms for Fenton reagent formation. Several workers have demonstrated the role of hydroxyl radical production in the degradation of wood by brown rot fungi (addressed in other chapters in this book). In white rots, the production of hydroxyl radicals may yet have a significant role in the degradation of wood. The ability to generate Fenton reagent is extremely advantageous as it results in a very strong oxidant, formed by the reaction of two substrates, Fe(II) and H_2O_2, neither of which are particularly strong oxidants. This is an interesting way to increase oxidation potential, in contrast to losing oxidation potential during mediation, as discussed previously.

Hydrogen Peroxide

Hydrogen peroxide is the oxidative substrate for the peroxidases. White-rot fungi can produce H_2O_2 enzymatically using the extracellular enzymes glyoxyl oxidase, an enzyme involved in the production of oxalate, and aryl oxidase. Both of these enzymes are produced during secondary metabolism along with the lignin degrading system. When grown on cellulose, wood-rot fungi produce CDH, which is able to directly reduce O_2 to H_2O_2. However, at low pH the rate of O_2 reduction by CDH is much slower than for other electron acceptors reduced by this enzyme. A nonenzymatic mechanism of H_2O_2 production is through the dismutation of $O_2^{\bullet-}$, which is formed by the reduction of O_2 by $CO_2^{\bullet-}$, Fe(II) or semiquinone radicals. The relative contribution of any of these systems on H_2O_2 production is not clear, but nonenzymatic evolution of H_2O_2 for sustained peroxidatic activity has been demonstrated (*36*).

Transition Metals

Cellobiose dehydrogenase has been shown to reduce Fe(III) (*4*) and nonenzymatic reduction of Fe(III) has been demonstrated using $CO_2^{\bullet-}$ and $O_2^{\bullet-}$ (*22, 25*). The formation of Fe(II) in the presence of H_2O_2 would yield the $^{\bullet}OH$ and promote the oxidation of chemicals that may not be directly oxidized by the peroxidases or laccases. The extracellular pH established by wood-rot fungi is generally quite low, often below four. At this pH iron reduction is favored over oxygen reduction by CDH and it allows oxalate-chelated Fe(II) to diffuse away into areas of higher pH before it reacts with H_2O_2 to produce $^{\bullet}OH$. This scheme has the advantage of forming the $^{\bullet}OH$ away from the fungal mycelia so the fungus will not be harmed by $^{\bullet}OH$, and the oxidation of lignin or cellulose can occur at a distance from the mycelium.

ADDITIONAL FUNGAL MECHANISMS FOR DEGRADATION

Plasma Membrane Redox Potential

Wood-degrading fungi have a plasma membrane redox potential (Figure 3) and the membrane redox potential of *P. chrysosporium* has been shown to reduce a number of chemicals (*37*). Although the plasma membrane potential has been shown to reduce and detoxify environmental pollutants such as TNT (*37*), its physiological function may be to protect the fungal hyphae from free-radicals produced by the extracellular enzymes during wood degradation. The membrane redox potential has also been shown to reduce quinones to hydroquinones, thus adding to the potential reservoir for semiquinone production by the peroxidases, or for methylation by methyl transferases leading to quinone degradation (*32*).

The reduction potential generated by wood-degrading fungi, in particular white-rot fungi, is maintained through the plasma membrane redox potential. This enables the fungi to maintain an extracellular environment at very low pH, for example *P. chrysosporium* maintains a physiological pH of 4.5. The ability to manipulate pH provides a more competitive environment for fungal growth and provides an optimal environment for the free-radical chemistry catalyzed by the extracellular enzymes. The low pH environment created by white-rot fungi increases the redox potential of the peroxidases enabling them to oxidize chemicals that other peroxidases are not able to oxidize.

Methylation by Wood-Rot Fungi

Methylation by a membrane bound methyltransferase system in wood-rot fungi may be important in denaturing fungicidal phenolic wood extractives and also aids in the degradation of phenolic pollutants and lignin (Figure 4A) (*38, 39*). In addition to limiting futile redox cycling of phenols, the methylation of phenolic lignins would make non-phenolic substrates more suitable for carbon-carbon bond cleavage, ring opening, and facilitates their oxidation. For example, the biotransformation of pentachlorophenol to its corresponding anisole permits other reactions such as oxidation by the peroxidases, nucleophilic attack by water, and reduction to occur (Figure 4B) (*40*). The chemical is then primed for further methylation and subsequent reactions leading to detoxification of the parent compound and its eventual degradation.

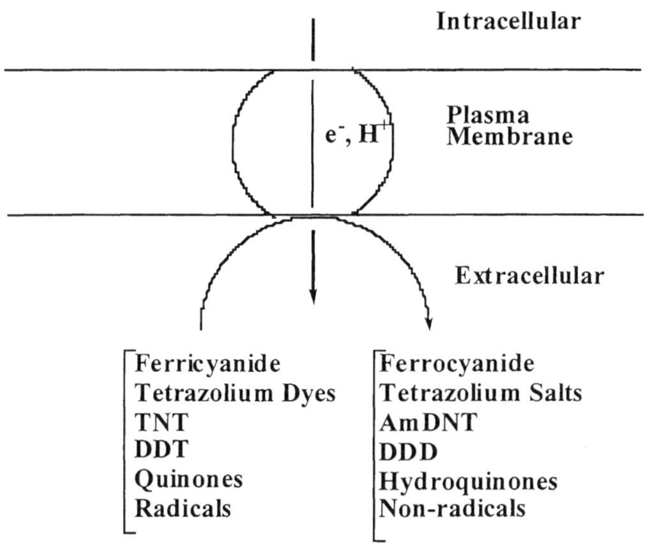

Figure 3. The fungal plasma membrane redox system.

HIERARCHY OF FREE RADICALS

Most free radicals undergo simple first- and second-order reactions due to their reactivity. Therefore, at least the elementary reactions of free radicals are often straightforward and predictable. Generally, the overall direction of a redox reaction is controlled by the feasibility of forming radical intermediates and in many instances, this feasibility also determines the kinetics of the reaction. To predict the direction of free radical processes, it is necessary to understand the thermodynamics of free radical reactions. A key thermodynamic property to consider is the reduction potential, from which predictions on reactions can be made (*41*). A comprehensive review on reduction potentials can be found in an article by Peter Wardman (*42*).

Predicting free radical reactions are generally straightforward. Each oxidized species is capable of abstracting an electron (or hydrogen atom) from any species with a lower reduction potential. Viewed another way, each reduced species can donate an electron (or hydrogen atom) to any species with a higher reduction potential. The reduction potential of LiP and MnP are higher than that of most other peroxidases. This allows for the oxidation of chemicals that are not easily oxidized, i.e., have a high reduction potential. The hierarchy of free radical electron transfer reactions allows us to predict the cascade of free radical reactions that will occur after initiation. For example, MnP can oxidize Mn(II) to Mn(III) which has a reduction potential of about 1.5 V relative to the standard hydrogen electrode. However, the oxalate-Mn(III) complex only has a reduction potential of about 0.9 V. Subsequently, the oxalate-Mn(III) complex can oxidize other compounds which have a reduction potential less than 0.9 V. Similarly, LiP can oxidize VA to a $VA^{\bullet+}$, which has a reduction potential near 1.2 V. Thus, $VA^{\bullet+}$ can oxidize numerous environmental compounds that have a reduction potential less than 1.2 V, such as the phenolic constituents of lignin. Most phenolic compounds have a reduction potential ranging from approximately 0.05 V to 0.8 V depending upon the substituted functional groups on the aromatic ring. These oxidized compounds can then in turn propagate free radical reactions by oxidizing compounds with a lower reduction potential.

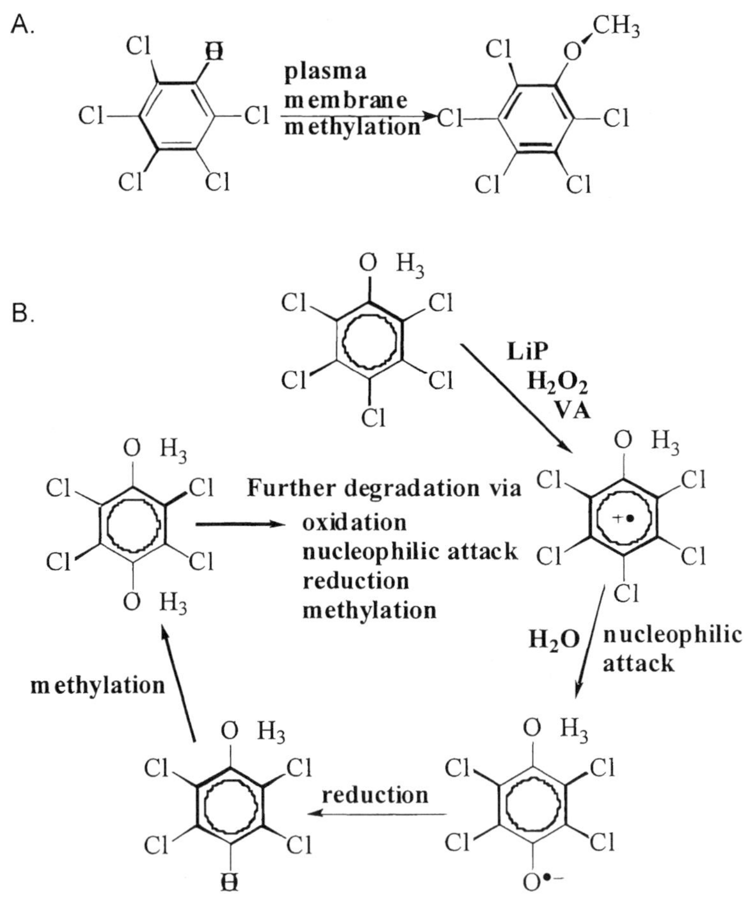

Figure 4. (A) Methylation of pentachlorophenol to pentachloroanisole by the membrane bound methylation system of the white-rot fungus *__Phanerochaete chrysosporium__*. *(B) Cycle of oxidative, reductive and methylation reactions leading to the degradation of pentachlorophenol.*

Interestingly, many of the fungal enzymes can produce an oxidant that is capable of oxidizing oxalate. One of the oxidation products of oxalate is the $CO_2^{\bullet -}$, which has a reduction potential of -1.9 V, and thus, is a powerful reductant. Subsequently, the $CO_2^{\bullet -}$ can reduce compounds with a higher reduction potential, e.g., O_2, which results in $O_2^{\bullet -}$. In general, each reaction in the sequence will occur according to their reduction potentials to generate less energetic radicals, with antioxidants producing the least energetic (most stable) radicals. However, the presence of O_2 and catalytic metals, such as iron, can result in reactions that produce more reactive radicals. For example, the Fenton reaction, $Fe(II) + H_2O_2 \rightarrow Fe(III) + {}^{-}OH + {}^{\bullet}OH$, produces the ${}^{\bullet}OH$, one of the most potent biological oxidants known, which has a reduction potential of 2.3 V.

For a spontaneous process, the change in Gibbs free energy must be negative, $\Delta G < 0$. For ΔG to be negative in the equation $\Delta G = -nF\Delta E$, the change in the reduction potential (ΔE) must be positive for a spontaneous reaction. However, thermodynamics are not the sole factor regulating reactions. Kinetics are also a large determining factor whether a reaction will occur in a reasonable time. Therefore, it must be kept in mind that a reaction that is thermodynamically possible may not be kinetically feasibility; i.e., the rate constant for the reaction may be too small for the reaction to be biologically significant. Reduction potentials can only predict the value of an equilibrium constant, not the rate. The feasibility of a reaction occurring therefore also depends on the concentrations of the reactants and the stability of the products. A reaction with a large positive ΔG can go to completion if one of the products is a radical that subsequently reacts at very high rates. Similarly, a reaction with a large negative ΔG may occur slowly, or not at all, due to kinetic constraints. It should be kept in mind that free radical reactions generally follow a trend to form less reactive, thus, more stable compounds. Additionally, due to kinetic restraints, a reaction that is thermodynamically favorable may not occur fast enough to be biologically significant.

SUMMARY

The hierarchy of free-radical reactions would predict that the most efficient wood degrading fungi, where efficiency is defined as ability to degrade chemicals with increasing reduction potential, are those that secrete LiP, followed by those having MnP and then laccases, because of their decreasing order of oxidation potentials. The ability to generate the ${}^{\bullet}OH$, by CDH for example, adds oxidation ability, primarily through the contribution of Fe(II). Mediation may increase the number of chemicals that can be oxidized but lowers the oxidation potential. Additional factors include the ability to generate a powerful reductant using oxalate. Futile redox cycling can be limited in a number of ways, one of which is by reduction and methylation reactions providing methoxyphenols which are also better substrates for aromatic ring opening and mineralization.

ACKNOWLEDGMENT

The investigations from the author's laboratory were supported in part by funding granted by the Stockhausen GmbH & Co. of Krefeld, Germany. We thank Terri Maughan for secretarial assistance in preparation of the manuscript.

REFERENCES

1. Cooke, R. C.; Rayner, A. D. M. In *Ecology of Saprotrophic Fung;,* Longman; New York, NY, 1984.
2. Singley, M. E. In *On-Farm Composting Handbook*; Rynk, R., Ed. Northeast Regional Agricultural Engineering Service; Ithica, NY, 1992; pp 179-180.
3. Crawford, R. L. In *Lignin Degradation and Transformation*; Wiley; New York,
NY, 1981.
4. Cameron, M. D.; Aust, S. D. *Enzyme Microb. Technol.* **2001,** *28*, 129-138.
5. Cameron, M. D.; Timofeevski, S.; Aust, S. D. *Appl. Microbiol. Biotechnol. **2001,** 54*, 751-758.
6. Khindaria, A.; Yamazaki, I.; Aust, S. D. *Biochemistry* **1996,** *35*, 6418-6424.
7. Millis, C. D.; Cai, D. Y.; Stankovich, M. T.; Tien, M. *Biochemistry* **1989,** *28*, 8484-8489.
8. Sutherland, G. R. J.; Haselbach, J.; Aust, S. D. *Environ. Sci. Pollut. Res.* **1997,** *4*, 16-20.
9. Timofeevski, S. L.; Nie, G.; Reading, N. S.; Aust, S. D. *Arch. Biochem. Biophys.* **2000,** *373*, 147-153.
10. Wariishi, H.; Dunford, H. B.; McDonald, I. D.; Gold, M. H. *J. Biol. Chem.* **1989,** *264*, 3335-3340.
11. Reading, N. S.; Aust, S. D. *Biochemistry* **2001,** *40*, 8161-8168.
12. Thurston, C. F. *Microbiology* **1994,** *140*, 19-26.
13. Kersten, P. J.; Kalyanaraman, B.; Hammel, K. E.; Reinhammar, B.; Kirk, T. K. *Biochem. J.* **1990,** *268*, 475-480.
14. Higuchi, T. *ACS Symp. Ser.* **1989,** *399*, 482-502.
15. Bourbonnais, R.; Paice, M. G. *FEBS Lett.* **1990,** *267*, 99-102.
16. Li, K.; Xu, F.; Eriksson, K-E. L. *Appl. Environ. Microbiol.* **1999,** *65*, 2654-2660.
17. Igarashi, K.; Samejima, M.; Eriksson, K-E. *Eur. J. Biochem.* **1998,** *253*, 101-106.
18. Ander, P. *FEMS Microbiol. Rev.* **1994,** *13*, 297-312.
19. Wood, J. D.; Wood, P. M. *Biochem. Biophys. Acta* **1992,** *1119*, 90-96.
20. Khindaria, A.; Grover, T. A.; Aust, S. D. *Biochemistry* **1995,** *34*, 6020-6025.
21. Goodwin, D. C.; Aust, S. D.; Grover, T. A. *Biochemistry* **1995,** *34*, 5060-5065.

22. Chung, N.; Aust, S. D. *Arch. Biochem. Biophys.* **1995,** *322*, 143-148.
23. Khindaria, A.; Barr, D. P.; Aust, S. D. *Biochemistry* **1995,** *34*, 7773-7779.
24. Paszczynski, A.; Huynh, V-B.; Crawford, R. L.. *Arch. Biochem. Biophys.* **1986,** *244*, 750-765.
25. Chung, N.; Shah, M. M.; Grover, T. A.; Aust, S. D. *Arch. Biochem. Biophys.* **1993,** *306*, 70-75.
26. Khindaria, A.; Grover, T. A.; Aust, S. D. *Environ. Sci. Technol.* **1995,** *29*, 719-725.
27. Shah, M. M.; Grover, T. A.; Barr, D. P.; Aust, S. D. *J. Biol. Chem.* **1992,** *267*, 21564-21569.
28. Dutton, M. V.; Evans C. S.; Atkey P. T.; Wood D. A. *Appl. Microbiol. Biotechnol.* **1993,** *39*, 5-10.
29. Schwarz, H. A.; Dodson, R. W. *J. Phys. Chem.* **1989,** *93*, 409-414.
30. Higuchi, T. *Wood Res.* **1986,** *73*, 58-81.
31. Westermark, U.; Ericksson, K-E. *Acta Chem. Scand. B.* **1975,** *29*, 419-424.
32. Rasmussen, S. J.; Chung, N.; Khindaria, A.; Grover, T. A.; Aust, S. D. *Arch. Biochem. Biophys.* **1995,** *320*, 243-249.
33. Barr, D. P.; Shah, M. M.; Grover, T. A.; Aust, S. D. *Arch. Biochem. Biophys.* **1992,** *298*, 480-485.
34. Barr, D. P.; Aust, S. D. *Environ. Sci. Technol.* **1994,** *28*, 78A-87A.
35. Faison, B. D.; Kirk, T. K. *Appl. Environ. Microbiol.* **1983,** *46*, 1140-1145.
36. Khindaria, A.; Grover, T. A.; Aust, S. D. *Arch. Biochem. Biophys.* **1995,** *314*, 301-306.
37. Stahl, J. D.; Aust, S. D.. *Biochem. Biophys. Res. Commun.* **1993,** *192*, 471-476.
38. Coulter, C.; Hamilton, J. T. G.; Harper, D. B. *Appl. Environ. Microbiol.* **1993,** *59*, 1461-1466.
39. Valli, K.; Brock, B. J.; Joshi, D. K.; Gold, M. H. *Appl. Environ. Microbiol.* **1992,** *58*, 221-228.
40. Mileski, G. J.; Bumpus, J. A.; Jurek, M. A.; Aust, S. D. *Appl. Environ. Microbiol.* **1988,** *54*, 2885-2889.
41. Buettner, G. R. *Arch. Biochem. Biophys.* **1993,** *300*, 535-543.
42. Wardman, P. *J. Phys. Chem. Ref. Data 18*, **1989,** 1637-1755.

Wood Deterioration Processes

Decay

Chapter 4

Microview of Wood under Degradation by Bacteria and Fungi

Geoffrey Daniel

Wood Ultrastructure Research Centre (WURC), Department of Wood Science, Swedish University of Agricultural Sciences, Box 7008, Uppsala, Sweden

Wood is colonized and degraded by a variety of microorganisms including bacteria, moulds, blue stain and wood rotting fungi (i.e. soft-, white and brown rots). The manner by which wood is degraded varies according to the organisms involved, type of wood substrate, nature of environment as well as interactive competition. Over the years microscopy (*e.g.* light and electron) has played a major role in improving our understanding of the morphological aspects of biological decay and has been useful as a tool for understanding fundamental aspects of the biochemical mechanisms of attack *in-situ*. This review outlines some of the major aspects in our knowledge of the micro- and ultrastructural changes which take place in wood during decay and where microscopy has aided this understanding.

The biodegradation of wood is a natural process of organic matter recycling and occurs in all natural environments (terrestrial, aquatic) assuming suitable conditions are available. Under natural conditions wood may be colonized quite rapidly by a variety microorganims which initiate the process of biodegradation whereby the wood structural components (*i.e.* polysaccharides, lignin) are solubilized to simpler molecules and finally CO_2 and water. When this process occurs on "man-made" structures it is considered of economic importance. However when we use the same biodeteriogens for producing novel proteins of commercial interest or to modify lignocellulose under appropriate experimental or industrial situations we consider them as "biological potential". The purpose of this chapter is to provide an update of our current understanding of the micro- and ultrastructural changes occurring in wood during various types of biological decay by different microorganisms. This information should provide a means to identify the different types of biodegradation to which wood is subjected in various environments as well as emphasizing the importance of a basic knowledge of wood structure (*i.e.* anatomy, micro- and ultrastructure) for understanding both wood decay and improving its protection by wood preservation. Several reviews on wood biodegradation have been written over the years *(e.g. 1-4, see also reviews in this book)* and the reader is advised to also consult these texts. Light and electron microscopy (TEM, SEM) have proven to be invaluable tools for studying aspects of wood decay by both fungi and bacteria, thereby improving our basic understanding of the decay mechanisms involved. Today microscopy is used routinely in a diverse range of applications from the simple diagnoses of microbial decay types in test stakes and building construction, to its use as an advanced research tool for studying aspects of wood cell wall biodeterioration, and wood preservative interactions at ultrastructural and nanoscales.

Wood as a Substrate for Biodegradation: Features of Importance

Wood anatomy, microstructure and ultimately cell wall ultrastructure (assumming suitable conditions prevail) has a governing role in regulating biodegradation. The anatomical and microstructural features of wood varies between different species with gymnosperms (softwoods) tending to have a comparatively simple structure in contrast to angiosperms (hardwoods)*(5-7)*. Representative morphological features of a hardwood are shown in Fig. 1. Up to 95 % of softwoods consists of tracheids with the remainder of cells non-lignified parenchyma cells (longitudinal or radial) residing in the rays or as epithelial cells surrounding resin canals *(6, 7)*. In contrast, hardwoods have a more complicated anatomy containing a large variety of cell types including vessels (characteristic only for hardwoods), fibres, parenchyma cells, fibre tracheids and libriform fibres and others *(6, 8)*. Chemically all wood cells are composed of various ratios of cellulose, hemicellulose and lignin *(9)* although the quantity of cellulose remains quite consistent (*i.e. ca* 40-50%) in both hard- and softwoods *(9)* with the lignin : hemicellulose ratio varying most. Normally hardwoods have a lower lignin content (*ca* 16-26 %, *9*) than softwoods (*ca* 28-32 %) which is reflected by a higher hemicellulose content, primarily of glucuronoxylan *(9)*. The higher lignin content of softwoods is normally reflected by a lower hemicellulose content primarily of galactoglucomannan *(9)*. A further major distinction between hard- and softwoods is the type of lignin monomer present with softwoods having guaiacyl lignin *(9)* and hardwoods both syringyl and guaiacyl lignins *(9)*. Lignin has a significant and governing role in natural wood protection with high lignin levels and presence of guaiacyl lignin considered to provide better natural protection than low lignin levels and syringyl lignin *(3, 10, 11)*. Since the distribution of syringyl and guaiacyl lignin can vary greatly in different cell types within hardwoods (see below) this can have a major influence on the manner and speed by which individual cells and tissues are degraded by different bacterial and fungal types. Both the variability and ratio of different cell types as well as presence and type of lignin in hardwoods and to a lesser extent softwoods plays a major role in regulating biodegradation. In broad terms all wood cells can be degraded by fungi and bacteria under suitable conditions, nevertheless the chemistry and ultrastructure of wood cell wall plays an overriding role.

While most fungi and bacteria colonize cut wood surfaces, there is a tendency for all species to initially colonize the rays and from there grow further into the wood structure via natural openings (*i.e.* pits: bordered, cross field) or directly from cell to cell by producing bore holes through the cell walls. Parenchyma cells in the rays tend to contain significant amounts of easily assimilated nutrient reserves and are non-lignified; thus any fungus or bacteria with cellulolytic and/or hemicellulolytic ability can attack these cells leading to subsequent increases in permeability of the wood substrate. Certain fungi (*e.g.* moulds, blue stain) and bacteria attack non-lignified cells but normally can not cause significant wood decay. Such fungi cause either superficial colonization (moulds) or deeper stain within the wood xylem substrate. In order to obtain a better understanding of wood decay it is first important to understand the nature of the wood cell wall micro- and ultrastructure.

Wood Cell Wall Ultrastructure: the Ultimate Factor Controlling Wood Biodegradation

Our understanding of the micro- and ultrastructure of wood cell walls has improved greatly over the last few decades as more refined techniques (*e.g.* confocal laser scanning microscopy (CLSM), field emission scanning electron microscopy (FE-SEM), cryo-SEM, transmission electron microscopy (TEM), nuclear magnetic resonance (NMR) and atomic force microscopy (AFM)) have been explicitly used to study wood fibre cell wall architecture, and as knowledge is derived from other areas of closely related research such as biosynthesis of wood cell walls *(12-14)* cell wall delignification and processing during chemical and mechanical pulping *(15-17)* as well as the use of fungi to reveal novel features of wood cell wall structure *(3, 18-21)*.

Numerous models are reported in the literature which outline the gross morphological architecture of wood cell walls *(22)*. While these models represent an over simplification of a highly complex structure and are general for fibres (tracheids) rather than representative for all cell types (*i.e.* reaction wood cells, juvenile cells, parenchyma, epithelia, vessels, fibre tracheids etc), they are useful for understanding some of the basic aspects of wood cell wall decay. A typical model proposed for a softwood tracheid by Cote (*23*) is shown in Fig. 1. Here an individual wood fibre is shown comprised of several discrete wall layers; *viz* a thin (*ca* 0.09 μm) primary wood cell wall (P) composed of multi-directional oriented cellulose microfibrils; and a secondary cell wall comprised of three cell wall layers S1 (*ca* 0.26-0.38 μm), S2 (*ca* 1.66-3.70 μm), S3 (*ca* 0.09-0.14 μm) which possess almost horizontal (*i.e.* S1, S3) or nearly vertical (*i.e.* S2) oriented cellulose microfibrils *(5)*. The thickness of the individual layers varies between early- and latewood and also with cell type. The S2 microfibrils also spiral helically in Z or S orientations along the fibre axis. The individual wood cells are bound together by an highly lignified middle lamella region (Fig. 1). Of the three secondary cell wall layers the S2 is of overriding importance because of its thickness and orientation of cellulose microfibrils giving fibres strength and ultimately imparting the major mechanical and physical characteristics to wood products. A further consideration of importance is the thickness and lignification of the S3 and S1 cell wall layers. The S3 layer in softwoods tends to be of variable thickness and well lignified *(24)*, whereas in certain hardwoods such as birch the S3 is normally very thin and composed of microfibrils showing similar orientations as the underlying S2 layer *(9, Daniel unpub. obs)*. While these two features have little effect on brown or white rot decay (see below) they may have a significant affect on soft rot attack. During soft rot attack of coniferous species, cavities tend not to be formed in the S3 layer because of its thickness and lignification, however, in soft rot attack of the birch cell wall, erosion decay is frequent (see below).

One of the major factors controlling the biodegradation of wood cells is the availability of the cellulose and hemicelluloses for decay within wood cell walls. The numerous studies conducted on a diverse range of fungi and bacteria have shown the majority to possess cellulase and/or hemicellulase activity. Since possession of cellulase and hemicellulase activity is not correlated with decay of wood (*e.g.* mould fungi like *Trichoderma* spp. are well known for pronounced cellulase production but tend not to degrade lignified wood cells efficiently), the major controlling factor is the availability of cellulose to decay in wood cell walls. In lignified cell walls, the cellulose is thought to be associated with the hemicelluloses which are in turn surrounded by lignin forming an impenetrable matrix to enzymes *(5, 9)*. Several models depicting this molecular arrangement have been put forward although there is no accepted model (Fig. 2).

Wood cells may also vary in degree and type of lignification; *e.g.* fibres in hardwoods are syringyl lignified while vessel walls tend to possess guaiacyl lignin *(1, 2)*. Similarly, ray parenchyma cells in rays from the sapwood of softwoods tend to be non-lignified and easily degraded, while adjacent ray tracheids are lignified and are not degraded by moulds and blue stain fungi. Rot fungi in particular use different mechanisms to overcome the "*lignin barrier*" which are manifested at the ultrastructural level. Soft rot fungi align themselves along the cellulose microfibrils of the cell wall and overcome the lignin barrier by being closely adpressed to both cellulose and lignin; the cellulolytic enzymes of soft rot fungi are thought to be associated with the hyphal surface. In contrast, white rot fungi are not restricted by the *lignin barrier* and are able to degrade all the wood components progressively from the cell lumen or preferentially attack the lignin and hemicelluloses (see below). Brown rot fungi, however, are able to degrade the cellulose and hemicelluloses by a so far poorly understood and highly effective chemical mechanism with lignin left over as a modified chemical entity. The basis of this decay is regulated at the cell wall level of wood cell walls. No fungi can apparently degrade and exist on lignin alone.

In recent years, more attention has been applied to understanding the ultrastructural architecture of wood cell walls particularly the biosynthesis, configuration and arrangement of the cellulose microfibrils *(25)*. While it is generally agreed that cellulose microfibrils (*ca* 2-4 nm) or sub-elementary fibrils (*ca* 1.5-2.0 nm; *26*) compose the basic armature and provide mechanical strength to wood cell walls, there is increasing evidence that there may be a greater order in which the cellulose microfibrils are complexed together to form what has been termed cellulose "aggregates" or "macrofibrils" *(27-30)* (Fig. 3).

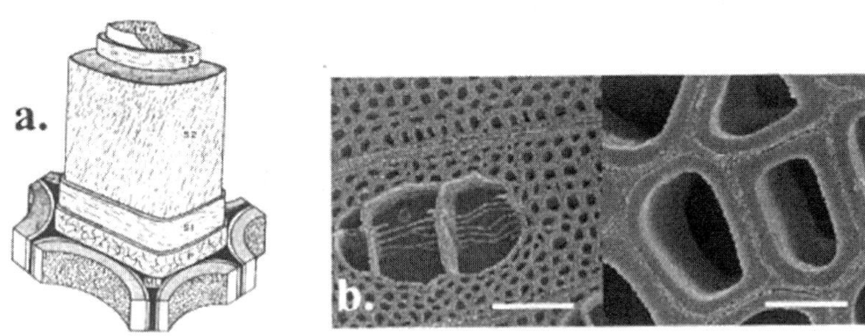

Figure 1. a) Softwood fibre (tracheid)(Côté, 23) showing the microfibrillar orientation of primary (p) and secondary cell wall layers (S1, S2 and S3) and position of the middle lamella (ML) layer (Reproduced with permission from reference 23. Copyright 1967 Syracuse University Press.) b) Cross-sections of birch fibres and vessels as seen using SEM. Bars: left, 30 μm; right, 10 μm.

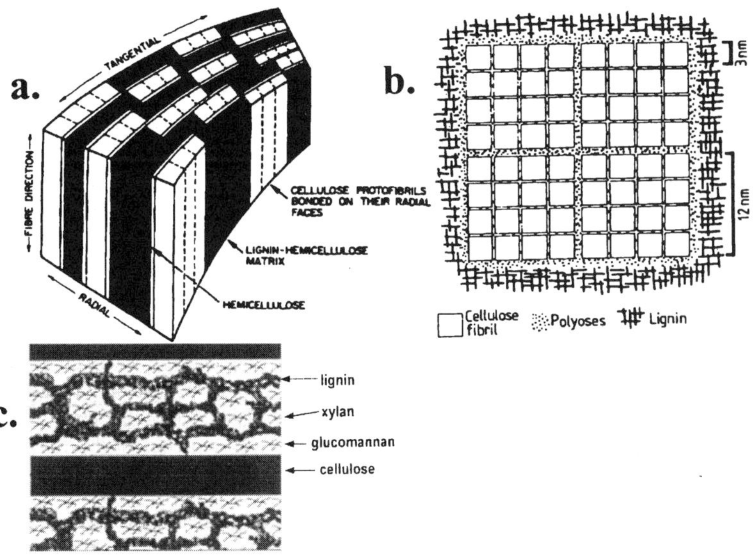

Figure 2. Proposed molecular arrangement of wood polymers in wood fibre cell walls: a) Kerr and Goring (31), with cellulose microfibrils forming interrupted lamellae embedded in a matrix of lignin and hemicellulose (Reproduced with permission from reference 31. Copyright 1975 Springer.) b) Fengel and Wegener (5) where elementary fibrils are surrounded by monolayers of hemicelluloses and larger units (macrofibrils) enclosed with hemicellulose and lignin (Reproduced with permission from reference 5. Copyright 1984 Walter de Gruyter.) c) Salmén and Olsson (32) where glucomannan is closely associated with cellulose and xylan to lignin. (Reproduced with permission from reference 32. Copyright 1998 Tappi.)

For example, these aggregates (10-60 nm) have been well documented in spruce kraft pulp fibres for all secondary cell wall layers using a variety of microscopical methods including FE-SEM, Cryo-FE-SEM and AFM *(16, 27, 28, 30)*. They have also been documented in spruce sulphite pulps *(33)* and in *Phlebia radiata* white rotted pine and birch woods (*34*, see later, Fig. 15) using Cryo-FE-SEM (see below). They have similarly been described and characterized in refined spruce pulps of variable molecular weight *(35)*. During the last decade there has also been considerable debate on the arrangement of cellulose microfibrils within the S2 cell wall and whether they follow a concentric or radial-concentric arrangement (*e.g. 17, 18, 28, 30, 37*). Recent studies *(30)* using AFM suggest that concentric lamellation is possibly the most common arrangement, but that radial alignment may develope when wood fibres are placed under tension forces. The arrangement of the cellulose aggregates is important in wood decay since it is frequently noted that decay radiates out in a concentric manner from sites of hyphal decay (*e.g.* cavities, troughs) of wood cell walls. Fundamentally, the arrangement of the wood components (cellulose, lignin) will also have marked bearing on the ability of wood preservatives to protect wood from decay. For example, copper-chrome-arsenic (CCA) is believed to fix primarily to lignin in wood and provide protection by both toxicity as well as changing the molecular structure of the wood *(38)*. Recent results regarding the cellulose aggregates described above would indicate that the cellulose is protected by a lignin-CCA complex. From *in-situ* toxicity studies with soft and brown rot fungi (*i.e.* the most well known Cu resistant wood degrading fungi; *39, 40)*, production of cavities and alignment of hyphae along the cellulose microfibrils may appear to be a more effective manner for causing decay than the brown rot decay mechanism, since soft rot fungi as a group tend to show greater Cu resistance than brown rotters. Other cell types like parenchyma cells which are not lignified are similarly constructed but the matrix surrounding the cellulose consists of hemicelluloses and pectins *(5)*.

Another important aspect concerns the compound middle lamella between wood fibres. This structure is well known to contain the highest lignin content (g/g) in wood cell layers *(41-43)*. Ultrastructural and spectroscopic studies, as well as the use of wood degrading organisms, has shown this layer not to be homogeneously lignified *(36)* and when subjected to biological decay from either brown- or white rot fungi shows variability in structure, although the layer is resistant to attack by soft rot fungi (see below).

Wood Degrading Organisms and Decay Types

Bacteria

The occurrence and negative effects of bacteria on wood have been known for a very long time *(44)*. However, these negative effects have primarily been recognized from the ability of bacteria (both anaerobic and aerobic) to cause a pronounced increase in the permeability of round wood (best known in softwoods), including refractory species like spruce, during ponding or water sprinkling. Economically, this still represents the most important effect of bacterial attack of wood. Over the last *ca* 20 years however, convincing evidence for bacterial degradation of lignified cell walls has been reported *(45-48)*. Such unequivocal evidence has primarily been obtained through the use of electron microscopy (TEM and SEM) on bacterially degraded wood samples produced under laboratory conditions or removed from either terrestrial or aquatic situations *(45, 46, 48, 49)*. Prior to the use of the electron microscopy it was difficult to distinuish and confirm bacterial decay patterns from fungal decay(s). It is now well established that *true wood degrading bacteria* are cosmopolitan in

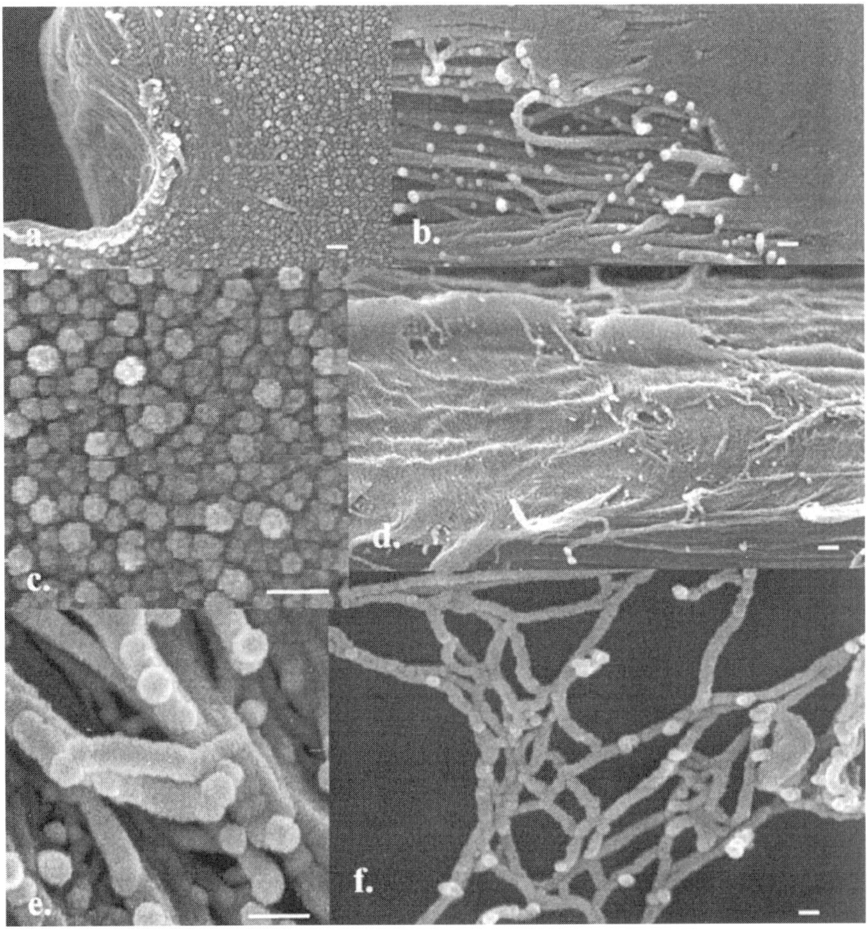

Figure 3. FE-SEM micrographs showing the typical cellulose macrofibrillar structure of spruce tracheid secondary (S1, S2, S3) and primary wall layers revealed following delignification using H_2O_2 and acetic acid or kraft pulping (e). a) S2 layer comprised of numerous macrofibrils in cross section; b) Macrofibrils comprising the S1 layer; c) High magnification of macrofibrils seen in cross section from a.; d) S3 layer macrofibrils of the lumen wall; e) Cylinder like appearance of macrofibrils from the S2 layer of spruce kraft pulp fibre; f) Macrofibrils comprising the primary cell wall. Bars: 100 nm.

the world, and have a tendency to be aerobic and exist in a diverse range of aquatic and terrestrial environments *(48)*. These bacteria can cause significant attack of lignified fibres and tracheids in both hard- and softwoods and have an ability to degrade preservative treated material (*e.g.* CCA treated; *46, 50, 51*),chemically modified timbers, and highly durable timbers containing high extractive levels like *Eusideroxylon zwageri (52)* or high lignin content (*e.g. Alstonia scholaris*)*(49)*. Decay by all currently known true wood degrading bacteria is carried out by single-celled, motile Gram negative, pleomorphic bacteria possessing considerable cell wall plasticity *(48)*.

Bacterial Decay Leading to Increased Permeability

This type of attack is known primarily from the "ponding" or "water spraying" (Fig. 4) of round woods which is still routinely employed in many countries to protect timber from decay fungi, blue stain and insect attack. Bacteria direct their attack to the non-lignified tissues such as the ray parenchyma cells and pit membranes of the sapwood; similar tissues and structures in the heartwood being less affected due to partial lignification or the encrusting of extractives. Typical examples of bacterial attack of *Pinus sylvestris* bordered pits and ray tissues after 18 weeks water sprinkling at a Swedish saw mill are shown in Figure 4. The ray parenchyma cells and pit tori are the sites of most rapid attack (Fig. 4). Both rod-shaped (*ca* 1-2 μm long, Fig. 4) and cocci forms have been reported involved in decay *(53)*. Degradation generally results in an increase in the permeability of sapwood and subsequently an irregular and often overabsorption of preservative solutions, varnishes, stains (Fig. 4), paints etc depending on the final use of the wood material *(54-56)*. While sawn wood from such material is unacceptable for joinery (*i.e.* can not be used for house facades), long-term water sprinkling and ponding of poles has been attempted as a possible method for reducing bleeding of cresote impregnated poles *(57, 58)* and increased penetration of preservatives into refractory wood species such as spruce. Transport of round wood via rivers is still a major means of transport in some tropical counties. The effects on permeabilty are similar as described above but when low lignified trees are involved, actual decay of wood fibres may result with time.

Bacterial Decay of Lignified Tissues

Currently two major forms of bacterial degradation, *viz* tunnelling (TB) and erosion (EB) decay of lignified tissues, have been reported *(1, 48)*. These two forms of attack are based on microscopic classification of the decay patterns produced in lignified fibres and are not based on taxonomic features of the

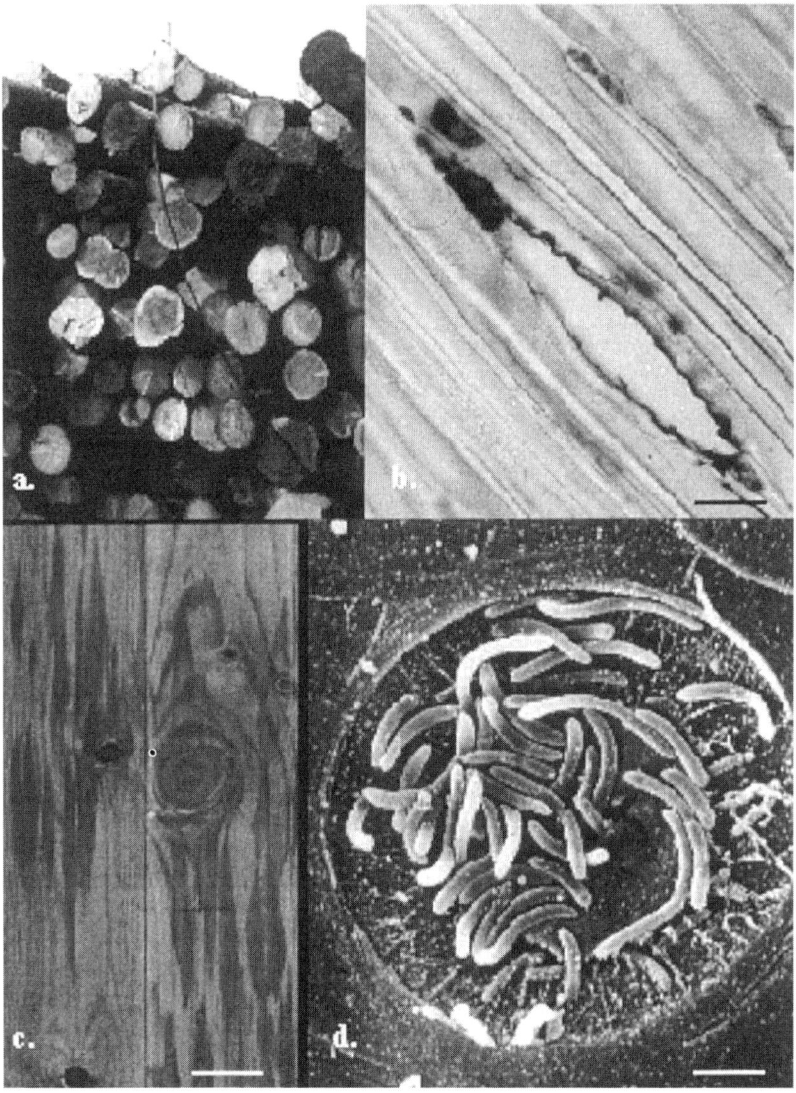

Figure 4. Bacteria attack of water sprayed pine wood. a) Stacked pine logs subjected to water spraying; b) LM tangential longitudinal section showing almost total decay of ray parenchyma cells; c) Sapwood plank after varnishing showing uneven uptake of varnish due to increased permeability; d) Typical attack of a bordered pit membrane by rod-shaped bacteria leading to increases in permeability (SEM photo). Bars: b, 50 μm; c, 50 mm; d, 5.0 μm.

organisms. Indeed, even at year 2002, no pure cultures of active wood degrading bacteria have been reported and thus the true taxonomic origin of the bacteria are unknown. Nevertheless mixed cultures from bacterial degraded wood from different habitats have been produced and under laboratory conditions shown to reproduce the decay patterns recognized in wood under natural conditions *(48, 59)*. Both electron microscopic and biochemical studies with mixed laboratory cultures (using labelled lignins, NMR) have confirmed the ability of wood degrading bacteria to attack lignin as well as wood polysaccharides *(59)*. Tunnelling bacterial decay of wood cells received considerable notoriety after it was unequivocally confirmed by TEM *(46, 60)* that the very peculiar, but thin branching and radiating decay patterns found in the S2 layers of wood cells was caused by single-celled pleomorphic bacteria (Figs 5, 6). Detailed TEM studies have shown a number of characteristic features concerning this type of bacterial decay including: direct cell wall penetration from the cell lumen (Figs 5, 6), and of the highly lignified middle lamella regions (Fig. 6). Only one bacterium is found per tunnel and the bacteria leave behind characteristic concentrically oriented slime secretions (Figs 5, 6) during progressive cell wall attack and motility. Evidence points to the polysaccharide nature of these secretions as well as negative charge. TEM-X-ray microanalytical studies have shown the slime secretions to bind heavy metals such as CCA during tunnelling bacterial attack of preservative treated wood (Fig. 6; *50*). Whether this feature represents indiscriminate binding of metals or a means of selective detoxification is currently unknown. Degradation of cell wall middle lamella regions and highly lignified wood species such as *Homalium foetium* and *Alstonia scholaris (49)* and species containing high extractive levels (*e.g. Eusideroxylon zwageri*)*(52)* have confirmed the ligninolytic and aromatic degrading ability of these bacteria. TB and EB (see below) also produce extracellular vesicles which are thought to carry cellulolytic and possibly peroxidase enzymes to sites of wood attack. This has been suggested by positive cytochemical staining with Benedicts reagent for cellulases and 3',3'-diaminobenzidine (DAB) for peroxidases (Daniel unpublish obs.). The tunnelling nature of TB would appear similar to the manner certain fungal hyphae ramify and penetrate through wood cell walls.

The degradative pattern produced by EB has morphological similarities to both white rot (simultaneous) and soft rot (*i.e.* Type 2) erosion decay in that the bacteria tend to initiate decay from the cell lumen through the S3 layer (Figs 7, 8). Unlike white rot fungi, EB tend not to actively degrade middle lamella regions (Fig. 7) although bleaching of this region during advanced stages of decay suggesting the presence of a low molecular weight redox system has been reported *(61)*. Cell wall degradation results from the combined activity of numerous bacteria progressively eroding the underlying cell wall, often in a uniform manner producing discrete, but characteristic, channels along which the bacteria move and divide while the colony progressively expands from its ends

(Figs 7, 8). This gives rise to a "stripped" decay pattern which is readily seen in longitudinal sections using polarized light; the loss in birefringence indicative of the attack of the underlying cellulose and associated hemicellulose. In TEM sections, EB often appears closely adpressed to the underlying cell wall encapsulated in slime. Studies so far have shown EB to have a rod-shaped appearance (*ca* 1-2 μm long, Fig. 8) often with pointed ends (Figs 7, 8)*(3, 48, 61, 62)*. Detailed observations indicate that EB, like TB, also produce extracellular secretions in the form of vesicles (*ca* 0.003-0.06 μm) *(48, 61)* or complexes, a feature consistent with that reported for some cellulolytic bacteria such as *Bacteroides succinogenes (63)* and for the multicellulase "cellulosomes" found on the outer cell wall of anaerobic bacteria such as *Clostridium thermocellulum (64)*. These structures have been noted on both the outer cell wall of EB and associated with the decay channels (Fig. 8).

In addition to cell wall erosion, the decay pattern is similar to soft rot attack in that the bacteria normally show alignment like soft rot cavity hyphae (see below) with the underlying cellulose microfibrils (Fig. 8). The bacteria do not actively degrade the high lignin containing middle lamella regions of hard- and softwoods; have difficulties with the lignified S3 layers of softwoods and may on occasions produce angular cavities reminiscent of soft rot cavities within the S2 layer of wood fibres after penetration via pit chambers *(61)*. Like soft fungi, the alignment of EB with the cellulose microfibrils may similarly represent a method to overcome the lignin barrier protecting the cellulose. A characteristic feature of EB attack is the irregularity by which cells are degraded with both highly attacked and non-degraded tracheids often visible in the same section (Fig. 7).

The interest in bacterial degradation of wood has varied over the years but despite the electron microscopic evidence *(48)*, and biotechnological potential these single-celled organisms have for the decay of recalcitrant waste chemicals, the isolation of pure cultures remains a major stumbling block preventing application and further study. Under terrestrial conditions wood degrading bacteria may be found in association with soft- , brown- and white rot decay, and under aquatic situations -where brown- and white rot tend to be absent- together with soft rot. However, these bacteria cannot apparently compete successfully with fungal attack unless fungal development is suppressed in some way to produce a comparatively competitive-free environment for the bacteria. Under oxygen limiting conditions such as buried shipwrecks *(1, 65, 66)* archaeological artefacts *(62, 67, 68)* and building foundations *(69)* fungal attack is suppressed and EB proliferate and may be a serious problem. Similarly, under conditions where woody materials are more resistant to fungal attack through wood preservation with high loadings of chemicals such as CCA and CCB *(51, 61)* or

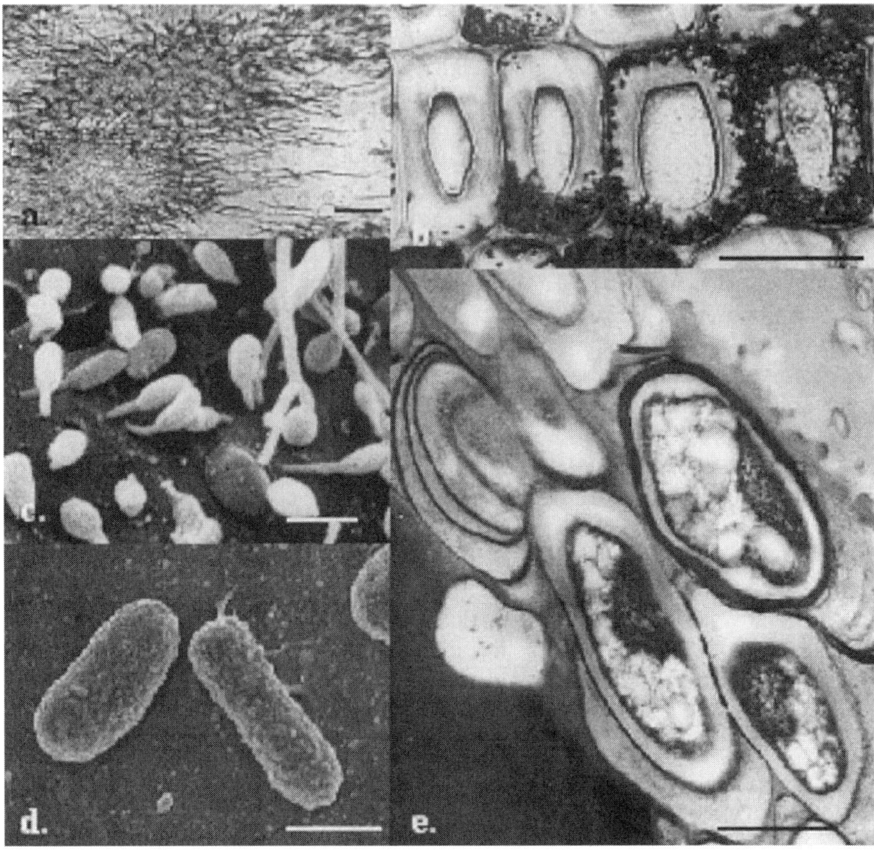

Figure 5. Aspects of tunnelling bacteria. a) Typical decay pattern in wood cells as seen with light microscopy; b) Cross section of spruce tracheids with TB attacking all wood cell wall layers including the middle lamella joining cells; c) TB enclosed in extracellular slime material (tubes) and attached to the lumen wood cell wall; d) Morphology of typical TB cells (SEM); e) Characteristic nature of TB decay of wood cell walls with concentric slime layers left behind during attack. Bars: a, 5.0 µm; b, 10.0 µm; c-e, 0.5 µm.

creosote *(70)* or are naturably durable through high lignin *(49)* or extractive content *(52, 70)*, a competitive free environment is created where bacteria, particularly tunnelling bacteria can proliferate. Conditions favourable for bacterial attack can also be produced by hydrofluoric acid treatment of stakes which are then placed in terrestrial environments. The acid treatment removes the cellulose and hemicelluloses leading to a higher lignin : carbohydrate ratio thereby suppressing fungal and encouraging bacterial attack. As described above, bacteria compete poorly with fungi, but where favourable conditions exist

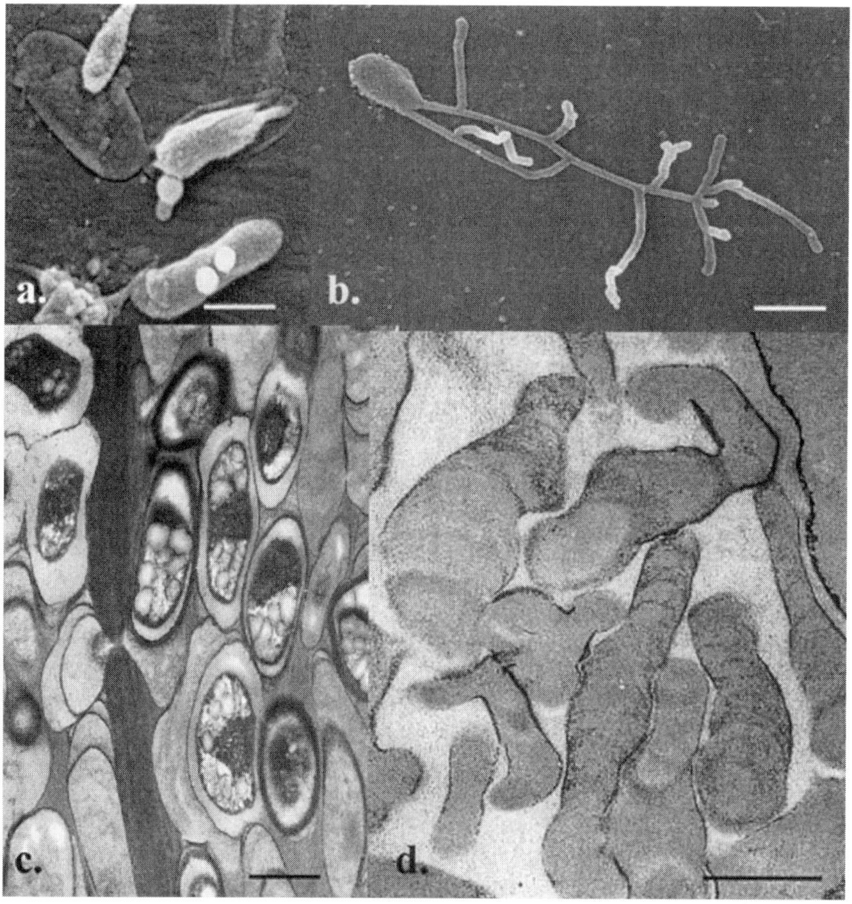

Figure 6. SEM and TEM micrographs of TB; a) Typical slime secretions which secure the entrance of TB tunnels into wood cell walls; b) TB frequently appear encapsulated in extracellular slime secretions (i.e. tubes); c) TB degrading the secondary cell wall and middle lamella regions of a softwood (Reproduced with permission from Daniel, G; Nilsson, T.; Forest Products Biotechnology; *Bruce, A.; Palfreyman, J. W., editors; page 304. Copyright 1998 Taylor & Francis.)*

d) TB tunnels formed in CCA treated radiata pine with CCA concentrated on tunnel walls (electron dense areas) and the concentric slime secretions within the tunnels. Bars: a, b, 1.0 μm; c, d, 0.5 μm.

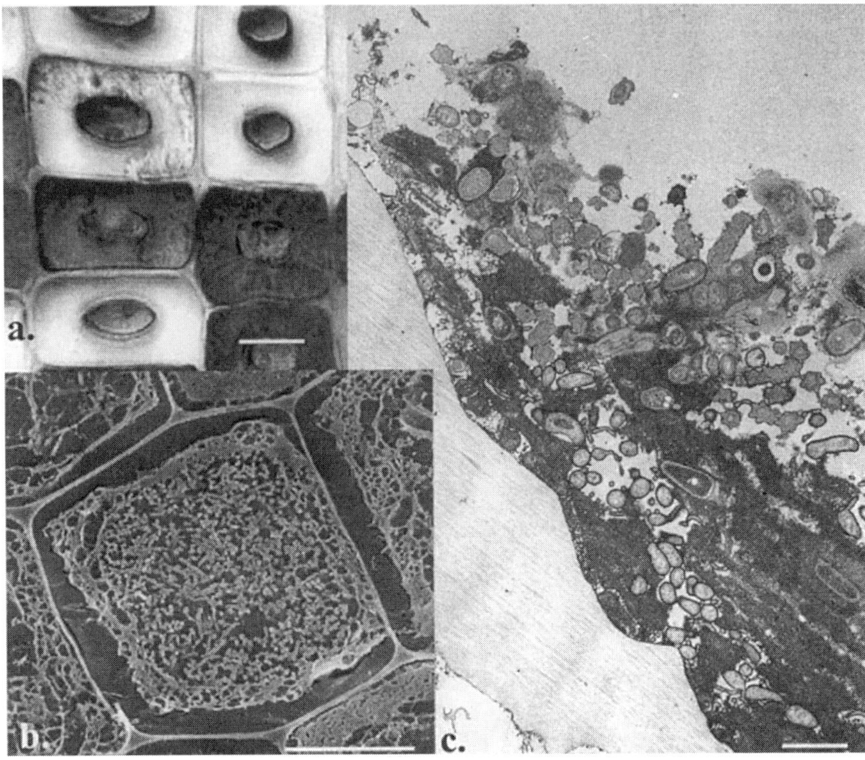

Figure 7. Aspects of erosion bacteria decay of wood cells. a) Light micrograph of a cross section of pine cells from an old shipwreck with adjacent tracheids showing severe and absence of attack; b) Cryo-SEM micrograph showing advanced EB decay of pine cells with only the middle lamella cell wall layers remaining with numerous EB concentrated together with slime in the central regions of the cells; c) TEM longitudinal section showing characteristic cell wall erosion. Bars: a, b, 10 µm; c, 1.0 µm.

they may represent a major form of decay of man-made structures. Examples include the decay of the durable wood greenheart used for moorings in lock gates *(70)* and preservative treated support pilings in New Zealand kiwi orchards *(51)*.

A feature of both EB and TB is the apparent encapsulation of the bacteria in extracellular slime materials (Figs 5-7) *(46-48)*. These materials presumably provide the organism with protection (*e.g.* from wood preservatives (Fig. 6), dessication, etc.) as well as providing a media for motility and passage of enzymes (Fig. 6) to the wood substrate. For a more detailed account on bacterial decay the reader should consult a recent review (Daniel and Nilsson, *48*).

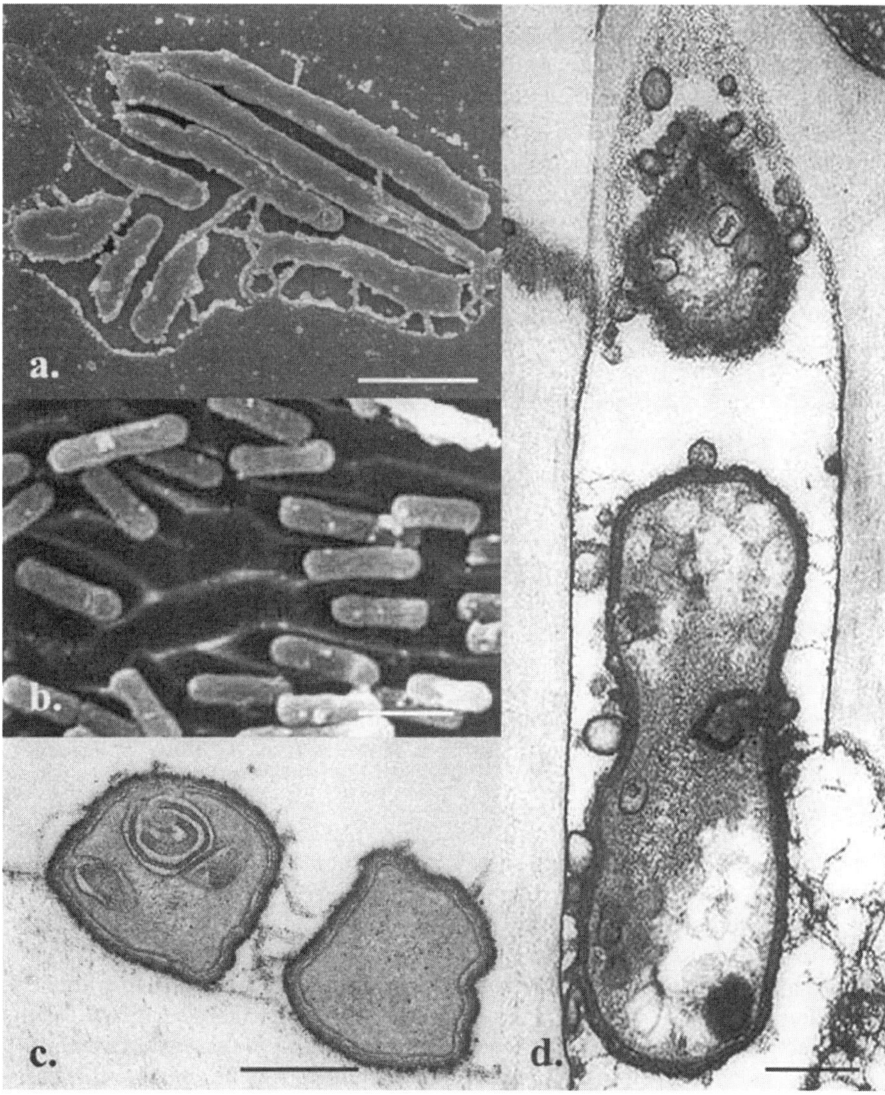

Figure 8. SEM and TEM micrographs showing erosion bacteria attack of wood. a) Initial stage in decay with rod-shaped EB encapsulated in slime attached to the lumen cell wall; b) Later stage of attack with EB forming characteristic erosion channels in the S2 layers; c) TEM micrograph showing a cross section of EB adpressed to the S2 cell wall; d) Oblique longitudinal section of EB cells within an erosion channel with presence of small extracellular vesicles. Bars: a, 0.4 μm; b, c, 1.0 μm; d, 0.2 μm.

A variety of other bacteria forms have also been shown present in woody tissues. These bacteria probably exist as secondary feeders and probably can cause only slight cell wall modification *(45, 48, 71, 72)*.

Filamentous bacteria (*i.e.* actinomycetes) have been variously reported over the years as important colonizers of wood in ground contact although few reports showing degradation of lignified wood fibres are available. These organisms are cosmopolitan in soils and are readily isolated from decaying plant and lignocellulose materials *(73, 74)* thus the general acceptance that they are impor-

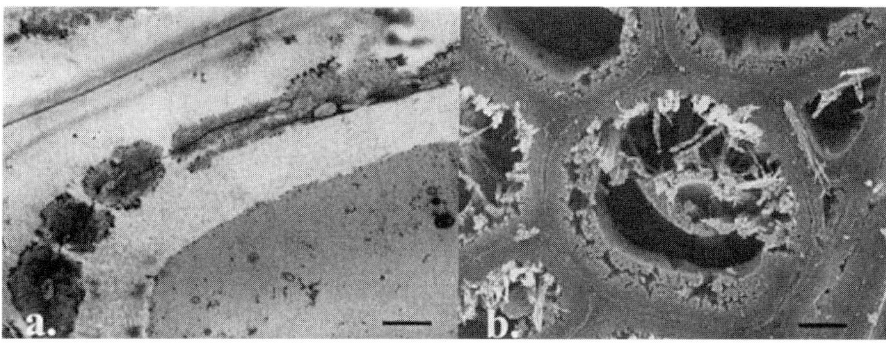

Figure 9. Decay of birch wood fibres by myceliated bacteria. a) TEM micrograph showing concentric orientated decay pattern within the S2 cell wall; b) SEM micrograph showing a similar form of attack. Bars. a, 1.0 μm; b, 4.0 μm.

-tant degraders of woody tissues. Apart from a report indicating soft rot of lime wood by an actinomycete *(75)* and more recent ultrastructural evidence for minor etching of wood cells by *Streptomyces* spp. *(76)*, the only microscopic evidence showing pronounced attack is by a so far unidentified species (Fig. 9)*(3)*. Decay is rather pronounced in that large areas of birch cell walls are degraded with the myceliated bacteria following a concentric pattern within the S2 cell wall (3). The fact that actinomycetes have not been shown as important degraders of lignified wood cells strongly suggests they are unlikely to be important in the degradation of preservative treated wood. Despite the few reports on wood degradation, actinomycetes have been shown to produce a wide variety of enzymes including cellulases, hemicellulases, pectinases and even lignin peroxidases *(77, 78)*. Their biotechnological potential may thus be better understood than for true wood degrading bacteria.

Moulds and Blue Stain Fungi

Numerous studies have been carried out on the health and discolouration problems associated with moulds of sawn timber and their prevention *(79)*. This has resulted in numerous physiological, and biochemical studies to understand toxin production by moulds such as *Aspergillus* and *Penicillium (80)*. Similarly over the years, considerable work has been directed towards preventing the invasion of round wood by blue stain fungi. The number of microstructural studies on moulds and blue stain fungi over the last few decades has been limited, although in the last few years interest has been revived as the staining of round- and sawn wood cut from tropical softwood forests *(81, 82)* and the moulding and staining of modern paint systems *(83)* on houses is beginning to present major problems. Microstructurally, mould and blue stain (*i.e.* various phycomycetes, ascomycetes and fungi imperfect) fungi colonizing wood have been reported to have very little effect on the strength of wood *(84)*, but may degrade non-lignified wood cells such as axial and ray parenchyma cells, pit membranes leading to increased permeability, and tension wood G-layers *(84)*. A strict delineation between moulds, blue stain and wood decay is not possible since many fungi traditional implicated in the discolouration of wood can cause soft rot if suitable conditions are prevailing (*e.g. Alternaria alternata, Cladosporium herbarum, Aspergillus fumigatus (150, 151)* and many of the most important soft rot fungi are highly melanized (*e.g. Phialophora* spp., see below). Blue stain fungi such as *Lasiodiplodia theobromae,* however, have been shown to cause cell wall decay (up to 20 % weight loss) of birch and rubber wood but only to have limited effects on softwoods like Caribbean pine *(81)*. Studies have shown *L. theobromae* to produce erosion and cavity decay of birch cell walls and behave like weak soft rot fungi *(81, 84)*. Microstructural studies on the blue stain fungi *Aureobasidium pullulans* and *Sclerophoma pityophila* using SEM *(85)* has shown some interesting results concerning a possible mechanism for micropore penetration in paint films with these fungi producing specialized microhyphae of very small dimensions (*i.e.* 0.2 μm) (Fig. 10). Such observations are also of interest with respect to the microhyphae produced during T-branching, and L bending of soft rot fungi and basidiomycete attack (see below).

Soft Rot Fungi

The term "soft rot" is used to describe a very specific type of decay caused by large numbers of Ascomycetes and Fungi Imperfect which typically produce chains of cavities with conical ends within the S2 cell wall layers of soft- and hardwoods exposed under terrestrial and aquatic environments (Fig. 11)*(86-89)*. The term was originally used by Savory *(89)* to describe the characteristic

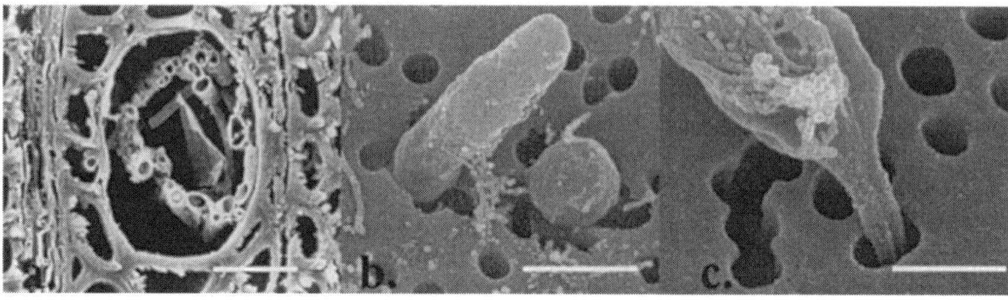

Figure 10. (left) SEM micrographs showing typical staining and colonization of vessels in Carribean pine by L. theobromae hyphae (a) and penetration of micropores by specialized microhyphae of Sclerophoma pityophila (b, c). Bars: a, 30 μm; b, c, 1.0 μm.

softening of wood by cellulose-destroying microfungi to distinuish it from that caused by basidiomycete white- and brown rot fungi. Although softening of wet wood is typical, studies on the soft rot attack of CCA treated timber has shown degraded wood to be hard, nevertheless the term is wildly accepted. Economically soft rot attack is of considerable importance since it is effective in degrading wooden constructions and preservative treated wood, especially in ground contact *(48, 89)*. Although cavity formation remains the most characteristic feature for microscopic detection of soft rot decay, many of the fungi involved also produce cell wall erosion similar to simultaneous white rot particularly in hardwoods (Fig. 12)*(87)*. Cavity formation and cell wall erosion are normally referred as Types 1 and 2, respectively *(90)*.

Considerable research has been directed towards understanding the manner of cavity formation within the S2 cell walls of fibres *(90-94)*. Studies on the mechanisms of soft rot attack have shown the involvement of specialized microhyphae (*ca* 0.3-0.4 μm thick)(*87, 88, 90-94*) which traverse the S2 layer from the wood cell lumen and then reorientate along the cellulose microfibrils by characteristic T-branching (growth in two directions) or L-bending (growth in one direction). In fact the characteristic alignment of soft rot cavities along the cellulose microfibrils is so well defined that it may be used as a precise means for determining the microfibril angle of individual cell wall layers (S1, S2, S3) in wood fibres *(18, 19, 95)*. Direct penetration of cell walls as described for blue stain fungi may also occur. Following growth of the microhyphae, T-branching or L-bending occurs over a finite distance, and cavities are then developed around the hyphae by the release of enzymes (putatively endocellulases) along what is then described as the proboscis hyphae *(48, 91, 92, 94, 96)*. After cavity formation – size, shape and form dependant on wood species and fungal enzymes – the hyphae produce further microhyphae and the sequence is repeated *(91, 93, 94)*. Particularly, the mechanism of cavity formation in relation to cellulose structure of the cell wall and T-branching/L-bending has fascinated

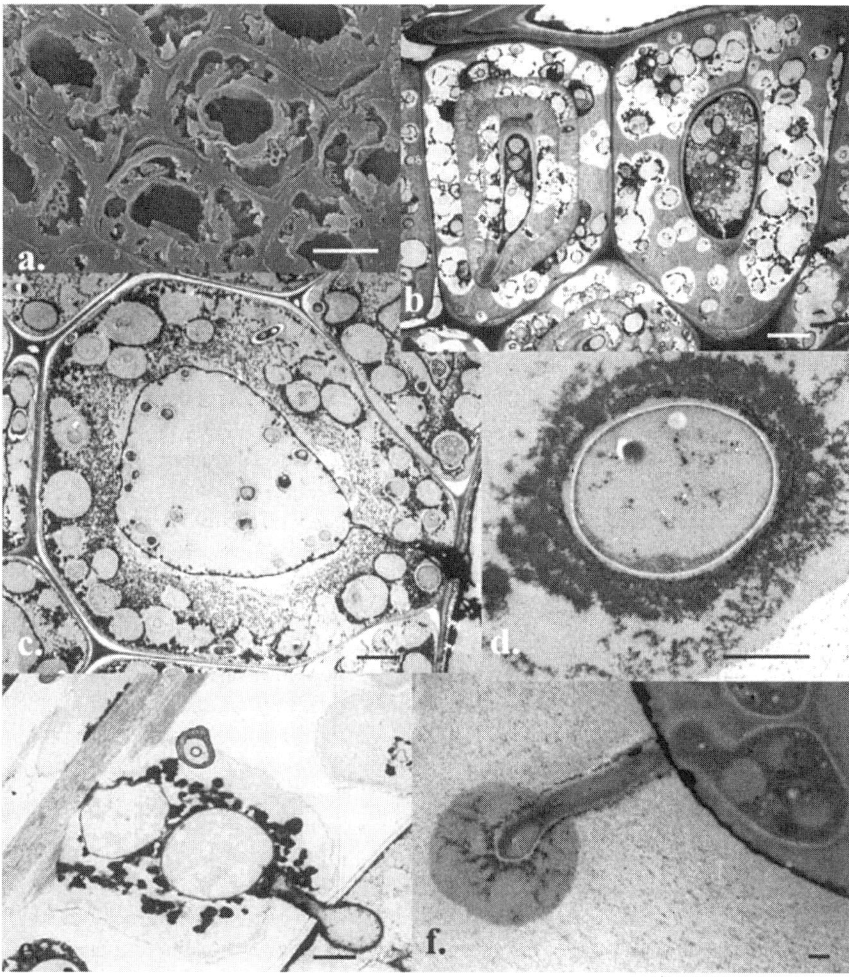

Figure 11. SEM and TEM micrographs showing aspects of soft rot cavity formation within soft- and hardwoods. a) Cross section of pine latewood tracheids with characteristic cavities containing hyphae within the S2 cell wall; b) Cavities within the hardwood Homalium foetium and restricting affect of lignin in concentric layers in the S2 cell wall (left cell); c) Highly degraded CCA treated birch fibre. S2 comprised of remaining CCA, lignin and lysed hyphae; d) Phialophora mutablis hypha encapsulated in melanin, CCA and cell wall breakdown products; e, f) S2 cell wall penetration by lumena hyphae and cavity formation in H. foetium (left) and birch fibres (right). Bars: a, 10.0 µm; b, 2.5 µm; c-f, 1.0 µm.

scientists for years although the process remains poorly understood. An understanding of the mechanism(s) of T-branching/L bending could lead to methods to inhibit this process, preventing cavity formation and resulting in simple cell wall penetration similar to that of blue stain fungi with less economic consequence. Continuous enlargement of existing cavities and the formation of new cavities normally leads to total destruction of the S2 layer (Fig. 11).

While the middle lamella and S3 layers appear easily penetrated by microhyphae, the former is not degraded by soft rot fungi and the S3 is only poorly attacked in softwoods. Both experience this limited attack because of the thickness of the layer and lignin content. SEM and TEM observations on cavity hyphae show the hyphae normally associated with a variety of granular and fibrillar materials including extracellular slime, melanin (*i.e.* pigments) and lignin breakdown products (Fig. 11)*(11, 88)*. These residual materials often remain even after death and lysis of cavity hyphae forming a skeleton in a highly degraded S2 cell wall matrix (Fig. 11). TEM-X-ray microanalytical observations on CCA treated wood degraded by preservative resistant soft rot species such as *Phialophora mutabilis (88)* have shown concentration of CCA in the granular materials surrounding the hyphae. Like TB, whether this reflects direct detoxification by the fungus or concentration in residual cell wall materials (*i.e.* lignin) following cellulose and hemicellulose removal is unknown.

Soft rot erosion (Type 2) (Fig. 12) decay causes a characteristic thinning of wood fibre walls from the cell lumen similar to that caused by white rot decay and higher ascomycete fungi like *Hypoxylon, Xylaria* and *Daldinia (3, 97)*. Decay occurs simultaneously with cavity formation in hardwoods *(1, 3, 87)* and is more frequently described for wood under high moisture conditions or in aquatic situations. Microstructural observations indicate that at advanced stages of decay only remnants of the S1 layer close to the middle lamella remain, forming a skeletal type wall structure *(3)* (Fig. 12). Observations on thin sections of wood using light microscopy have shown differences in white- and soft rot erosion of the lumen wall, the former producing rounded and the latter angular cavities *(150, 151)*.

Possibly the most important factor affecting the micromorphology of soft rot attack is the type and level of lignin. Soft rot fungi generally produce greater decay in hard- than softwoods which is thought to be related to the higher lignin content and guaiacyl type of lignin in softwoods compared with the lower and syringyl-guaiacyl nature of hardwood lignin *(48)*. Soft rot susceptibility appears inversely correlated with lignin level *(48)*. As described above, the architecture of wood cell walls and particularly the molecular arrangement of the cellulose and hemicelluloses and its protection (*i.e.* coating) by lignin probably "regulates" the decay process. Just as the high lignin containing middle lamella layers are not degraded by soft rot fungi (Figs. 11, 12), high lignin containing thin concentric layers within the S2 layers of polylaminate cell walls are initially also poorly attacked (Fig. 11)*(98)*. Possibly one of the main mechanisms for protection of wood by CCA from soft rot is its fixation to lignin which is also inversly correlated with lignin type and level; softwoods retain CCA better and are therefore more protected. When wood cell walls are delignified the remaining wall structure will consist of a cellulose aggregate structure rather like that shown in Figure 3 after chemical delignification or preferential white rot degradation. Such delignified wood cell walls tend to be easily degraded by all cellulolytic fungi, including moulds *(99, 100)* and bacteria *(101, 102)*, while soft rot fungi loose their ability to produce cavities; both features reflecting not only

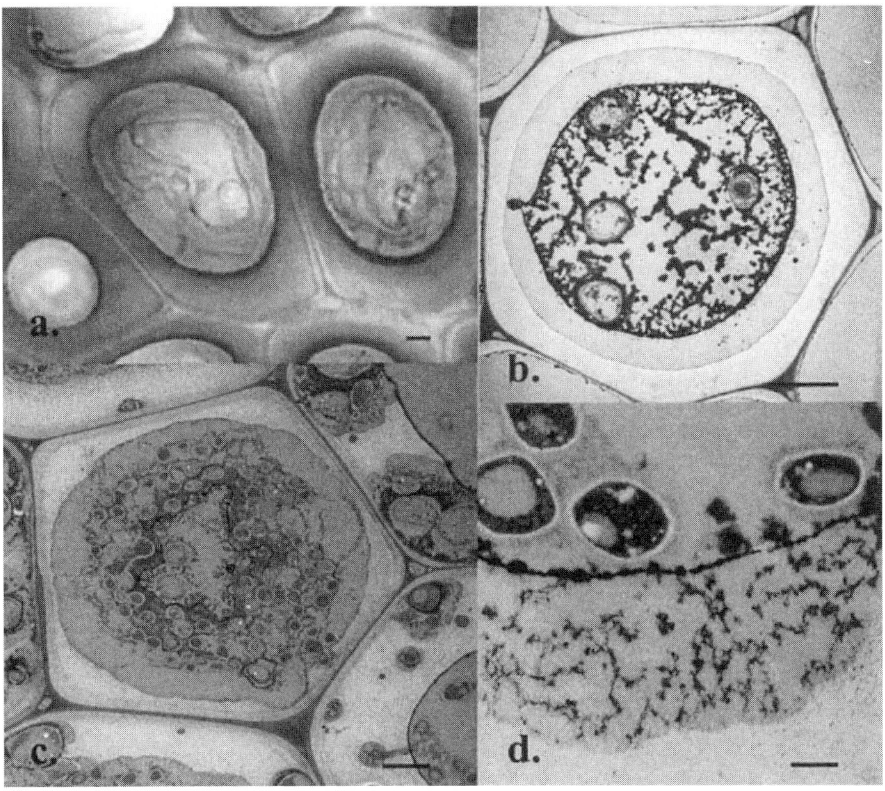

Figure 12. Light (a) and TEM micrographs showing aspects of soft rot erosion decay of pine and birch. a) Erosion of pine S3 and S2 cell wall layers by Chaetomium globosum; b, c) Advanced cell wall erosion of CCA treated birch fibres by hyphal complexes of Phialophora mutablis. Note melanin deposits associated with lumenal hyphae in b. (Reproduced with permission from Daniel, G; Nilsson, T.; Forest Products Biotechnology; *Bruce, A.; Palfreyman, J. W., editors; page 304. Copyright 1998 Taylor & Fran-cis.) d) Soft rot erosion of birch S2 cell wall through a layer of CCA precipitates (black layer) lining the lumen wall. Bars: a-c, 2.0 μm; d, 0.5 μm.*

the importance of lignin type and level but also its molecular location with respect to the cellulose.

White Rot Fungi

Of all the wood decay types, white rot by basidiomycete fungi has been studied at the microstructural level with the greatest intensity over the last 25 years. In addition to pure morphological studies to characterise features of wood cell wall decay, several studies have also been carried out to determine the

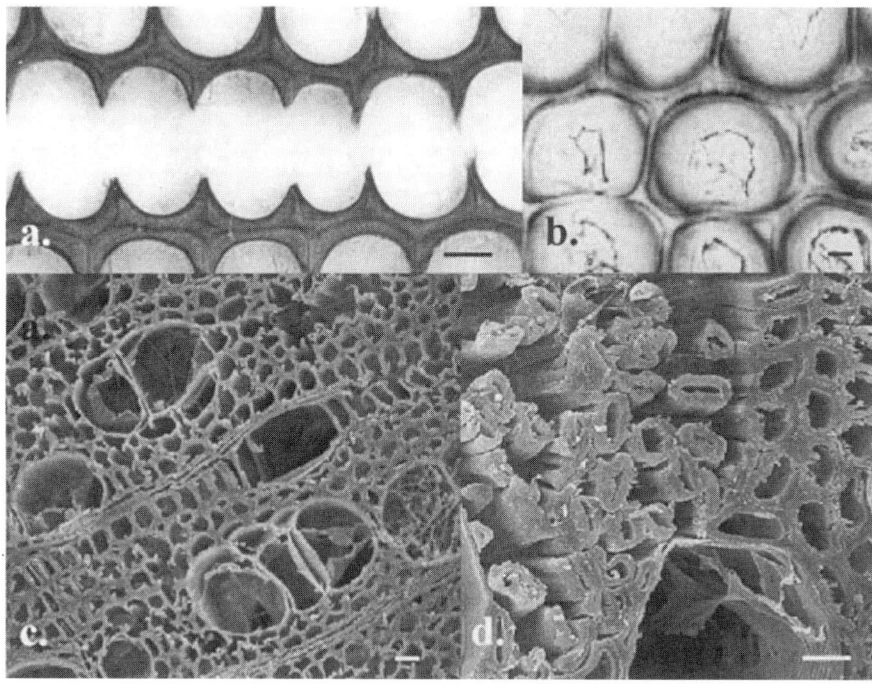

Figure 13. Aspects of simultaneous (a, c) and preferential white rot of pine (b,d) and birch. a) Advanced cell wall thinning of pine leads to early rupture of the tangential cell walls as these are thinner than the radial cell walls; b) Preferential decay of pine tracheids by Phlebia radiata Cel 26, the dark rings indicating "in time and space" sites of lignin and hemicellulose attack; c) Advanced simultaneous attack of birch resulting in total decay and of areas lacking fibres; d) Preferential white rot of birch by Heterobasidium annosum resulting in defibration. Bars: a- c, 5.0 μm; d, 10.0 μm.

in-situ spatial distribution of white rot fungal produced wood cell wall degrading enzymes (*i.e.* ligninolytic, cellulolytic and hemicellolytic) and their fungal origin *(3, 103-111)*. Developments to understand white rot decay have been spurred on by the biotechnological potential of using these fungi or their enzymes in the pulp and paper industry in such processes as biopulping, bleaching and fibre modification in addition to all the other non-lignocellulose possibilities (*see Chp. by Konealey and Jefferies in this book*). In nature, white rot fungi are normally found colonizing hardwoods in terrestrial environments and on the whole, apart from a few species show poor tolerance to salt preservatives under laboratory test conditions *(79)*. Nevertheless, white rot attack of wood has been reported under marine conditions *(112)* and has been recorded on CCA treated *Syncarpia* mooring poles (Daniel and Cookson, unpublished obs), in cooling towers under service conditions *(113)* and test stakes treated with TnBTO-AAC and AAC *(114)*. Our understanding of the microstructural aspects of white rot decay therefore has been undertaken primarily for biotechnological potential rather than for protection of wood.

Two main morphological decay types described almost 100 years ago by Schacht (1863, *115*) as simultaneous and preferential (*i.e.* selective) attack, are the principle ways in which wood cell walls in both hard- and softwoods are degraded by white rot fungi (Fig. 13). Colonization of the wood substrate occurs via rays and penetration between cells via pits and by development of bore holes (Figs 13, 14; *79*). Fungal hyphae are typically localized in the wood cell lumena during the decay process. With simultaneous white rot (*e.g. Phlebia radiata, Phanerochaete chrysosporium,* wild types) all the wood components (cellulose, lignin hemicellulose) are degraded simultaneously from the cell lumen outwards (Figs 13, 14), although decay progresses more rapidly along the comparatively thinner tangential in contrast to the thicker radial- cell walls. As decay progresses outwards, even the lignin rich middle lamella regions are degraded with decay often progressing into adjacent wood cell walls. Decay *in-situ* is usually distinguished at both micro- and ultrastructural levels by a thin, advancing zone (Figs. 13, 14) in which wood cell wall mineralization takes place. This zone is easily demonstrated using stains (*e.g.* safranin) at the light microscopic level but is recognized as only a thin electron lucent layer at the TEM level (Fig. 14). Using immunocytochemical techniques and antibodies, particularly with gold-labelled probes *(3 for review),* it has been possible to localise ligninolytic (*e.g.* lignin peroxidase, Mn-peroxidase, laccase), cellulolytic *(3, 103, 104, 106-111)* hemicellulolytic *(107)* and glucose oxidizing *(105)* enzymes in these zones in direct contact with the degrading wood cell wall (Fig. 14). Similar labelling techniques have also shown the fungal origin of the enzymes and their location (or the location of antigenic subunit components) in the fungal hyphal periplasmic space and intracellularly in association with other cell membranes including double membrane structures *(3)*. A major problem however remains in determining how enzymes move from the lumena-based

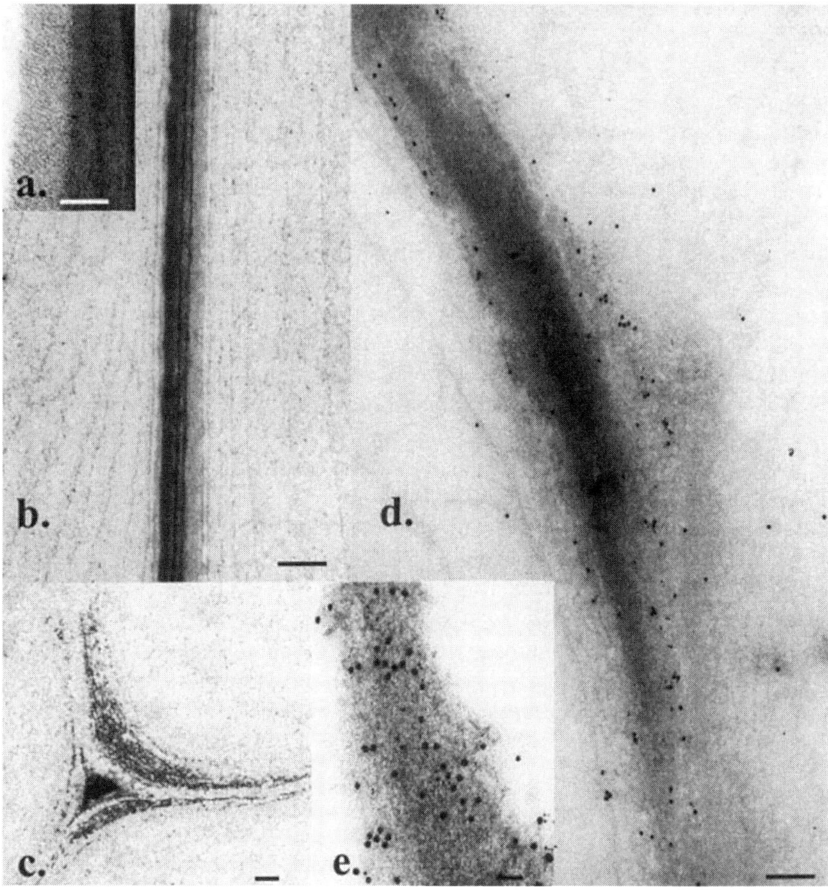

Figure 14. TEM micrographs showing aspects of simultaneous white rot of birch. a) Initial attack of the lumen cell wall and opening up of the wood structure; b) Advanced stage of attack with extracellular slime (seen as layers) filling the cell lumen; c) Remaining middle lamella cell corner encapsulated in slime; d, e) Remaining middle lamella and cell corner regions labelling positively for the presence of the lignin degrading enzymes using gold labelling and antibodies to MnP (large black dots) and laccase (small black dots). Bars: a-d,1.0 μm; e, 0.1 μm.

hyphae to the degrading wood cell wall. TEM immunolabelling and Cryo-FE-SEM with *Phlebia radiata* and *P. chrysosporium* suggest that this may be achieved via extracelluar slime materials which fill the region between the hyphae and the degrading wood cell wall (Fig. 14) and possibly tripartite membranes (*3*). Tripartite membranes have been implicated in this process earlier *(116, 117)*.

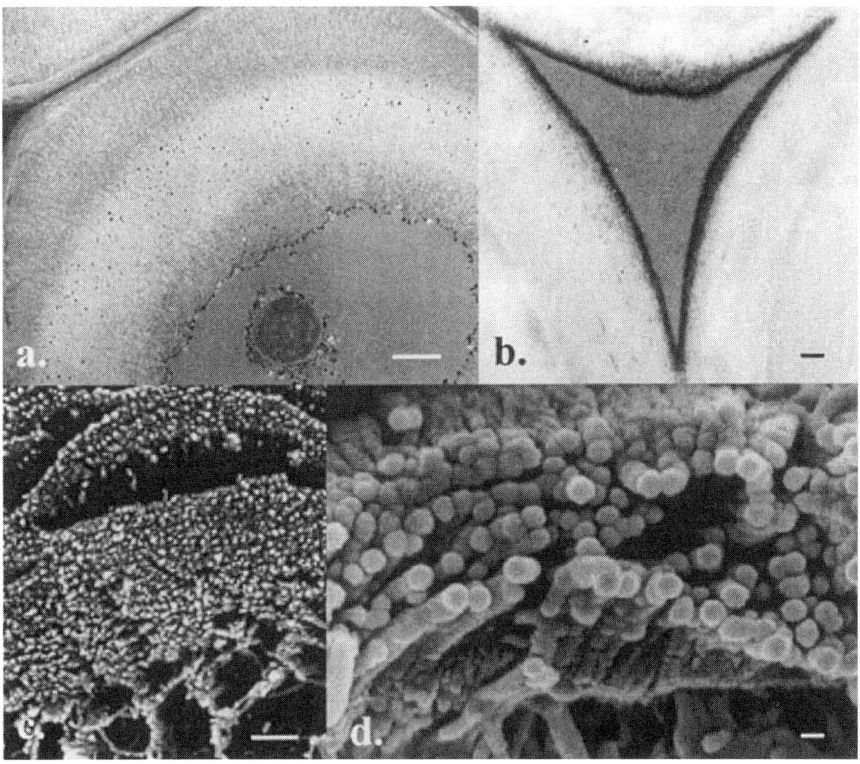

Figure 15. TEM and Cryo-SEM micrographs showing preferential white rot decay of pine and birch secondary cell walls. a) Cross section of pine latewood tracheid showing the characteristic zones (electron lucent regions) produced in the S2 wall during preferential removal of lignin and hemicelluloses by lumenal hyphae; b) Remaining middle lamella corner in birch showing attack of peripheral regions; c, d) Cryo-SEM micrographs showing the remaining macrofibrillar structure comprising the S2 cell wall after lignin and hemicellulose removal during preferential white rot. Bars: a, c, 1.0 μm; b, d, 100 nm.

How products of decay are absorbed by the hyphae is also currently unknown. In advanced stages of decay, the only wood substance remaining is the middle lamella cell corners (normally surrounded by slime)(Fig. 14). These areas are also degraded peripherally (like secondary cell walls) and show evidence for the presence of ligninolytic enzymes involved in decay *(3)*(Fig. 14). Microstructural and immunolabelling observations on simultaneous rot have lead to two recent developments in our understanding of the simulaneous decay process *viz*: that the enzymes involved in decay appear only to be present at sites of attack in which the cell wall is "open", and that they are unable to penetrate into the "unmodified" wood cell wall matrix *(103, 106)*. This is consistent with the known molecular weights (*ca* 40-70,000 Kda) of cellulolytic and ligninolytic enzymes and studies carried out on the porosity changes in wood cell walls during decay *(118)*. Secondly, immuno-studies proved for the first time that enzymes isolated under laboratory culture conditions were actually involved in true wood degradation *in-situ* and have shown that of the two peroxidases (LiP and MnP), MnP has been more frequently and more easily detected than LiP during white rot of wood (*e.g.* in *Phanerochaete chrysosporium, Phlebia radiata*), despite the fact that LiP is able to degrade non-phenolic lignin structures *(119)*.

The microstructural events during preferential white rot decay when lignin and hemicelluloses are selectively and progressively removed across wood cell walls has also received considerable attention over the years, both with native and laboratory developed cellulase mutants (*e.g. P. chrysosporium*; *P. radiata* Cel 26). This type of attack is best observed in the latewood cells of softwoods where the zones of decay are easily recognized as distinct rings progressing across the wood cell walls *(3)*(Fig. 13). Like simultaneous attack, the zones pass through middle lamella regions into adjacent wood cell walls. The location of the rings mark the sites in "*space and time*" with regard to the mineralisation of lignin and hemicelluloses. Unlike simultaneous white rot and despite the fact that similar ligninolytic enzymes have been proven to be involved, they have not been localized in the zones of attack at sites of cell wall attack. They have, however, been found associated with hyphae localized in the wood cell lumen (*3*, Daniel unpublished obs). Staining with $KMnO_4$ (providing some specificity for lignin) and chemically labelling of lignin *(120, 121)* has indicated that lignin decay takes place in the outer regions of the zones and apparent lignin breakdown products have been localized in the decay zones (Figs 13, 15). As decay progresses and the lignin and hemicelluloses are removed, the remaining fibre cell wall comprised of primarily cellulose swells into the cell lumen.

Preferential white rot decay and the inability of known ligninolytic and hemicellulolytic enzymes to penetrate wood fibre cell walls and produce the characteristic decay zones has puzzled scientists for years. The explanation would appear to the involvement of non-enzymatic processes similar to that proposed for brown rot attack (*see below and other chapters in this book*) and involve the action and diffusion of low molecular weight oxidating agents (*e.g.* metals) into the wood cell walls. In particular metals such as Mn, Fe and possibly Cu have variously been reported as being being implicated in the oxidative mechanisms of non-enzymatic attack by both white- and brown rot fungi *(122-125)*. From a microstructural point of view, proof for the involvement of such oxidating agents *in-situ* comes primarily from indirect evidence of their action on fibre cell wall structure rather than evidence for presence of the metal; apart from Mn. The presence of dark flecks (Mn as MnO_2) was shown associated with white rot decay in wood under natural conditions as early as 1878 by Hartig *(126)*. This was confirmed by more recent studies *(3, 127-129)*. Using TEM-X-ray microanalysis, the presence of Mn has been shown to be associated with fungal hyphae (*e.g. Phlebia radiata)*, extracellular slime and to penetrate into the characteristic decay zones during decay by various white rot fungi *(113, 128, 129)*. Manganese within fibre cell walls has also been shown to be associated with bleaching (presumably lignin attack) of the cell wall structure and delamination of the cellulose-lignin matrix *(128, 129)* in both the S2 and middle lamella layers. This decay morphology is similar in appearance to that produced by the oxidative effects of MnII + H_2O_2 and manganic organic complexes *(130)*. Currently it is not known what minimum levels of Mn and which cofactors (*e.g.* H_2O_2) and stabilizers (*i.e.* organic acid, chelators, oxalate, malonate, lactate) may be required to produce preferential cell wall modification. Possibly the source of H_2O_2 may arise enzymatically from H_2O_2 producing oxidases such as pyranose 2-oxidase (P2O) *(105, 131)*. For example, *in-situ* immunolabelling of *P. chrysosporium* P2O, has shown the presence of the enzyme in the periplasmic space of luminal hyphae *(105, 131)*. Biochemical studies on the purified enzyme have shown the ability of the enzyme to carry out double oxidation of sugars and products to produce two molecules of H_2O_2 per molecule of glucose *(132)*. TEM observations of decay zones has often shown a marked orientation of the lamellae in wood cell walls during preferential white rot; an orientation possibly reflecting the underlying wood cell wall ultrastructure. This suggests that decay may follow the original pathway of fibre cell wall biosynthetic development. More detailed ultrastructural work is required to understand non-enzymatic attack of wood cell walls by white rots.

Unfortunately in typical TEM sections the appearance of white rotted wood cell walls gives no indication of the cellulose macrofibrils or aggregates named above (*e.g.* see Fig. 15), but rather of a homogeneous wood structure. Such aggregates have now been detected in a wide range of fibres using different preparation processes. Rapid freezing in liquid nitrogen slush *(i.e.* -210 C) or propane with observations made under cryo-conditions is possibly the best way of visualizing these structures without artifacts. These structures indicate that the matrix materials *(i.e.* lignin/hemicelluloses) between the aggregates are removed. Cellulose aggregates have also been demonstrated in resin sections of pulp fibres *(i.e.* lignin removed) using AFM and using SEM of undegraded wood samples fractured under tension *(28, 30).*

Brown Rot Fungi

Brown rot fungi are characterized by their extensive and rapid depolymerization of cellulose leading to a rapid loss in wood strength at early stages of the decay process *(1-3, 122-124).* Of all wood decay types, brown rot fungi are generally considered the most important rot type for wood *in-service,* whether in ground contact such as transmission poles or in dwellings, particularly in house cellars where *Serpula lacrymans,* the causal agent of dry rot is well documented for its destructive effects *(133).* Brown rot fungal attack is normally characterized by an excessive removal of cellulose and hemicelluloses and although demethylation of lignin occurs (degree dependant on fungal species) the lignin is normally left over as a weak amorphous residual skeleton which easily fractures cubically and crumbles when dry *(1, 2).* A major characteristic of brown rot attack is that both the level and type of lignin has very little effect on decay and since the levels of cellulose are similar in hard- and softwoods, similar weight losses are normally achieved during laboratory decay tests with these fungi *(134).*

Brown rot-, like white rot fungi, typically colonize wood tissues via the rays, (Figure 16) from where the hyphae ramify out into the axial wood structure penetrating through cells using pits or by producing bore holes. In softwoods *(e.g.* pine), the characteristic degradation of the window pit membranes by brown rot fungi such as *Gloeophyllum trabeum* produces a diamond decay pattern reminiscent of soft rot erosion *(3).* As similar attack profiles can be produced using purified cellulases, this further indicates that the cross-field pit membranes are non-lignified.

Microstructural and ultrastructural studies on brown rotted wood samples have provided unique evidence to support that: i) close proximity between hyphae and the wood cell wall is not necessary for cellulose/hemicellulose depolymerization; decay radiating out across wood cells in a manner similar to that of preferential white rot attack, the loss in cell wall crystallinity indicated by a loss in birefringence using polarized light (Figure 16); ii) that the agent(s) responsible for decay are able to diffuse through the S3 and S2 layers causing initial morphological changes at the S2-S1 layer interface (Figure 16); iii) that the depolymerization agent is likely to be a low molecular agent (probably oxidizing

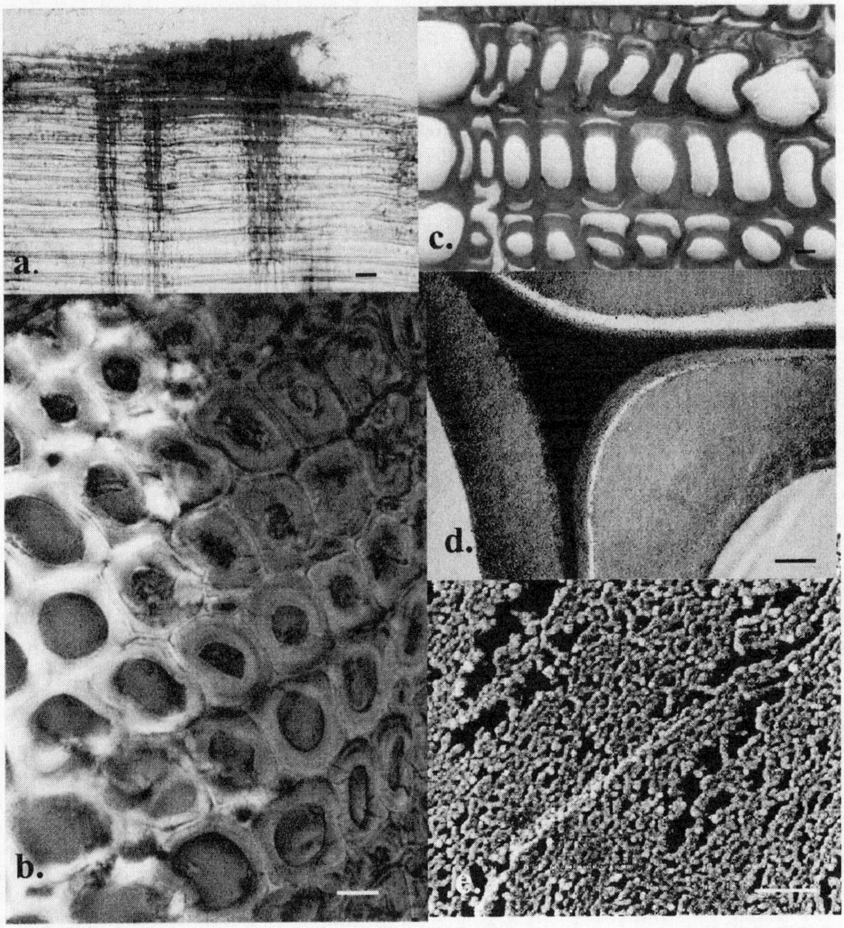

Figure 16. Light and electron micrographs showing aspects of wood cell wall degradation by the brown rot fungus G. trabeum. a) Colonization of rays by G. trabeum: b) Attack of pine latewood tracheids and loss of loss of birefringence due depolymerization of crystalline cellulose; c) Advanced decay of pine with almost total decay of fibres which irreversibly shrinks after drying; d) Early decay of pine latewood with first signs of attack at the S1/S2 interface; e) Cryo FE-SEM observations on birch fibre degraded by G. trabeum leaving a lignin skeleton without "macrofibrillar" structures (cf Fig. 13 showing preferential white rot). Bars: a, b, c, *5.0 μm; d, 1.0 μm; e, 100 nm.* **(This figure is also in color insert.)**

agent), at least in the initial stages of degradation, due to the rapidity of the decay process and the apparent tight and complex structure of the S2 layer in softwoods (Plate. 16); iv) the demethylation of lignin as shown by the reaction of dihydroxyphenol substances in brown rotted cell walls with OsO_4 forming electron dense precipitates *(135);* v) the selective removal of polysaccharides from middle lamella regions (*e.g.* cell corners) *(36)*; a feature also consistent with a non-homogeneous distribution of components in the middle lamella; and finally, vi) the association of extracellular membrane structures (sheaths, mycofibrils, tripartite membranes) like white rot fungi with hyphae during decay processes *(116, 117)*.

According to the evidence put forward at the beginning of this chapter on the importance of wood ultrastructure in regulating decay, TEM observations on rapidly frozen, freeze-substituted and resin embedded samples and cryo-FE-SEM observations have confirmed cellulose/hemicellulose removal from wood cell walls and an apparent localized and concomitant increase in cell wall porosity (Plate 16). TEM observations of S2 layers at high magnification show remaining lignin aggregated around electron-lucent regions previously containing cellulose and hemicellulose; while at low magnifications the lignin often appears organized into lamellae complementing the original organization of the cellulose (Daniel, unpublished obs). Observations of rapidly frozen and fractured brown rotted wood cells observed under cryo-conditions (samples retained at –120 C) provides further evidence for an increase in porosity and the organization of remaining lignin into globular structures (Plate 16). Degraded wood cell walls also lack the characteristic cellulose macrofibrillar structure shown with cryo observations on preferential white rotted wood samples (*cf* Fig. 15; Daniel unpublished obs) discussed above. A perplexing microstructural problem hindering the understanding of brown rot decay is the demonstration *in-situ* of the non-enzymatic agents *viz* oxalic acid *(123, 136),* low molecular weight chelators *(137, 139)*, Fenton reagent *(122),* glycoproteins *(138),* and enzymes (cellulases, xylanases) variously implicated in decay. Currently, how these agents function *in-situ* to degrade amorphous and crystalline cellulose is poorly understood. Demonstration of the agents *in-situ* is surrounded with technical difficulties such as loss and movement (*i.e.* diffusion) of the low molecular agents from sites of activity during the preparation of samples for electron microscopy and difficulties of assuming that the agents are initially present in the wood tissue. Both variability in the physiological traits of different brown rot fungi (*e.g.* some brown rot fungi produce exoglucanases (*Coniophoroid* spp.) others do not, others strongly reduce the pH of the substrate (*Postia* spp.) while still other species (*e.g. G. trabeum*) retain the wood at a slightly acid pH. A further complication is the wood structure itself.

Ultrastructural observations show distinct dissimilarities in S2 structure following attack by the same fungus. For example, birch fibres often show an open structure after brown rot attack while softwoods tend to retain a rather dense and compact cell wall matrix *(3)*. This difference is probably related to the degree and type of lignification as described above. The open structure of the birch cell walls after decay would not seem to present a barrier to enzymatic diffusion. Possibly, the lignified and compatively thicker S3 layer in softwood tracheids, in contrast to that found in hardwood fibres like birch and poplar and which remains during decay, also influences enzymatic diffusion.

Immunological studes using antibodies to brown rot fungal extracellular metabolites have nevertheless confirmed their penetration into wood cell walls *(140, 141)* and their association with membrane structures in fungal hyphae. Studies with Fenton reagents and metals have further given morphological indications for attack in wood cell walls at remote sites (*i.e.* S1-S2 interphase) similar to that recognized during fungal attack of wood (*6-8, 141*).

The mechanism(s) of brown rot decay and their visualization *in-situ* remains a major challenge for ultrastructural studies. While considerable effort is being concentrated on the perplexing aspects of depolymerization agents (types, diffusion etc) (*147-143*), possibly more effort should be given to the involvement of enzymes as a possible indirect source of depolymerization agents (*e.g.* H_2O_2 and radicals). For example as described above, certain brown- like white rot fungi, are known to produce H_2O_2 via oxidation of glucose and other sugar oxidoreductases like pyranose oxidase *(144)*. Possibly this enzyme could represent a source of H_2O_2 for Fentons reagent and the hydroxyl radicals postulated over the years as involved in brown rot depolymerization *(122)*.

Cavity Formation by Basidiomycete Fungi

In addition to the usual patterns of basidiomycete white- and brown rot attack where the hyphae are normally restricted to the cell lumen (apart from bore holes), evidence is mounting for certain of these fungal types to produce cavities (reminiscent of soft rot cavities) within the S2 layer of wood fibres *(144-152)*. This type of decay is reflected by the development (at least initially) of very thin hyphae (*ca* 0.3-0.5 μm) within the S2 wall layers of wood tissues (*e.g. Oudemansiella mucida*; *147*). These hyphae subsequently cause decay, which for white rot typically results in the development of small cavities, which may or may not be aligned with the cellulose microfibrils as indicated above for soft rot fungi *(147)*. With certain species (*e.g. O. mucida*) cavities are produced following L/T-branching of the hyphae like soft rot fungi within the S2 layer *(147)*. With brown rot fungi such as *G. trabeum,* hyphal development and cavities can also be occasionally observed within the S2 layer of wood fibres, but here the remaining wood (primarily lignin) substance may have direct contact with hyphae but typical angular cavities like soft rot have so far not been observed. More recent work by Kleist and Schmitt *(149)* have shown characteristic soft rot cavities produced by *Coniophora puteana* in *Sapelli* wood. Cavity development by white and brown rot fungi has also been noted for basidiomycete pathogens attacking trees *(148)* and for pure cultures attacking treated wood materials (*145*; Daniel unpublished). An interesting feature of this type of attack is the development of very fine hyphae in the S2 layer, not dissimilar to that produced by soft rot fungi during T/L branching and that by blue stain fungi and basidiomycetes during transwall and bore hole hyphal development. The development of such hyphae by wildly different fungal species from different taxonomic groups suggests a common feature for the attack and penetration into wood cell walls. Possibly it is a mode of attack which develops in wood at high moisture content, such as seen with cavity formation by soft rot fungi. Further studies are needed to develop an understanding of this decay process.

Summary

Microstructural studies on morphological features of the wood cell wall undergoing degradation by fungi and bacteria over the last 20 years have been invaluable in confirming decay patterns and the finer details of the processes involved. Electron microscopy, for example, has been used to: confirm the pathways of fungal colonization into wood; determine that single celled bacteria are capable of wood decay; show that fungal hyphae need not have close proximity to the wood cell walls to cause decay; elucidate stages in soft rot cavity development and T-branching; demonstrate the interactions between wood preservatives in cell walls and fungi/bacteria; and the *in-situ* visualization of the involvement of ligninolytic and cellulolytic *in-situ* at sites of wood decay. Microstructural results emphasize the important need for an improved understanding of wood cell wall ultrastructure (*i.e.* morphology, chemistry, physical properties). A better understanding of wood ultrastructure will allow increased understanding of: i) mechanisms of wood decay *in-situ*; ii) development of better protective measures; and iii) biotechnological exploitation of ligno-cellulolytic fungi and bacteria.

References

1. Blanchette, R.A.; Nilsson, T.; Daniel, G; Abad, A. In *Archaeological Wood Properties, Chemistry and Preservation;* Rowell, R.M.; Barbour, R.J., Eds.; American Chemical Society, Los Angeles. **1990**, pp. 141-177.

2. Eriksson, K.-E.; Blanchette, R.A.; Ander, P. *Microbial and Enzymatic Degradation of Wood and Wood Components.* Springer Series, Berlin, **1990**, pp 407.
3. Daniel, G. *FEMS Microbiol. Rev.* **1993**. *13*, 199-23.
4. Liese, W. *Ann. Rev. Phytopathology*, **1970**, *8*, 231-258.
5. Fengel, D. Wegener, G. *Wood Chemistry, Ultrastructure, Reactions*, Walter de Gruyter, Berlin **1984**. pp. 613.
6. Ilvessalo-Pfäffli, M-S. *Fiber Atlas: Identification of Papermaking Fibers.* **1995**, Springer Series, Berlin, pp. 400.
7. Parham, R.A.; Gray, R.L. In *The Chemistry of Solid Wood*; Rowell, R. Ed.; American Chemical Society, **1984**, Seattle, pp. 540.
8. Panshin, A. J. Zeeuw, C. de. *Textbook of Wood Technology;* 4th Ed. McGraw-Hill Book Company, New York. **1980**, pp. 705.
9. Fujita, M.; Harada, H.: In *Wood and Cellulosic Chemistry*; Hon, D.N.-S.; Shiraishi, N.; Eds; Marcel Dekker, Inc, NY, **1991**, pp. 3-57.
10. Nilsson, T.,; Obst, J.; Daniel, G. Int. Res. Group Wood Preserv. **1988**, Document No. IRG/WP/1357.
11. Daniel, G.; Nilsson, T. Int. Res. Group Wood Preserv. **1987**. Document No. IRG/WP/1338.
12. Abe, H.; Ohtani, J.; Fukazawa, K. *IAWA Bull.* **1991**, *12*, 431-438.
13. Hafrén, J.; Fujino, T.; Itoh, T. *Plant Cell Physiol.* **1999**, *40*, 532-541.
14. Fujino, T.; Itoh, T. *Holzforschung*, **1998**, *52*, 111-116.
15. Hult, E.-L.; Larsson, P.T.; Iversen, T. *Holzforschung* **2001**, *2*, 56 -61.
16 Duchesne, I.; Daniel, G. *N. P. P. J.* **2000**, *15*, 54-61.
17. Fahlén, J.; Salmén, L. In *Proceedings of COST Action E20 Workshop, Interaction Between Cell Wall Components*, 26th-28th April **2001** at SLU, Uppsala, Sweden.
18. Daniel, G.; Nilsson, T. Department Forest Products Report Series, Swedish University of Agricultural Sciences; **1984**, 184, pp. 34.
19. Anagnost, S.E.; Mark, R.E.; Hanna, R. B. *Wood Fiber Sci.* **2000**, *32*, 81-87.
20. Bailey, I.W.; Vestal, M.G. *J. Arnold Arboretum* **1937**, *18*, 196-205.
21. Khalili, S.; Nilsson, T.; Daniel, G. *Holz Roh- Werkst.* **2000**, *58*, 439-437.
22. Brandström, J. Ph.D Thesis, Department of Wood Science, Swedish University of Agricultural Sciences, Uppsala, Sweden, **2002**.
23. Cote, W.A. Jr. *Cellular Ultrastructure of Woody Plants*, Syracuse; Syracuse University Press, **1967**, pp. 603.
24. Donaldson, L. Wood Sci.Technol. **1987**, *21*, 227-234.
25. Brown, R.M. Jr. *Pure Appl. Chem.* **1999**, *71*, 767-775.
26. Hanna, R.B.; Côte, W.A. *Cytobiol.* **1974**, *10*, 102-116.27.

27. Daniel, G.; Duchesne, I. In Proceedings of the 7th Int. Conf. On *Biotechnology in the Pulp and Paper Industry, Vancouver, Canada,* **1998,** pp. 81-85.
28. Sell, J.; Zimmerman, T. *Holz Roh- Werkst.* **1993**, 51, 384.
29. Hult, E.-L.; Larsson, P.T.; Iversen, T. *Cellulose* **2000**, 7, 33-55.
30. Fahlén, J. Licentiate Thesis, Department of Pulp and Paper Chemistry and Technology, Royal Institute of Technology, Stockholm, **2002**.
31. Kerr, A.J.; Goring, D.A.I. *Cellulose Chem. Technol.* **1975**, 9, 563-573.
32. Salmén, L.; Olsson, A.-M. *J. P. P. S.* **1998**, *24*, 99-103.
33. Duchesne, I.; Hult, E.-L., Molin, U.; Iversen, I.; Lennholm, H. *Cellulose*, **2001**, *8*, 103-111.
34. Daniel., G.; Volc, J.; Niku-Paavola, M.-L. *Abstracts 8th International Conference on Biotechnology in the Pulp and Paper Industry, Finland.* **2001**, pp. 42-44.
35. Molin, U.; Daniel, G. *Holzforschung* **2002** *(Submitted).*
36. Daniel, G.; Nilsson, T.; Pettersson, B. *IAWA Bull.* **1991**, *12*, 70-83.
37. Bailey, I.W. *Indust. Engineer. Chem.* **1938**, *30*, 40-47.
38. Pizzi, A. *J. Polymer Sci.*, **1982**, *20*, 739-764.
39. Daniel, G.; Nilsson, T. *Int. Biodet.* **1988**, 24, 327-335.
40. Nilsson, T.; Henningsson, B. *Mater. Org.* **1978**, *13*, 297-313.
41. Saka, S.; Thomas, R.J. *Wood Sci. Technol.* **1982**, *16*, 1-118.
42. Saka, S.; Goring, D.A.I. In *Biosynthesis and Biodegradation of Wood Components*; Higuchi, T.; Ed.; Academic Press, Orlando, 1985, pp. 51-62.
43. Saka, S. In *Wood and Cellulosic Chemistry*; Hon, D.N.-S.; Shiraishi, N.; Eds; Marcel Dekker, Inc, New York, **1991**, pp. 59-88.
44. Liese, J. 1950. In *Handbuch der Holzkanservierun. 3rd edit.* Mahike, F.; Troschel, R.; Liese, J.; Eds; Berlin, **1950**; pp. 44-111.
45. Holt, D. M. *J. Inst. Wood Sci.* **1983**, *9*, 212-223.
46. Daniel, G.; Nilsson, T.; Singh, A. *Can. J. Microbiol.* **1987**, *33*, 943-948.
47. Singh, A.P.; Butcher, J.A. *J. Inst. Wood Sci.* **1991**, *12*, 143-157
48. Daniel, G.; Nilsson, T. In *Biotechnology in Study of Soft Rot and Bacterial Decay;* Bruce, A.; Palfreyman, W.; Eds Taylor and Francis publishers, Salisbury, UK.; **1997**, pp. 37-63.
49. Singh, A.; Nilsson, T.; Daniel, G. *J. Inst. Wood Science,* **1987**, *15*, 361-378.
50. Daniel, G.; Nilsson, T. *Int. Res. Group Wood Preserv.* **1985**, IRG/WP/1283.
51. Nilsson, T. *Int. Res. Group Wood Preserv.* **1984**, IRG/WP/1234.
52. Nilsson, T.; Singh, A.; Daniel, G. *Holzforschung* **1992**, *46*, 361-367.

53. Daniel, G.; Elowson, T.; Nilsson, T.; Singh, A.; Liukko, K. *Int. Res Group Wood Preserv.* **1993**, IRG/WP/93, 10033.
54 Boutelje, J.; Jonsson, U. *STFI – Meddelande Serie A*, **1976**, nr 376 (TT52).
55 Bauch, J.; Liese, W.; Berndt, H. *Holzforschung* **1990**, *24*, 199-205.
56. Boutelje, J.; Henningsson, B.; Lundström, H. *STFI Meddelande Serie A*, **1977**, Nr 435 (TT A:70).
57. Bergman, Ö.; Henningsson, B.; Persson, E. *Svenska Träskyddsinstitutet Meddelande*, **1975**, Report No, 132.
58. Bergman, Ö.; Martinsson, S. *Svenska Träskyddinstitutet Meddelande*, **1979**, 119.
59. Nilsson, T.; Daniel, G. In *3rd Int. Conf. Biotechn. in the Pulp and Paper Ind., Proceedings,* **1986**, 54-57.
60 Nilsson, T.; Daniel, G. *Int. Res. Group Wood Preserv.* **1983**, IRG/WP/1186.
61. Daniel, G.F.; Nilsson, T.; *Int. Res. Group Wood Preserv.* **1986**, IRG/WP/1283.
62. Björdal, C.G.; Daniel, G.; Nilsson, T. *Int. Biodeter. Biodegrad.* 2000, *45*, 15-26.
63. Forsberg, C.W.; Beveridge, T.J.; Hellström, A. *Appl. Environ. Microbiol.,* **1981**, *42*, 886-896.
64. Lamed, R.; Setter, E.; Bayer, E. *J. Bacter.* **1983**, *156*, 828-836.
65. Kim, Y.S.; Singh, A. *Mokuzai Gakkaishi*, **1994**, *40*, 554-562.
66. Riess, W.; Daniel, G. *Int. J. Nautical Arch.* **1997**, *26*, 330-338.
67. Kim, Y.S. *Mater. Org.* **1989**, *24*, 271-286.
68. Björdal, C.G.; Daniel, G.; Nilsson, T. *Int. Biodeter. Biodegrad.* **1999**, *43*, 63-71.
69. Boutelje, J.B.; Bravery, A.F. *J. Inst. Wood Sci.*; **1968**, *20*, 47-57
70. Pitman, A.; Cragg, S.M.; Daniel, G. *Mater. Org.* **1997**, *31*, 281-293.
71. Powell, K.L.; Pedley, S.; Daniel, G.; Corfield, M. *Int. Biodeter. Biodegrad.* **2001**, *47*, 165-173.
72. Holt, D.M.; Jones, E.B.G.; Furtado, S.E.I. *Rec. Ann. British Wood Pres. Assoc.* **1979**, pp. 13-24.
73. Calvalcante, M.S. Ph.D thesis, Dept. Biological Sciences, Portsmouth University, England, **1981**.
74. Safo-Sampah, S. Ph.D Thesis, University of California, Berkeley, **1985**.
75. Baecker, A.A.W.; King, B. *J. Inst. Wood Sci.* **1981**, *9*, 65-71.
76. Nilsson, T.; Daniel, G.; Bardage, S. *Int. Res Group Wood Preserv.* **1990**, IRG/WP/1444.
77. Adhi, T.P.; Korus, R.A.; Crawford, D.L. *Appl. Environ. Microbiol.* **1989**, 55, 1165-1168.

78. Ramachandra, M.; Crawford, D.L.; Hertel, G. *Appl. Environ. Microbiol.* **1988,** 54, 3057-3063.

79. Eaton, R.A.; Hale, M.D. *Wood Decay Pests and Protection*, Chapman and Hall, London, **1993**, pp. 546.

80. Hendry, K.M.; Cole, E.C. *J. Toxicol. Environ. Health* **1993**, *38*, 183-198.

81. Encinas, O.; Daniel, G. *Mater. Org.* **1997**, *30*, 239-259.

82. Encinas, O.; Henningsson, B.; Daniel, G. *Holzforschung* **1998**, *52*, 82-88.

83. Bardage, S.L. *Holz Roh- Werkst.* **1998**, *56*, 359-364.

84. Encinas, O.; Daniel, G. *Mater. Org.* **1995**, *29*, 255-272.

85. Bardage, S.; Daniel, G. *Mater. Org.* **1997**, *31*, 233-247.

86. Nilsson, T. *Stud. For. Suec.* **1973**, 104.

87. Mouzouras, R. *J. Inst. Wood Sci.* **1989**, *11*, 193-201.

88. Daniel, G.; Nilsson, T. *J. Inst. Wood Sci.* **1989**, *11*, 162-171.

89. Savory, J.G. *Ann. Appl. Biol.* **1954**, *44*, 336-347.

90. Corbett, N.H. *J. Inst. Wood Sci.* **1965**, *14*, 18-29.

91. Leightley, L.E. Ph.D. Thesis, Department of Biological Sciences, Portsmouth, University, Porstmouth, England, **1977**.

92. Hale, M.D. Eaton, R.A. *Int. Res. Group Wood Preserv.* **1984**, IRG/WP/1227.

93. Hale, M.D.; Eaton, R.A. *Mycologia*, **1985**, *77*, 447-463.

94. Hale, M.D.; Eaton, R.A. *Mycologia*, **1985**, *77*, 594-605.

95 Bergander, A.; Bränström, J.; Daniel, G.; Salmén, L. *Japan J. Wood Science*, **2001** *(In Press)*.

96. Khalili, S.; Nilsson, T.; Daniel, G. *Holz als Roh-und Werkstoff* **2001**, *58*, 439-447.

97. Nilsson, T.; Daniel, G.; Kirk, K.; Obst, J. *Holzforschung* **1989**, *43*, 11-18.

98. Daniel, G.; Nilsson, T. In *Recent Advances in Wood Anatomy;* Donaldson, L.A.;.Singh, A.P., Butterfield, B.G.; Whitehouse, L.J.; Eds; **1994**, pp 369-373.

99. Zainal, A.S. *Mater. Org.* **1975**, *3*, 121-127.

100. Zainal, A.S. *Mater. Org.* **1976**, *11*, 295-301.

101. Holt, D.M.; Jones, E.B.G. *Mater. Org.* **1978**, *13*, 13-30.

102. Singh, A.; Nilsson, T.; Daniel, G. *Mater. Org.* **1987**, *22*, 257- 269.

103. Daniel, G.; Nilsson, T.; Pettersson, B. *Appl. Environ. Microbiol.* **1989**, *55*, 871-881.

104. Daniel, G.; Pettersson, B.; Nilsson, T.; Volc, J. *Can. J. Bot.* **1990**, *68*, 920-933.

105. Daniel, G.; Volc, J.; Kubatova, K.; Nilsson, T. *Appl. Environ. Microbiol.* **1992**, 58, 3667-3676.

106. Srebotnik, E.; Messner, K.; Foisner, R. *Appl. Environ Microbiol.* **1988,** *54*, 2608-2614.
107. Blanchette, R.A.; Abad, A.R.; Farrell, R..L.; Leathers, T.D. *Appl. Environ. Microbiol.* **1989,** *55*, 1457-1465.
108. Ruel, K.; Odier, E.; Joseleau, J.P. In *Proc. of Biotechnology in Pulp and Paper Manufacture;* Kirk, T.K.; Chang, H.M.; Eds.; Butterworths, Heinemann, Boston, **1990,** pp. 83-99.
109. Daniel, G.; Jellison, J.; Goodell, B.; Paszczynski, A.; Crawford, R. *Appl. Microbiol. Biotech.* **1991,** *35*, 674-680.
110. Daniel, G.; Pettersson, B.; Volc, J.; Nilsson, T. In *Proceedings of Biotechnology in Pulp and Paper Manufacture*; Kirk, T.K.; Chang, H.M.; Eds. Butterworth-Heinemann, Boston, **1990** pp. 99-111.
111. Evans, C.; Gallagher, I.M.; Atkey, P.T.; Wood, D.A. *Biodegradation* **1991,** *2*, 93-106.
112. Leightley, L.E. *Bot. Marina,* **1980,** *23*, 387-395.
113. Schmidt, O.; Schmitt, U.; Moreth, U.; Potsch, T. *Holzforschung* **1997,** *51*, 193-200.
114. Daniel, G.; Bergman, Ö. *Holz als Roh. Werkstoff* **1997,** 55, 197-201.
115. Schacht, H. *Jahrb. Wiss Bot.* **1863,** *3*, 442-483.
116. Foisner, R.; Messner, K.; Stachelberger, H.; Röhr, M. *Trans. Br. Mycol. Soc.* **1985,** *85*, 257-266.
117. Palmer, J.G.; Murmanis, L.; Highley, T.L. *Mycologia* **1983,** 75, 1005 1010.
118. Flournoy, D. S.; Kirk, T.K.; Highley, T. *Holzforschung* **1991,** 45, 383-388
119. Kirk, T.K.; Tien, M.; Kersten, P.J.; Mozuch, M.D.; Kalyanaraman, B. *Biochem. J.* **1985,** *226*, 455-460.
120. Blanchette, R.A.; Otjen, L.; Carlson, M.C. *Phytopathology* **1987,** 77, 684-690.
121. Otjen, L.; Blanchette, R.A. *Holzforschung* **1988,** *42*, 281-288.
122. Koenigs, J.W. *Wood Fiber* **1974,** *6*, 66-80.
123. Shmidt, C.J.; Whitter, B.K.; Nicholas, D.D. In *Proceedings Am. Wood Preserv. Assoc.*, **1981,** *77*, 157-164.
124. Ilman, B.L.; Meinholtz, D.C.; Highley, T.L. In *Biodeterioration Research II.* O′ Rear, C.E.; Llewellyn, D.C.; Eds. Plenum Press, New York, pp. 485-496, **1988.**
125. Hammel, K.E.; Tardone, P.J.; Moen, M.A.; Price, L.A. *Arch. Biochem. Biophys* **1989,** *270*, 404-496.
126. Hartig, R.; *Die Zersetzungserscheinungen des Holzes der Nadelholzbäume und der Eiche in Forstlicher, Botanisher und Chemischer Richtung*, Springer, Berlin, Heidelberg, New York, 1878.
127. Blanchette, R.A.; *Phytopathology*, **1984,** *74*, 725-730.

128. Daniel, G.; Bergman, Ö. *Holz Roh- Werkst.* **1997**, *55*, 197-201.
129. Daniel, G.; Nilsson, T.; Volc, J. *Can. J. Microbiol.* **1997**, *43*, 663-671.
130. Joseleau, J.P.; Gharibian, S.; Ruel K. *FEMS Microbiol Rev.* **1993**, *13*, 255-264.
131. Daniel, G.; Volc, J.; Kubatova, K. *Appl.Environ. Microbiol.* **1994**, *60*, 2524-2532.
132. Volc, J.; Sedmera, P.; Havlicek, V.; Prikrylova, V.; Daniel, G. *Carb. Res.* **1995**, *278*, 59-70.
133. Bech-Andersen, J. In *The Dry Rot Fungus and Other Fungi in Houses*; Hussvamp Laboratoriet ApS, **1980**.
134. Nilsson,T.; Daniel, G. 1987 *Int. Res Group Wood Preserv.* **1987**, IRG/WP/1320.
135. Messner, K.; Srebotnik, E.; Stachelberger, H. In *Proceedings 3rd International Conference on Biotechnology in the Pulp and Paper Industry*; Eriksson, K.-E.; Ander, P.; Eds. pp.17-19. Pub. Swedish Forest Products Research Lab. Stockholm, **1986**.
136. Green, F.; Hackney, J.M.; Clausen, C.A.; Larsen, M.J.; Highley, T.L. *Int. Res. Group Wood Preserv.* **1993**, IRG/WP/93-10028.
137. Jellison, J.; Chandhoke, V.; Goodell, B.; Fekete, F.A. *Appl. Microbiol. Biotechnol.* **1991**, *35*, 805-809.
138. Enoki, A.; Tanaka, H.; Fuse, G. *Wood Sci. Technol.* **1989**, *23*, 1-12.
139. Fekete, F. A.; Chandhoke, V.; Jellison, J. *Applied Environ. Microbiol.* **1989**, *55*, 2720-2722.
140. Jellison, J.; Chandhoke, V.; Goodell, B.; Fekete, F. A. *Applied Microbiol. Biotech.,* **1991**, *35*, 805-809.
141. Goodell, B.; Jellison, J.; Liu, J.; Daniel, G.; Paszczynski, A.; Fekete, F. Krishnamurthy, S.; Jun, L.; Xu, G.; *J. Biotechnol.* **1997**, *53*, 133-162.
142. Xu, G.; Goodell, B. *J. Biotechnol.* **2001**, *87*, 43-57.
143. Goodell, B.; Qian, Y.; Jellison, J.; Richard, M.; Qi, W. *Biotechnology in the Pulp and Paper Industries,* **2002**.
144. Izumi, Y.; Furuya, Y.; Yamada, H. *Agric. Biol. Chem.* **1990**, *54*, 799-801.
145. Duncan, C. G. *For. Serv. Rep. FPL-2173*, **1960**, 70 pp.
146. Liese, W.; Schmid, R. *Angew. Bot.* **1962**, *36*, 291-298.
147. Daniel, G.; Volc, J.; Nilsson, T. *Mycol. Res.*, **1992**, 96, 49- 54.
148. Schwarze, F.W.M.R.; Lonsdale, D.; Fink, S. *Mycol. Res.* **1995**, 99, 813-820.
149. Kleist, G.; Schmitt, U. *Holzforschung*, **2001**, *55*, 573-578.
150. Worrall, J.J.; Anagnost, S.E.; Zabel, R.A. *Mycologia* **1997**, *89*, 199-219.
151. Worrall, J.J.; Wang, C.J.K. Electric Power Research Institute Final Report No TR-101244. Electric Power Research Institute, Palo Alto, California 1992.
152. Anagnost, S.E. *IAWA J.* **1998**, *19*, 141-167.

Chapter 5

Overview of White-Rot Research: Where We are Today

Kurt Messner[1], Karin Fackler[1], Pongsak Lamaipis[1], Wolfgang Gindl[2], Ewald Srebotnik[1], and Takashi Watanabe[3]

[1]Institute for Biochemical Technology and Microbiology, University of Technology Vienna, Getreidema 125rkt 9, A–1060 Vienna, Austria
[2]Wood Research Institute, University of Agricultural Science, Gregor Mendel-Strasse 33, 1180 Vienna, Austria
[3]Wood Research Institute, Kyoto University, Gokasho, JP–611–0011 Kyoto, Uji, Japan

White-rot fungi are the most active lignin degrading organisms and thus, they play a key role in the carbon cycle on earth. Some species are known to cause heavy damage to wood construction and building materials, requiring that this damage be prevented by wood preservatives. Chemical products to increase wood durability currently in use are very effective but in addition, a strong demand to develop new products with less environmental impact cannot be overlooked. Biotechnological processes have been successfully implemented in the pulp and paper industry during the last decade. In the past, developments were driven mainly by environmental considerations. In the future however, the main driving force for research and development will be reductions in manufacturing costs using new, low investment delignification processes. The application of white-rot fungi, or their ligninolytic systems, is one option for this. Understanding the microbial mechanisms leading to wood- and especially lignin degradation is a prerequisite for understanding both the development of new wood preservatives as well as wood biotechnology processes.

This paper briefly reviews what is known about the effects of simultaneous and selective white-rot fungi with regard to the morphological and ultrastructural changes in wood. An overview is given on reactions mediated by ligninolytic enzymes. It is generally accepted that low molecular weight compounds, smaller than enzymes, are the agents responsible for selective white-rot delignification. A new mechanism involving a powerful lignin degrading system based on coordinated Cu and peroxide, either hydrogen peroxide or organic peroxides, is proposed to be the agent involved at least in the initial depolymerization and degradation of lignin. Hypothetical pathways for both reaction types, involving H_2O_2 or organic hydroperoxides, respectively, are presented. The capacity of the Cu system to degrade native wood lignin was evaluated by two methods: 1) a newly developed method employing section staining, and 2) UV-microscopy. It was shown that the cell walls of a hardwood species were almost completely delignified and the middle lamella was degraded after treatment with coordinated Cu and organic hydroperoxides. The treatment matches the effect of selective white-rot fungal degradation. With H_2O_2 only, the cell walls of the hardwood were degraded to some extent, while the middle lamella was not attacked. The likelihood of the coordinated Cu system functioning as the agent primarily responsible for selective white-rot is discussed and a hypothetical scheme of wood degradation is presented.

The Cu system seems to be related to the brown-rot mechanism, both of which employ reactions catalysed by transition metals. Inhibiting this mechanism may offer a chance to develop new wood preservatives. The possibility of applying the Cu system in pulp and paper production is discussed.

Introduction

The defining character of white-rot fungi in the conversion of lignocellulose in nature is their strong capacity to degrade lignin, the second most abundant biopolymer in nature. In principle both softwoods and hardwoods are colonized by white-rot fungi but generally these fungi degrade the hardwoods, whereas the brown-rot fungi preferentially attack softwood species. Although white-rot fungi may play a secondary role in the decay of wood used for construction and building materials, wood preservatives designed to prevent fungal decay must be effective against both types of fungi. The wood preservatives currently in use are very effective; however, the trend to products with improved environmental performance cannot be overlooked. Understanding the biochemical mechanisms involved in decay may enable the development of products acting in a more targeted manner to control decay fungi.

Besides its use as construction and building material, wood is also a primary substrate for the pulp and paper industry. In the past decade, biotechnological processes have been successfully implemented that contribute to the constant improvement of technology of this industrial sector. Several new applications of enzymes have reached, or are approaching, the stage of commercial use. These include enzyme-aided bleaching with xylanases, direct delignification with oxidative enzymes, energy saving refining with cellulases, pitch removal with lipases, freeness enhancement with cellulases and hemicellulases as well as enzymatic slime control in the paper machine (*1*). Besides enzymes, biopulping -the use of white-rot fungi to treat wood chips - is close to mill application.

While environmental aspects have been a major driving force in the past, the need for significant reduction in manufacturing costs has become a primary goal of the pulp and paper industry. This is expected to be reached only by a major redesign of the industry's core technologies. New revolutionary breakthrough technologies may be developed in bio-manufacturing and involve lignin degradation processes such as bio-pulping and bio-bleaching. Implementation of these technologies could result in an estimated 50% reduction in capital cost requirements (*2*). Reduction of investment and energy costs by low temperature cooking processes allowed by implementation of biopulping processes is just one of the technologies that offers this potential.

To reach this goal, new approaches in research are needed and the known concepts of delignification by white-rot fungi must be re-evaluated. Ligninolytic enzymes and their catalytic cycle, and the degradation of lignin model compounds and pulp lignin have been studied in detail. However, this work has not been able to explain the degradation of native wood lignin, and we still do not know how lignin is depolymerized by white-rot fungi in nature.

This chapter is divided into two sections. The first section provides a brief overview of white-rot degradation processes whereas the second section focuses on a coordinated Cu system for the oxidation of lignin that the authors have researched. The coordinated Cu system includes components of naturally occurring white-rot biochemical degradation mechanisms and its study offers insight into how certain white rot fungi may degrade lignocellulosic materials.

White-rot

The literature reviewed in this paper focuses mainly on decay mechanisms. A comprehensive, general description of degradative fungi is presented elsewhere (*3*). The primary path of colonisation of wood by all types of fungi is usually the rays where easy access to non-structural nutrients contained in parenchyma cells is provided. From there, hyphae spread into fibres and vessels, and further colonization takes place via pits or less often by bore hyphae. Electron microscopic studies have demonstrated that an extracellular mucilage or slime layer (4-9) covers the hyphae. This extracellular slime produced by *Phanerochaete chrysosporium* has been characterized as a glucan.

When wood degraded by white- and brown-rot fungi was investigated in TEM after ruthenium red staining, extracellular tripartite membranous structures were found covering the hyphae and slime layers throughout all stages of wood decay (*10*). The extracellular structures of *Phanerochaete chrysosporium* were isolated by enzymatic digestion of the fungal cell walls. They were found to be composed of equal amounts of carbohydrates, lipids (none of these were phospholipids), and proteins, including five fractions with molecular weights between 30,000 and 200,000 (*11*).

The extracellular slime layer may cover the entire surface of the wood cell wall. As the optimum moisture content of wood for white-rot fungal activity is far above the fibre saturation point, the slime layer permits a film of liquid water to surround the wood cell wall. This water film presumably contains the slime components and the watery matrix they form may decrease evaporation when wood dries. Furthermore it is very likely that this layer is involved in regulating cell wall degradation by regulating the glucose level in the medium (*12*). Production of glucane polymers may either occur directly via the involvement of extracellular enzymes localised in the extracellular membrane, or by metabolism of carbohydrates and *de novo* synthesis of glucanses. The slime layer may also form a microenvironment where H_2O_2 needed for lignin degradation is maintained. As the glucane polymers are also depolymerised by extracellular laminarinase (12) it is most likely that slime fractions of a molecular weight, low enough to penetrate the pores of the wood cell wall are created. The chemical structure of the extracellular slime layer and its viscosity might also have an effect on diffusion of degrading agents into the cell wall as well as affecting the passage of degradation compounds to and from the hyphae. Unfortunately, the functions of the slime layer and of the proteins localized in the extracellular membrane are not understood and further study must confirm the concepts discussed.

Taxonomically, most white-rot fungi belong to the order of *Aphyllophorales* within the basidiomycetes and comprise a large number of species. Interestingly, some species causing white-rot and brown-rot are taxonomically closely related to each other. One example is the genus *Tyromyces,* with *T. chioneus* and *T.*

placenta causing white-rot and most other species causing brown-rot. This might lead to the assumption that although white- and brown-rot decay are very different from each other in appearance, and they attack different chemical components of the wood cell wall, the mechanisms of decay of both types might be closely related to each other. The mechanism of white-rot with its preference for lignin degradation may be based on Cu reactions while Fenton reactions based on Fe have been suggested to be involved in brown-rot. The types of wood cell wall attack caused by white-rot fungi can be diverse, and macroscopic as well as microscopic differences have been reported. The unifying feature of most white- rots is the extensive degradation of lignin resulting in a bleached appearance. Several types of white-rots were characterised by macroscopic and microscopic differences as early as 1878 by Robert Hartig (*3*). White- rots have been classified by macroscopic characteristics into different sub-categories such as white-pocket, white-mottled, and white-stringy. The types of decay produced are affected by the fungal species, wood species, and ecological conditions among other things.

From the microscopic and ultrastructural observation, two main types of white-rot have been distinguished. These types are classified with regard to the order in which different amounts of components are degraded (*13,14*) as follows:

- Simultaneous white-rot: lignin, cellulose and hemicellulose are lost more or less simultaneously.
- Selective white-rot: preferential removal of lignin and hemicellulose. In most cases, cellulose is also degraded to some extent.

Simultaneous White-Rot

Degradation of the wood cell wall starts by erosion of the lumen surface, sometimes around hyphae, and progresses from the cell lumen to the middle lamella. All components are degraded, the middle lamella is degraded last and only at exposed areas after complete cell wall degradation. Some fungi like *Fomes fomentarius* are not able to degrade the middle lamella. Walls of the surrounding cells remain lignified, and the lignin degrading agent does not diffuse into adjacent cell walls. The cell corners remain in an undegraded state for a long period and sometimes are left after decay. Losses of crystalline cellulose have been examined using polarized light microscopy (*15,16*). Distribution of lignin was determined by bromination and X-ray microanalysis by Saka and Thomas (*17*) as well as by UV-microscopy (*18*). It was demonstrated that lignin was removed from the inner circumference of the secondary wall near the lumen before cellulose was degraded, and lignin was removed continuously in advance of cellulose degradation.

Selective White-Rot

Selective white-rot fungi have the ability to remove large quantities of lignin from the cell wall without destruction of cellulose. Lignin is preferentially removed from the cell wall causing a loosening of cells. Hemicellulose is degraded concomitantly with lignin and cellulose is left unmodified to a large extent (*3,13,14*). After incubation of wood chips for 6 weeks with *Ceriporiopsis subvermispora,* or *Dichomitus squalens,* TEM observations demonstrated disintegration of the wood tissue by complete dissolution of the middle lamella. Nevertheless, the wood fibres showed no sign of visible damage (*19*). By employing selective staining of lignin and cellulose with safranin and astra blue, it was obvious that the middle lamellae and the wood cell walls had both been delignified (*20*). Greater birefringence was observed when wood was viewed in polarized light (*21*). The crystalline nature of cellulose was not destroyed which is in contrast to the removal of cellulose by simultaneous white-rot.

Blanchette and Reid (*8*) demonstrated progressive stages of selective delignification by *Phlebia tremellosa* by fixation of wood with OsO_4-glutaraldehyde and post-staining with uranyl acetate. Delignification of the cell wall started directly adjacent to the hyphae and progressed through the S_2 layer to the middle lamella, which was gradually lost. This occurred around the entire circumference of the cell wall. Remnants of cell corners persisted.

The ability to degrade hardwood vessels varies among selective white-rot fungi: e.g. *Dichomitus squalens* does not delignify the vessels while *Phlebia tremellosa* removes lignin selectively from vessels (*8*). This might be related to the ability to cope with the higher guaiacyl content of vessels (*18,22*). Syringyl lignin is degraded more rapidly than guaiacyl lignin (*23-26*).

Both types of attack, simultaneous as well as selective decay, can be produced by the same white-rot fungus and even in the same piece of wood. Factors that determine different modes of fungal degradation remain obscure. Unpublished results of the authors have lead us to the conclusion that the moisture content of wood is one of the determining factors.

Penetration of Enzymes

Selective white-rot fungi are used in so called "biopulping processes" to treat wood chips prior to mechanical or chemical pulping. After 2 - 3 weeks of incubation, weight loss of the wood chips is only about 2%, but cell wall components are already modified, leading to energy savings of approximately 30% in mechanical pulping (*27*), or lower kappa numbers in sulphite (*19*) or Kraft pulping (*28*). The chemical nature of these effects was found recently to be depolymerization of cell wall lignin (*28*) by fungal metabolites. This demonstrates that lignin is attacked by fungal metabolites in the cell wall and middle lamella in the early phases of colonisation.

Based on immunogold labelling experiments carried out by Srebotnik et al. (*29*) and confirmed later by Daniel (*30*) and Blanchette (*31*), it is now generally accepted that the enzymes secreted by fungi into the lumen – their molecular weights are usually above 20 kD - are too large to penetrate the cell wall until relatively late stages of decay. Even when insulin (5.7 kD) was used as a size marker molecule after 2 weeks of colonisation of the wood samples by *Ceriporiopsis subvermispora,* it could be identified only in the inner area of the wood cell wall, close to the lumen (*30*). Consequently, the depolymerization of lignin leading to biopulping effects must be caused by molecules smaller than 5.7 kD which clearly excludes the involvement of any direct enzymatic activity in the cell wall. These low molecular weight agents involved in selective white-rot have become one of the main scientific targets in white-rot research.

Low Molecular Weight Agents

Various reactive low molecular weight agents may be formed directly or indirectly by oxidative fungal enzymes and have been proposed to participate in delignification starting at the cell lumen surface and penetrating deeper into the wood cell wall. Several of the systems proposed to generate low molecular weight agents are reviewed below:

Manganese Peroxidase / Mn(II) / Oxalate

Manganese peroxidase (MnP) is a very common extracellular enzyme produced by simultaneous as well as selective white-rot fungi. Its catalytic cycle includes two one-electron reducing stepsby Mn(II) (*32, 33*). The Mn(III) formed is then chelated and released from the enzyme by the fungal metabolite oxalate *in vivo*. The relatively stable Mn(III) oxalate is able to oxidize phenolic lignin compounds and has been discussed as a diffusible agent acting in the wood cell wall, distant from MnP located in the fibre lumen. Since phenolic lignin makes up only a small portion of the cell wall lignin, total delignification as observed in selective white-rot is considered unlikely to be caused by Mn(III) alone. However, Hofrichter et al. (*34*) have recently succeeded in creating conditions favourable for the efficient depolymerisation of native lignin by the MnP / Mn(II) couple. Furthermore, MnO_2-deposits accumulate during white-rot decay, and are probably formed by disproportionation of MnP-generated Mn(III). This may then be "reactivated" by reduction to Mn(II) (*35*) or the formation of reactive oxalate complexes, which may contribute to the selective delignification of the wood cell wall (*36*).

Manganese Peroxidase / Mn(II) / Oxalate / Cellobiose Dehydrogenase (CDH)

CDH is oxidized by molecular oxygen and metal ions such as Fe(III) and Cu(II) to form hydrogen peroxide and reduced metal ions. Fe(II) and Cu(I) react with hydrogen peroxide to generate hydroxyl radicals which in turn are proposed to demethoxylate and hydroxylate non-phenolic lignin. Thus, non-phenolic lignin is converted to phenolic lignin which can then be attacked by MnP-generated Mn(III) (*37*). The hydroxyl radical may also be formed by other pathways, e.g. via hydroquinone redox cycling involving semiquinones produced by peroxidase or laccase, which reduce both Fe(III) and O_2 to provide the ingredients for Fenton-type hydroxyl radical formation (*38*).

Manganese Peroxidase / Mn(II) / Oxalate / Lipids

The oxidative potential of MnP can be extended by the presence of lipids. Peroxidation of unsaturated fatty acids is promoted by Mn(III) resulting in the formation of peroxyl radicals which are diffusible, potentially ligninolytic agents (*39*). Watanabe et al. (*40*) suggested that Mn(III) directly abstracts hydrogen from fatty acids to form acyl radicals. This system has been shown to depolymerise both phenolic as well as non-phenolic lignin (*41*). Substantial amounts of fatty acids and hydroperoxides have recently been detected in cultures of *Ceriporiopsis subvermispora*, providing further support that lipid-derived radicals may in fact be involved in selective white-rot degradation (*42*).

Lignin Peroxidase / Veratryl Alcohol

Lignin peroxidase (LiP) has the highest redox potential of all enzymes believed to be involved in lignin degradation, and in principle is able to oxidise phenolic as well as non-phenolic lignin. However, for steric reasons discussed above its action on lignin is restricted to surface areas of the wood fibre. The veratryl alcohol radical, generated during turnover of LiP when compound II is reduced to the resting state by veratryl alcohol, was proposed by Harvey et al. (*43*) to act as a charge transfer system in wood. However, due to its short lifetime (*44*) this radical is not expected to diffuse into deeper areas of the cell wall. A very recent study proposes self-propagation of chemical reactions initiated by lignin peroxidase, involving lignin-derived peroxyl radicals (*45*). However, it is not known whether such radicals would be able to depolymerize lignin inside the wood cell wall.

Laccase / Mediators

Laccase, a phenol oxidase which is produced by most white-rot fungi, oxidizes the phenolic moieties in lignin which primarily leads to the polymerisation of lignin. However, when laccases are combined with low molecular weight charge transfer agents, so-called mediators, they can delignify wood pulps. A significant kappa reduction can be achieved in pulp bleaching when using synthetic mediators carrying N-OH groups (*46*). The effect of the laccase/mediator-system is probably based on ketone formation by the mediator radical, which makes the lignin molecule more susceptible to alkaline hydrolysis during extraction. In addition to pulp bleaching, the laccase/mediator system is also able to depolymerise non-phenolic guaiacyl lignin (*47*). However, the occurrence of natural laccase mediators during white-rot decay has not been demonstrated so far.

Low Molecular Weight Peptides

Tanaka et al. (*48*) discovered peptides, produced by white-rot fungi, which are of the molecular weight range of 1000-5000 D. These peptides have been proposed to catalyse a redox reaction between molecular oxygen and an electron donor to produce hydroxyl radicals via reaction with hydrogen peroxide. This may lead to ligninolytic reactions similar to those described above for CDH.

Conclusion

Current research suggests that lignin depolymerization is a highly complex process. Numerous pathways have been suggested and it is difficult to assess their individual contributions to ligninolysis. However, it can be assumed that radical reactions involving Mn(III) and reduced oxygen species play a key role in selective lignin degradation by white-rot fungi. In this regard, we propose another mechanism, the coordinated Cu-system, which may also play a key role in white-rot decay and which is discussed in the second section of this chapter.

The Coordinated Copper System

Introduction

Hypothetical Pathway for the Generation of Active Oxidants from Hydrogen Peroxide by Coordinated Copper

The transition metals Fe and Cu are the active centres of all the enzymes assumed to be involved in lignin degradation. These metals are coordinated by amino acids and/or N-atoms of protoporphyrins. The concentrations measured for these metal ions in wood are 9.2 ppm for Fe(III) and 0.7 ppm for Cu(II) (*49*). There is strong evidence that Fe is involved in the mechanism of brown-rot (*50*) resulting mainly in the degradation of cellulose and hemicellulose. Recently, siderophores containing catechol groups that are able to reduce and solubilise the Fe attached to cellulose have been isolated (*51, 52*). In Fenton chemistry reactions, Fe(II) generates hydroxyl radicals from hydrogen peroxide produced by both brown-rot and white-rot fungi (*50, 38*). These radicals are strong oxidants and degrade cellulose and hemicellulose and can alter lignin chemistry.

Besides its function as the active centre in laccases, almost no attention has been paid in the past to the possible involvement of Cu in cell wall degrading reactions; although Cu(I) would also lead to Fenton reactions. One of the reasons might be that the main focus has been on Fenton chemistry and Fe is the more likely reaction partner as its concentration in wood is higher than that of Cu.

It was demonstrated by Watanabe et al. (*53*) that Cu coordinated by low molecular weight organic compounds like pyridine together with peroxides – either hydrogen peroxide or lipid hydroperoxides - leads to very strong oxidative reactions. These reactions are able to oxidize various dyes, pulp lignin and even a non-phenolic synthetic lignin model compound (Figure 1). Compared to uncoordinated Cu, the coordinated reaction was much more effective.

The products obtained after the reaction of the non-phenolic lignin dimer with Cu/ pyridine/ H_2O_2 were similar to those after oxidation of similar substrates with LiP/ H_2O_2 (*54*). The Cu/ pyridine/ H_2O_2 products were not consistent with oxidation by free hydroxyl radicals, which would instead lead to aromatic hydroxylation or hydrogen abstraction (*55*). The nature of the active oxidant is not known, but since the substrate was oxidized by a single electron transfer reaction it may be speculated that a Cu-centered oxo or peroxo complex (*56*) is the one-electron oxidant produced by the system (Figure 2).

Like the other models for lignin degradation, one can only conclude the existence of the coordinated-Cu system in nature from the detection of the reaction components in treated wood and comparing the results achieved to wood degraded naturally by selective white-rot fungi. As discussed earlier, Cu and hydrogen peroxide are present in white-rotted wood. Evidence for the production of compounds containing a pyridine nucleus by fungi is provided in the literature (*57*) and includes, homarine, pycolic acid, dipycolic acid, fusaric acid, dehydrofusaric acid, nicotinamide, PQQ etc.

In the first step of the reaction Cu(II) is reduced to Cu(I) by either H_2O_2 or reducing groups contained in wood or pulp lignin. Cu(I) together with H_2O_2

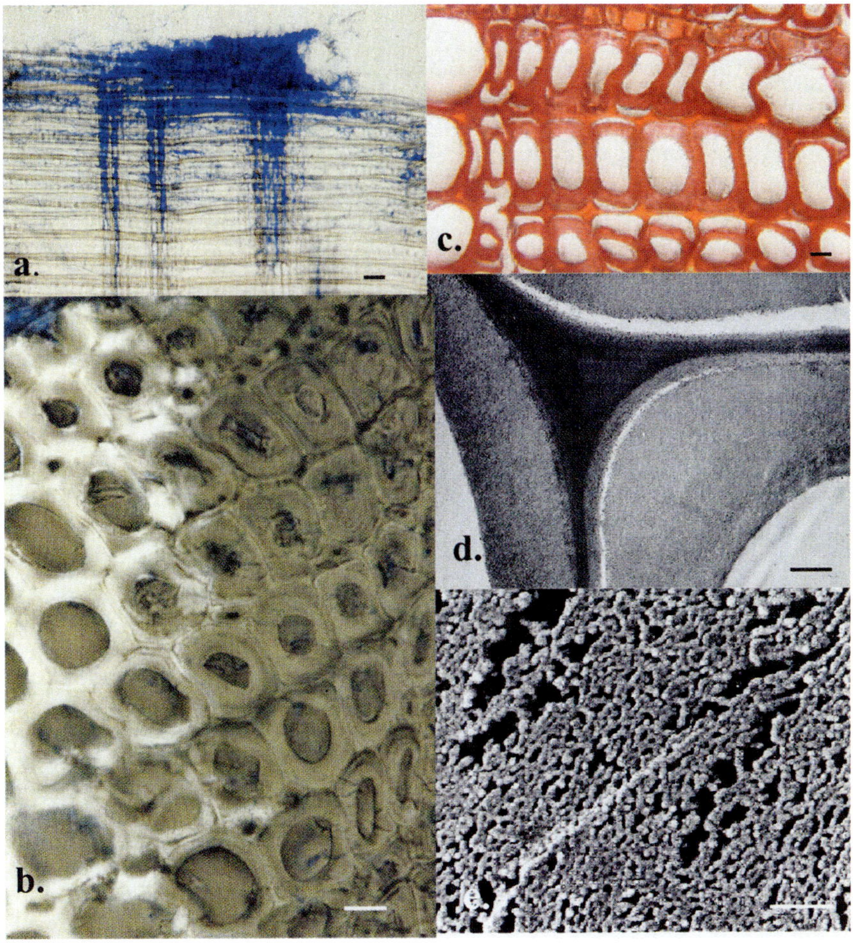

Figure 16. Light and electron micrographs showing aspects of wood cell wall degradation by the brown rot fungus G. trabeum. a) Colonization of rays by G. trabeum: b) Attack of pine latewood tracheids and loss of loss of birefringence due depolymerization of crystalline cellulose; c) Advanced decay of pine with almost total decay of fibres which irreversibly shrinks after drying; d) Early decay of pine latewood with first signs of attack at the S1/S2 interface; e) Cryo FE-SEM observations on birch fibre degraded by G. trabeum leaving a lignin skeleton without "macrofibrillar" structures (cf Fig. 13 showing preferential white rot). Bars: a, b, c, *5.0 μm; d, 1.0 μm; e, 100 nm.*

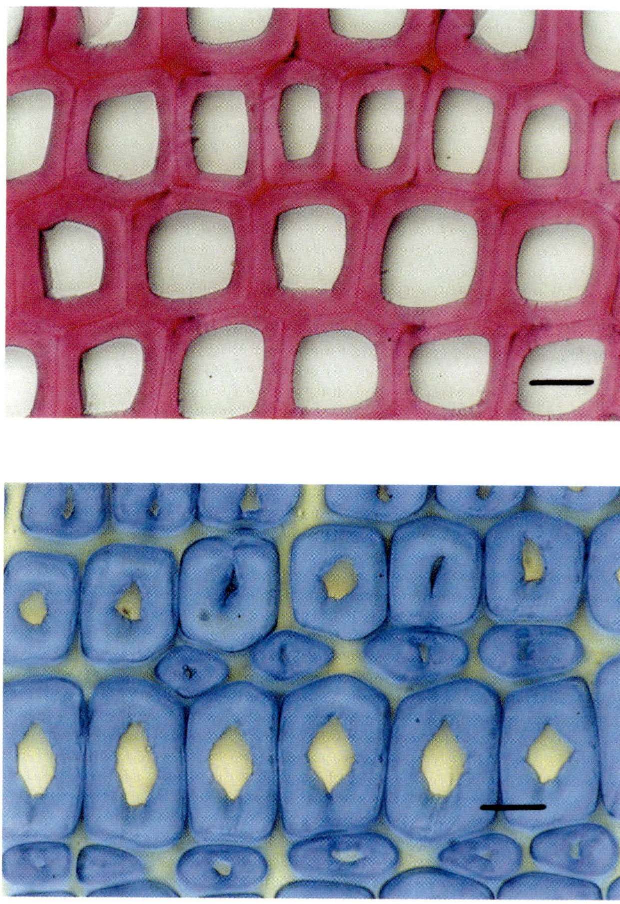

Plate 1. Transverse section of softwood stained with safranin/astra blue (bar: 20µm). ***Above***: *softwood; control (red).* ***Below***: *softwood; 0.1 mM $CuSO_4$, 70 mM 4-AP and 50 mM CHP (blue).*

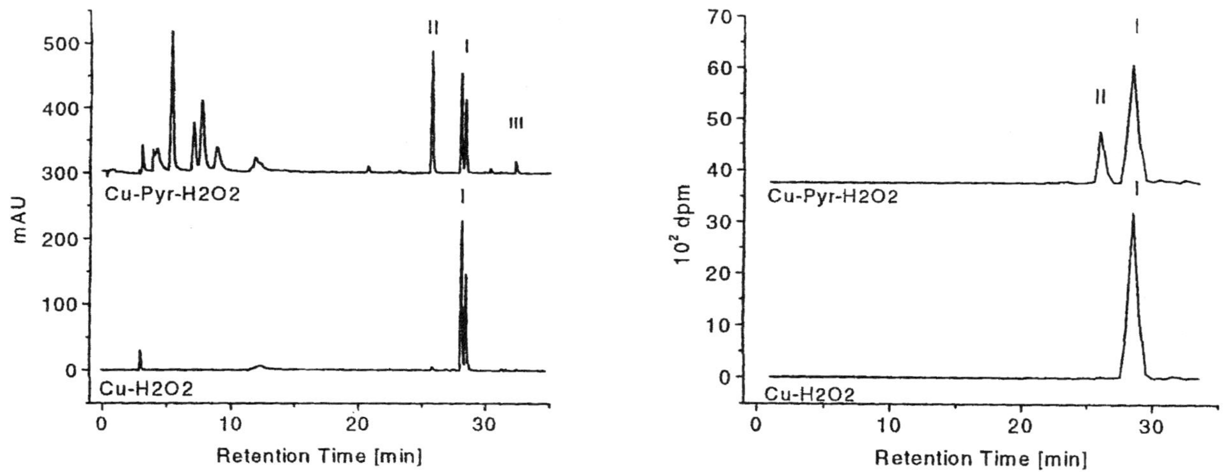

Figure 1. HPLC analysis and radioactive products obtained after incubation of a [^{14}C]-labelled non-phenolic lignin dimer (I) with Cu/ H_2O_2 in the presence and absence of pyridine. Veratraldehyde (II) was the only main oxidation product in the presence of the coordinator pyridine.

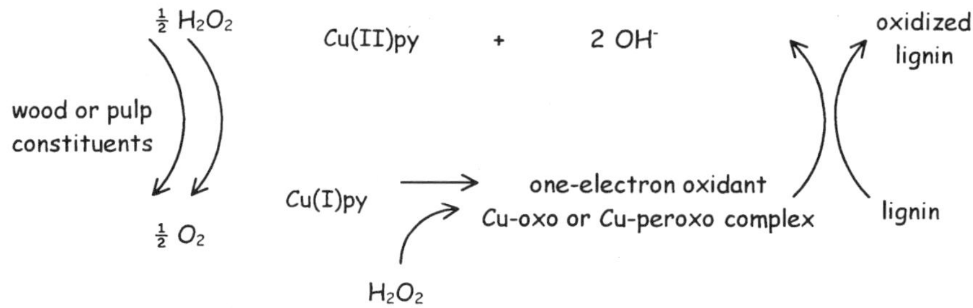

Figure 2. Hypothetical pathway for the generation of active oxidants from hydrogen peroxide by coordinated Cu.

forms a highly reactive one-electron oxidant that can oxidize both phenolic and non-phenolic lignin.

Hypothetical Pathway for the Generation of Free Radicals from Lipid Hydroperoxide by Coordinated Copper

Additional oxidants besides hydrogen peroxide include lipid hydroperoxides which are produced from unsaturated fatty acids. Enoki et al. (*42*) found that unsaturated fatty acids are secreted by white-rot fungi when grown on extracted wood. Furthermore, they are a component of resin which is a nutrient source for fungi in the incipient stages of colonisation in some softwoods. Lipid hydroperoxides were also found to be produced mainly in early stages of decay. Watanabe et al. (*58*) found that these peroxides can be generated from linoleic acid by Mn(III). Mn(II) is a chain breaking agent for lipid peroxidation, hence MnP controls the generation of lipid hydroperoxides by oxidizing Mn(II). Both the Mn(III) chelate and unsaturated fatty acids can diffuse freely within the wood cell wall, enabling the production of lipid hydroperoxides at any site in the cell wall. Their conversion to free lipid radicals is expected to be catalysed by coordinated transition metals and this is supported by the fact that Fe and Cu have been found to be evenly distributed in wood cell walls (*49*). If a hypothetical, pyridine-related coordination compound is produced by selective white-rot fungi, it would be diffusible due to its low molecular weight and would also coordinate transition metals within the entire wood cell wall (Figure 3).

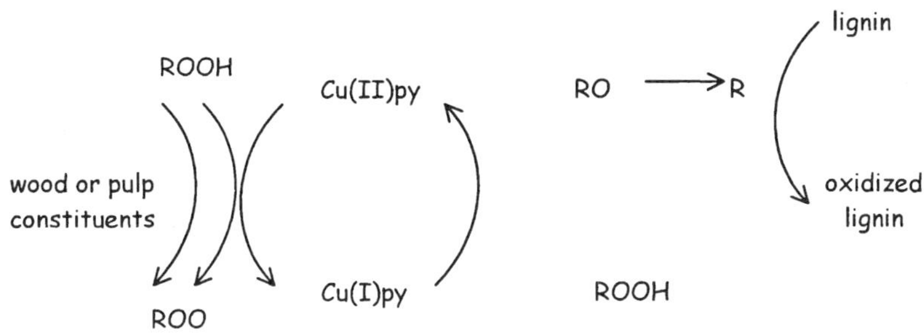

Figure 3. Hypothetical pathway for the generation of free radicals from lipid hydroperoxides by Cu/pyridine.

In the first step of the reaction, Cu(II) is reduced by the hydroperoxide or by reducing groups contained in wood or pulp lignin with concomitant formation of a peroxyl radical ($ROO^{\bullet}$). Cu(I) is then reoxidized by the lipid hydroperoxide. The alkoxyl radical ($RO^{\bullet}$) that is formed during this reaction decays quickly via β-scission and forms a carbon-centered radical ($R^{\bullet}$).

Figure 4 summarizes the results described above and explains the morphological appearance of selective white-rot, where lignin degradation is not restricted to the lumen surface but seems to take place simultaneously throughout the whole wood cell wall and middle lamella.

Semi-Quantitative and Quantitative Assay of the Capacity of the Coordinated Copper / Peroxide System to Degrade Native Wood Lignin

The potential of the Cu system to delignify native wood was evaluated by two tests (59): 1) Microtome sections of wood samples were prepared and immersed into solutions containing the Cu system. After incubation, the sections were selectively stained with safranin-O and astra-blue. Red staining of the cell wall indicated residual lignin, blue staining indicated that lignin has been degraded. In addition to providing a semiquantitative estimation of the residual lignin content in the cell wall, the microscopic test also allowed a morphological evaluation to be made in assessing the remaining cellulose of the cell wall. 2) To quantify the residual lignin content, small wood blocks were treated with the Cu system. By scanning the sections under the UV-microscope the lignin content in wood cell wall and middle lamella was quantified. The combination of these methods provided excellent information on the delignification capacity of the Cu system, but it also may prove useful in the evaluation of other ligninolytic systems that may degrade native lignin.

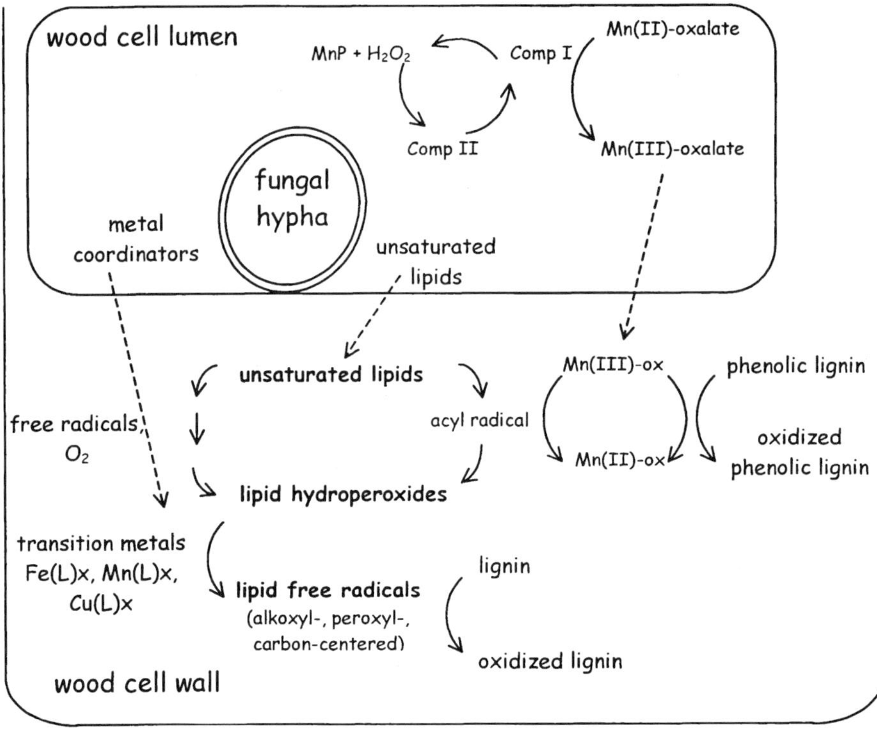

Figure 4. Hypothetical pathway of the delignification of native wood by the coordinated Cu system.

Materials and Methods

Semiquantitative Determination of Native Wood Lignin on Wood Sections

Small wood blocks (2×2×6 mm) of a hardwood (*Betula pendula*) and a softwood (*Pinus sylvestris*) were soaked in water under vacuum. Cross cryosections of 20 μm thickness were obtained with a Leitz cryostat microtome at -20°C. Softwood thermomechanical pulp, TMP (Steyrermühl AG Ges.m.b.H., Austria) was used as supplied. The presence of transition metals, other than the added Cu, was reduced by extraction of most of the metals contained in wood sections and TMP fibers before the delignifying step. This extraction was performed using 25 mM ethylenedinitrilo tetraacetic acid disodium salt dihydrate (EDTA) at 90°C, for 1 hour. The delignification step was then carried out on wood sections by immersing the sections in a solution containing 0.1 mM $CuSO_4$, 70 mM 4-amino pyridine (4-AP) and 50 mM of either H_2O_2 or the organic hydroperoxides cumene hydroperoxide (CHP) or *tert*-butyl hydroperoxide (TBHP). CHP and TBHP were used as models for lipid hydroperoxides. The reaction was allowed to proceed for 3 hours at 90°C, then terminated by washing the sections with water. Washed sections were transferred to glass slides. Staining with 1% aqueous safranin-O and subsequently with 1% aqueous astra-blue was carried out at room temperature for 3-5 minutes. The samples were rinsed with water after each staining step and then immersed in water. Finally, they were examined under light microscopy and the images captured as colour prints.

Quantitative Determination of Native Wood Lignin

Small (2×2×6 mm) hardwood birch blocks (*Betula pendula*) were prepared and treated with the Cu system in a solution containing 0.1 mM $CuSO_4$, 70 mM 4-AP and 0.7 M of either H_2O_2, CHP or TBHP. Subsequently the blocks were embedded in Spurr's resin (*60*). Cross sections (1 μm thick) were cut from these samples on an ultramicrotome by means of a diamond knife. The samples were floated off on water and mounted on quartz slides with quartz cover slips. UV absorption spectra were determined using a ZEISS MPM 800 UV-microscope at a wavelength of 280 nm. At each position the lignin content of cell walls was determined from lumen to lumen across the cell wall/middle lamella/cell wall (CW/ML/CW). Three measurements were taken along the S_3/S_2-layer, one in the compound middle lamella region and again 3 measurements along the adjacent S_2/S_3 area. The diameter of the measuring spot was 1 μm, the monochromator bandwidth was 5 μm, and the magnification 1000×. One series of measurements was taken across each CW/ML/CW area of 10 adjacent cell walls. Lignin concentration was estimated according to Fergus and Goring (*18*) and Okuyama (*61*) using the Beer-Lambert law,

$$A = \varepsilon \times C \times d$$

where A, ε, C and d are the UV absorption, an absorption coefficient ($cm^{-1}lg^{-1}$), the lignin concentration (g/l), and the thickness of the section, respectively. The lignin concentration (g/g) can be obtained after converting the unit grams per liter to grams per cubic centimeter and multiplying by the specific volume of 1.07 cm^3/g for the water swollen cell wall (*62*). The embedding procedure using gradual solvent-exchange with Spurr's resin preserves the initial volume of the cell wall and therefore the value of 1.07 proposed by Stone and Scallan (*62*) instead of the value of 0.67 valid for the dry cell wall was chosen. Sections to determine the lignin content from wood blocks were taken from the outer part of the blocks.

Delignification of Wood Sections by Different Coordinated Metal Peroxide System

After extraction of metals by EDTA, wood sections were treated with the coordinated metal peroxide system. The delignifying potential of the metal complex/peroxide couples was assayed by safranin-O/astra-blue staining. All coordinated metal/peroxide systems contained 70 mM 4-AP and 50 mM of either H_2O_2 or organic peroxides (CHP or TBHP). $CuSO_4$, $FeSO_4$, $MnSO_4$, and Mn(III)-tartrate, respectively were added in a concentration of 0.1 mM.

Delignification of TMP Fibers by the Coordinated Copper Peroxide Systems

TMP fibers (metal extracted with EDTA) 5 mg dry weight were incubated at 70°C in the solution containing 0.1 mM $CuSO_4$, 70 mM 4-AP and 50 mM of either H_2O_2 or organic peroxide (CHP or TBHP). The chemical delignifying potential was determined by employing the safranin-O/astra-blue staining method.

Results

Semiquantitative Assay

Examination of untreated wood sections and TMP fibers by light microscopy revealed that they were stained red by safranin-O which indicated the presence of lignin. The complex composed of coordinated Cu and organic hydroperoxides was able to depolymerize native lignin selectively at ambient temperature (Plate 1), demonstrating the high potential of this system. The result is comparable to the delignification activity of selectively delignifying white-rot fungi. The cell walls are stained blue, indicating that cellulose remains. The middle lamella is no longer present after treatment. The cell walls are swollen but still present, demonstrating the selectivity of the Cu system for lignin removal.

As shown in Table 1, hydrogen peroxide as the only oxidant showed no apparent delignification activity on either hardwood or softwood sections, or on TMP fibers, as indicated by red staining. When Cu was used as an uncoordinated metal, no delignification took place. If the transition metals were not first extracted from the wood, and no Cu was added, sections were also delignified but at a much slower rate. When either Fe(II), or Mn(II) or Mn(III) was used instead of Cu(II) with CHP and 4-AP, intensive delignification was observed only after incubation of the sections for one month at 30°C.

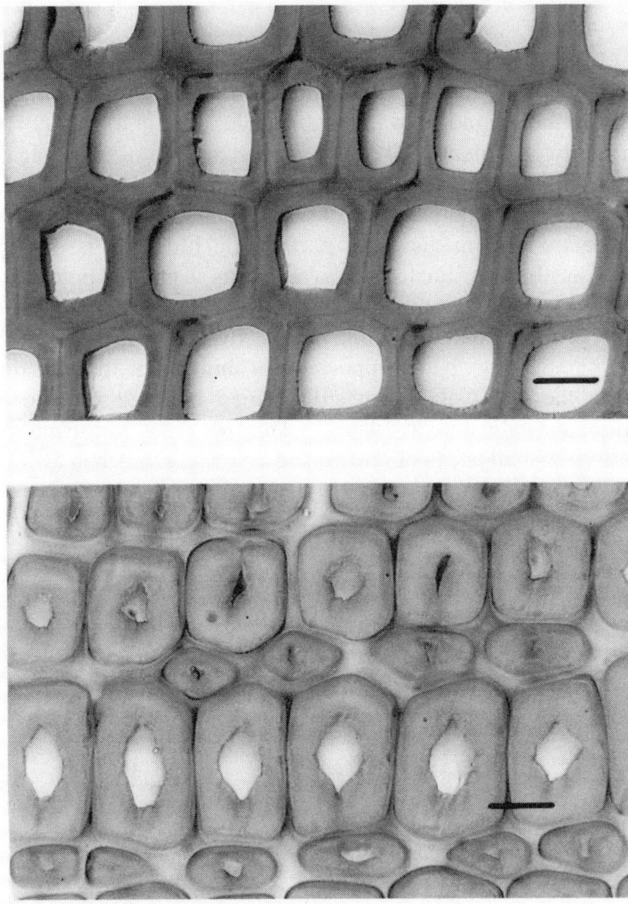

*Plate 1. Transverse section of softwood stained with safranin/astra blue (bar: 20μm). **Above**: softwood; control (red). **Below**: softwood; 0.1 mM $CuSO_4$, 70 mM 4-AP and 50 mM CHP (blue).* (This figure is also in color insert.)

Table I. Minimum incubation time (hours) of wood sections and TMP, respectively, to completely delignify native lignin with the coordinated Cu hydroperoxide system

Peroxides	Hardwood (30°C)	Hardwood (70°C)	Softwood (30°C)	TMP (70°C)
CHP	24	1	72	12
TBHP	48	1	120	12
H_2O_2	No apparent effect	No apparent effect	No apparent effect	No apparent effect

Quantitative Assay

UV-microscopy allowed the determination of lignin distribution and lignin concentration in the secondary walls and middle lamella of the untreated control and treated hardwood blocks. The highest concentration of lignin in untreated cell walls was found in the middle lamella, coinciding with values reported in the literature (18). The potential of the coordinated Cu peroxide system was assessed on sections from the outer part of the block. It is evident that with organic peroxide the coordinated Cu system has a greater potential to degrade lignin than the hydrogen peroxide system. However, as shown in Figure 5, the lignin content of the secondary wall layer was reduced by approximately 50% of the control with H_2O_2. It is important to highlight that at this stage of delignification the cell walls were still stained red and showed no apparent delignification, as indicated in Plate 1. This provides confirmation that blue staining of sections indicates delignification levels greater than 50%!

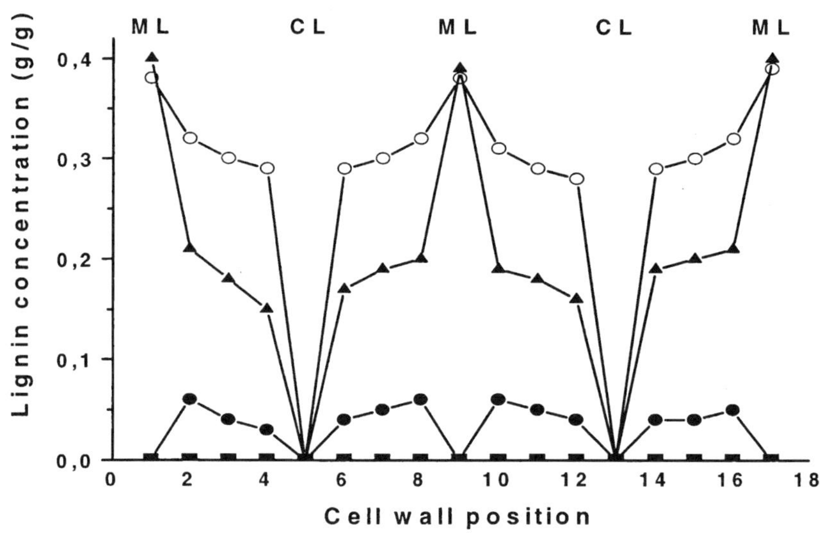

Figure 5. Lignin concentration of cell walls of birch wood at the outer part of the blocks: Untreated control (○); H2O2/Cu/4-AP (▲); CHP/Cu/4-AP (●); TBHP/Cu/4-AP (■). Position 9 is the position of the middle lamella, positions 8 and 10 are approximately the S1 layer, and positions 6 and 12 are the outer part of the S2 layer up to the S3 layer. Positions 5 and 13 demonstrate the lumen to the left and right side of the cell walls. ML=Middle lamella; CL=Cell lumen.

The cell walls of wood blocks that were treated with CHP and TBHP typically separated from their adjacent cells, indicating that the middle lamella was completely delignified. TBHP was the most efficient organic peroxide tested that could completely degrade both secondary wall and middle lamella lignin, while with CHP the lignin concentration of secondary wall layers remained at about 10% of the initial (Figure 5). When the two organic peroxides were used with the Cu system to delignify wood sections, CHP performed better. This is presumably related to a higher diffusion rate of TBHP into the wood block, while on thin sections the result is less affected by diffusion rate.

Conclusions

- The coordinated Cu system was highly efficient in removing native lignin from wood and TMP fibers.
- The Cu system removed lignin selectively from the cell wall, without major visual damage to the cellulose part of the cell wall, when organic hydroperoxides were used.
- Hardwood lignin was degraded faster than softwood lignin – as is also the case in natural decay by white-rot fungi.
- Delignification was achieved only when Cu was coordinated.
- The delignifying potential of Mn or Fe complexes with 4-aminopyridine was much weaker than that of Cu complexes.
- The time needed for complete delignification of softwood sections was reduced from 24 hours at 30°C to only 1 hour at 70°C.
- Although good delignification could be achieved with hydrogen peroxide on pulp (*63*), no visible effect was observed on wood sections using our microscopy techniques. The UV microscopic investigation revealed approximately 50% degradation of lignin in the wood cell wall but not in the middle lamella.
- The effect achieved is comparable to that of selectively delignifying white-rot fungi.
- Assays: Treating wood sections with ligninolytic systems followed by staining with safranin-O/astra-blue is a fast and very simple method to evaluate the delignification capacity of such systems. The technique indicates delignification only when lignin is almost completely depleted. By subsequent UV-microscopy, the rate of delignification can be determined quantitatively.

Summary

Ligninolytic enzymes, including their related low molecular weight compounds, are unable to degrade native wood lignin *in vitro.* To date, the coordinated Cu system is the only *in-vitro* system that permits the degradation of lignin in hardwood and softwood samples, and furthermore mimics the effect of selectively delignifying white-rot fungi. Based on the results presented here, and on the presence of compounds previously isolated from white-rot environments, it is conceivable that a similar system is involved in decay caused by selective white-rot. Unsaturated lipids are produced by white-rot fungi or are present in softwood resins. Mn(III) (to initiate lipid peroxidation) can be produced by MnP in almost all white-rot fungi. Potential Cu coordinating low molecular weight compounds with a pyridine nucleus were shown to be produced by some fungi. Cu, in small amounts, is available in all wood cell walls and could therefore potentially lead to the production of lipid radicals. Particularly with the organic model hydroperoxides, CHP or TBHP, the coordinated Cu system proved to possess extremely strong delignifying properties. The low molecular weight of all components would enable free diffusion within the wood cell wall and explain the depolymerization of lignin in early stages of decay. The system's selectivity for lignin has been demonstrated here. However, for a radical system, side reactions with carbohydrates cannot be excluded, perhaps explaining the concomitant degradation of hemicellulose which occurs in selective white-rot.

If such a system is responsible for early degradation of lignin, it may well be that this accounts for all of lignin degradation in the white-rot fungi. If so, the role of the ligninolytic enzymes would have to be re-evaluated. There is no doubt that they are somehow involved in the complex reaction pathway, but this could be a secondary role, e.g., modifying, detoxifying or polymerizing the low molecular weight depolymerization products diffusing into the lumen of the wood fiber. Since the primary activity, especially for laccase, is phenolic polymerization, a hypothesis that addresses the roles of the most common enzyme couple, MnP and laccase may be suggested. MnP may function only to initiate lipid peroxidation, creating lipid hydroperoxide for the Cu system, whereas laccase may detoxify or modify the degradation products of lignin. It must be clearly pointed out here though, that no specific pyridine coordinator for Cu has been identified in any fungal culture or decayed-wood extract, so additional work showing that this system actually does play a role in natural white-rot decay processes must still be performed.

An earlier observation made by bromination and X-ray microanalysis [Saka and Thomas] (*17*), as well as by UV-microscopy (*18*), showed that in simultaneous white-rot lignin was removed from the inner circumference of the secondary wall near the lumen before cellulose was degraded. The lignin was removed continuously in advance of cellulose. This suggests that a similar system functions not only in selective white-rot but also in simultaneous white-rot degradation. For reasons unknown, this system does not seem to be as diffusible as it is in selective white-rot. One possible reason could be the higher viscosity of the slime sheath decreasing the diffusion rate of the Cu-coordinating compound.

Based on the close taxonomic relationship of brown- and white-rot fungi one would expect that the physiological background of their decay mechanisms should also be similar, although the visual effect and chemistry of the resulting wood are very different from each other. A radical system based on Fe-Fenton reactions for brown-rot and a radical system based on coordinated Cu reactions for white-rot could be considered similar in regard to the basic reaction mechanisms of the decay fungi. From the standpoint of wood preservation this also suggests the development of radical scavenging systems could be developed to prevent both types of decay. Another option in this regard would be to inactivate the transition metal ions.

With respect to pulp and paper production, one of the conclusions from these experiments is that delignification of wood by the coordinated Cu/lipid hydroperoxide system can be achieved within a time frame acceptable for technical applications. The system can be applied in aqueous solution at temperatures below 100°C. These features demonstrate the potential of the system to become a new process in the pulp and paper industry, resulting in a reduction of capital requirements and energy costs. However, these developments have only started recently and the technology has to be optimised with respect to the right coordinator, generation of lipid hydroperoxides, viscosity losses, among other factors.

References

1. Viikari, L. *2000 Tappi.* 2000, Order number: PULP/PPQCD-00; ISBN: 0-89852-974-3.
2. Ragauskas, A. J. In 8th International Conference on Biotechnology in the Pulp and Paper Industry; Vahala, P; Lantto, R.; Eds.; VTT Biotechnology: Helsinki, FINLAND, 2001; pp 15-16.
3. *Microbial and Enzymatic Degradation of Wood and Wood Components*; Eriksson, K. -E. L.; Blanchette, R. A.; Ander, P., Eds.; Springer Series in Wood Science; Springer-Verlag: Berlin, GERMANY, 1990; pp 1-43.
4. Schmidt, R.; Liese, W. *Mater. Org.* **1966,** *1*, 252-261.
5. Highley, T. L.; Murmanis, L.; Palmer, J. G. *Holzforschung.* **1983,** *37*, 272-279.
6. Palmer, J. G.; Murmanis, L.; Highley, T. L. *Mycologia.* **1983,** *75*, 995-1004.

7. Palmer, J. G.; Murmanis, L.; Highley, T. L. *Mycologia.* **1983,** *75*, 1005-1010.
8. Blanchette, R. A.; Reid, I. D. *Appl. Environ. Microbiol.* **1986,** *52*, 239-245.
9. Bes, B.; Pettersson, B.; Lennholm, H.; Iversen, T.; Eriksson, K. -E. L. *Biotechnol Appl. Biochem.* **1987,** *9*, 310-318.
10. Foisner, R.; Messner, K.; Stachelberger, H.; Roehr, M. *Trans. Br. Mycol. Soc.* **1985,** *85*, 257-266.
11. Foisner, R.; Messner, K.; Stachelberger, H.; Roehr, M. *J. Ultrastruc. Res.* **1984,** *92*, 36-46.
12. Eriksson, K.-E.; Pettersson, B.; Volc, J.; Musilek, V. *Appl. Microbio. Biotechnol.* **1986,** *23*,257-262.
13. Blanchette, R. A. *Phytopathology* **1984,** *74*, 153-160.
14. Blanchette, R. A. *Appl. Environ. Microbiol.* **1984,** *48*, 647-653.
15. Greaves, H.; Levy, J. F. *J. Inst. Wood Sci.* **1965,** *15*, 55-63.
16. Wilcox, W. W. *US For. Serv. Res. Pap.* **1968,** *FPL-70*, p. 46.
17. Saka, S; Thomas, R. J. *Wood Sci. Technol.* **1982,** *16*, 1-18.
18. Fergus, B. J.; Goring, D. A. I. *Holzforschung* **1970,** *24*, 118-12.
19. Messner, K.; Srebotnik, E. *FEMS Microbiol. Rev.* **1994,** *13*, 351-364.
20. Srebotnik, E; Messner, K. *Appl. Environ. Microbiol.* **1994,** *60*, 1383-1386.
21. Peek, R. -D. von; Liese, W.; Parameswaran, N. *Eur. J. For. Path.* **1972,** *2*, 237-248.
22. Saka, S.; Goring, D. A. I. In Biosynthesis and biodegradation of wood components; Higuchi, T.; Ed.; Academic Press: Orlando, USA, 1985; pp 51-62.
23. Kirk, T. K.; Chang, H. M.; Lorenz, L. F. *Wood Sci. Technol.* **1975,** *9*, 81-86.
24. Highley, T. L. *Mater. Org.* **1982,** *17*, 205-214.
25. Ander, P.; Eriksson, K.-E. L.; Yu, H. S. *J. Gen. Microbiol.* **1984,** *130*, 63-68.
26. Faix, O.; Mozuch, M. D.; Kirk, T. K. *Holzforschung* **1985,** *39*, 203-208.
27. Akhtar, M.; Scott, G. M.; Houtman, C. J. In 8th International Conference on Biotechnology in the Pulp and Paper Industry; Vahala, P; Lantto, R.; Eds.; VTT Biotechnology: Helsinki, FINLAND, 2001; pp 39-41.
28. Ferraz, A..; Guerra, A.; Sousa Gruz, P. B.; Mendonça, R. In 8th International Conference on Biotechnology in the Pulp and Paper Industry; Vahala, P; Lantto, R.; Eds.; VTT Biotechnology: Helsinki, FINLAND, 2001; pp 44-45.
29. Srebotnik, E.; Messner, K.; Foisner, R. *Appl. Environ. Microbiol.* **1988,** *54*, 2608-26014
30. Daniel, G.; Nilsson, T.; Petterson, B. *Appl. Environ. Micrbiol.* **1989,** *55*, 871-881.
31. Blanchette, R. A.; Krueger, E. W.; Haight, J. E.; Akhtar, M.; Akin, D. E. *J. Biotechnol.* **1997,** *53*, 203-3013.
32. Wariishi, H.; Valli, K.; Gold, M. *J. Biol. Chem.* **1982,** *267(33)*, 23688-95.
33. Banci, L.; Ciofi-Baffoni, S.; Tien, M. *Biochemistry* **1999**, *38(10)*, 3205-10.

34. Hofrichter, M.; Vares, T.; Kalsi, M.; Galkin, S.; Scheiner, K.; Fritsche, W.; Hatakka, A. *Appl. Environ. Microbiol.* **1999,** *65*, 1864-1870.
35. Roy, B. P.; Paice, M. G.; Archibald, F. S.; Misra, S. K.; Misiak, L. E. *J. Biol. Chem.* ***1994,*** *269(31)*, 19745-19750.
36. Hames, B. R.; Kurek, B.; Pollet, B.; Lapierre, C.; Monties, B. *J. Agri. Food Chem.* **1998,** *46*, 5362-5367.
37. Henriksson, G.; Zhang, L.; Li, J.; Ljungquist, P.; Reitberger, T.; Pettersson, G.; Johansson, G. *Biochem Biophys. Acta.* **2000,** *1480*, 83-91.
38. Guillen, F.; Gomez-Toribio, V.; Martinez, M. J.; Martinez, A. T. *Arch. Biochem. Biophys.* **2000,** *383*, 142-147.
39. Kapich, A. N.; Jensen, K. A.; Hammel K. E. *FEBS Lett.* **1999,** *461*, 115-119.
40. Watanabe, T.; Katayama, S.; Enoki, M.; Honda, Y. H.; Kuwahara, M. *Eur. J. Biochem.* **2000**, *267*, 4222-4231.
41. Bao, W.; Fukushima, Y.; Jensen, K. A.; Moen, M. A.; Hammel, K. E. *FEBS Lett.* **1994,** *354*, 297-300.
42. Enoki, M.; Watanabe, T.; Nakagame, S.; Koller, K.; Messner, K.; Honda, Y.; Kuwahara, M. *FEMS Microbiol. Lett.* **1999,** *180*, 205-211.
43. Harvey, P. J.; Schoemaker, H. E.; Palmer, J. M. *FEBS Lett.* **1986,** *195*, 242-246.
44. Candeias, L. P.; Harvey, P. J. *J. Biol. Chem.* **1995,** *270*, 16745-16748.
45. Ten Have, R.; Frañssen, M. C. R.; Field, J. A. *Biochem. J.* **2000,** *347*, 585-591.
46. Call, H. P.; Mücke, I. *J. Biotechnol.* **1997,** *53*, 163-202.
47. Srebotnik, E., Hammel, K.E. *J. Biotechnol.* **2000,** *81*, 179-188.
48. Tanaka, H.; Itakura, S.; Enoki, A. J. Biotechnol. **1999,** *75*, 57-70.
49. Berglund, A.; Brelid, H.; Rindby, A.; Engström, P. *Holzforschung* **1999,** *53*, 474-480.
50. Kerem, Z.; Jensen, K.; Hammel, K. E. *FEBS Lett.* **1999**, *446(1)*, 49-54.
51. Goodell, B.; Jellison, J.; Liu, J.; Daniel, G.; Paszczynski, A.; Fekete, F.; Krishnamurthy, S.; Jun, L.; Xu, G. *J. Biotechnol.* **1997,** *53*, 133-162.
52. Xu, G.; Goodell. B. *J. Biotechnol.* **2001,** *87*, 43-57.
53. Watanabe, T.; Koller, K.; Messner, K. *J. Biotechnol.* **1998,** *62*, 221-230.
54. Kirk, T. K.; Mozuch, M. D.;Tien, M. *Biochem. J.* **1985,** *226*, 455-460.
55. Gierer, J. *Holzforschung* **1997,** *51*, 34-46.
56. Johnson, G. R. A.; Nazhat, N. B.; Saadalla-Nazhat, R. A. *J. Chem. Soc., Faraday Trans. 1.* **1988,** *84*, 501-510.
57. Gallois, A.; Gross, B.; Langlois, D.; Spinnler, H. -E.; Brunerie, P. *Mycol. Res.* **1990,** *94*, 494-504.
58. Watanabe, T.; Enoki, M.; Sato, S.; Honda, Y.; Kuwahara, M.; Shirai, N.; Messner, K. *IRG/WP.* **2001,** Doc. O1-10412.
59. Lamaipis, P.; Fackler, K.; Watanabe, T.; Gindl, W; Messner, K. In 8th International Conference on Biotechnology in the Pulp and Paper Industry;

Vahala, P; Lantto, R.; Eds.; VTT Biotechnology: Helsinki, FINLAND, 2001; pp 236-237.
60. Spurr, A. J.; *Ultrastructure Research* **1969**, *26*, 31-43.
61. Okuyama, T.; Takeda, H.; Yamamoto, H. *J. Wood Sci.* **1998,** *44*, 83-89.
62. Stone, J. E.; Scallan, A. M. *Tappi* **1967**, *50,* 496-501.,
63. Fackler, K.; Srebotnik, E.; Watanabe, T.; Lamaipis, P.; Humar, M.; Tavzes, C.; Sentjurc, M.; Messner, K. In 8th International Conference on Biotechnology in the Pulp and Paper Industry; Vahala, P; Lantto, R.; Eds.; VTT Biotechnology: Helsinki, FINLAND, 2001; pp 61-62.

Chapter 6

Brown-Rot Fungal Degradation of Wood: Our Evolving View

Barry Goodell

Wood Science and Technology Department, University of Maine, Orono, ME 04469–5755 (email: Goodell@UMIT.Maine.edu)

This overview focuses on fungal attack and degradation of wood by the brown rot fungi. These fungi are perhaps the most important agents involved in the degradation of wood products and in the degradation of dead wood in coniferous ecosystems. In the decay process, brown rot fungi actively metabolize the carbohydrate fraction as well as a more minor portion of lignin, leaving behind a chemically modified lignin residue. The wood loses strength very rapidly in the early stages of degradation because of the rapid depolymerization of the cellulosic fraction. More heavily degraded wood typically displays a cubical, crumbly, brown appearance with little residual strength. Mechanisms employed by brown rot fungi in the biodegradation of wood are both enzymatic and non-enzymatic. These fungi produce no lignin degrading enzymes, but they do have a mechanism that results in lignin modification and slow lignin depletion from wood undergoing decay. Recent work by several groups has suggested that non-enzymatic, low molecular agents produced by the brown rot fungi are responsible for early stages of wood cell wall depolymerization through the production of free radical species. Similarities and differences between brown rot fungi, white rot fungi, and molds are discussed with regard to pH contorl in wood, oxalate production, and metal metabolism. Development of several hypotheses for low molecular weight metabolite function and systems postulated for non-enzymatic degradative activities are reviewed.

Introduction

Researchers have made significant advances in understanding brown rot degradation of wood over the last 30 years, particularly in understanding the components involved in the enzymatic and non-enzymatic systems which initiate brown rot attack. Brown rot degradation processes are of particular importance because these fungi preferentially attack coniferous wood products, and coniferous woody debris which is abundant in forests, especially in the northern hemisphere. Residues, largely composed of the modified lignin produced during brown rot decay, form a large part of the humic material found in coniferous ecosystems (*1*). It has been estimated that the equivalent of 10% of the timber cut in the United States decays in service each year (*2*). Since the majority of timber used in construction in the northern hemisphere is from coniferous species, and brown rot fungi preferentially degrade softwood timber, a large part of this destruction is due to the action of brown rot fungi. Interestingly, although brown rot fungi are more the more common causal agent of decay in coniferous wood products, white rot fungi, including those that cause white 'pocket rots', 'stringy rots', and amorphous white rot decays, are found at least in equal numbers to brown rot fungi as the agents producing heart- or sap-rot in living coniferous trees (*3,4*).

Early or incipient stages of brown rot decay display little outward evidence of attack. When mass losses are between five and 10 percent in brown rotted wood, in its hydrated state the incipiently decayed wood may only appear to be water stained. As decay progresses, the wood darkens, turning brown because of the preferential accumulation of modified lignin residues with removal of the holocellulose fraction. After drying, wood in advanced stages of degradation undergoes extreme shrinkage. In advanced decay stages, volume loss with drying in both longitudinal and transverse directions is much greater than in sound wood, and volume losses exceed 30% in some cases. The wood is friable and, in its weakened state, readily fractures across the grain due simply to drying stresses. The wood takes on a checked appearance with cubical pieces of the degraded wood prevalent.

In general, the degradative systems of white rot fungi have received far more attention over the last 50 years than those of brown rot fungi. Thus, we have a better understanding of the methods white rot fungi employ for woody tissue breakdown. The reason for this has been at least in part due to the economic promise held out for the application of white rot systems in industrial processing. Brown rots were originally thought not to have the capacity to degrade phenolic compounds such as lignin. Because white rot fungi are well equipped with lignin degrading enzymes, and some even degrade lignin selectively (*5,6*), their potential as bio-agents in paper bleaching and pulping processes was first explored. Later work examined their aromatic ring degrading properties for use in bioremediation of a variety of xenobiotic organic compounds found as pollutants in the environment (*7,8*). As a result, our understanding of the lignin degrading systems active in selected white rot fungi has, until recently, been significantly ahead of our understanding of brown rot mechanisms.

This began to change in the 1990's as white rot researchers and the funding agencies supporting them began to realize that significant obstacles remained before the aromatic degrading systems involved in the white rots could be effectively employed in industrial bio-processing. Concurrently, reports on the use of brown rot fungi to degrade aromatic pollutants (*9*) appeared, with the capacity of some brown rot species such as *Gloeophyllum* spp. to degrade aromatic compounds now well documented (*10,11*). The advancement of hypotheses concerning non-enzymatic systems in brown rot fungi capable of generating oxygen radicals (*11-15*) also occurred in the 1990's. It was recognized that low molecular weight systems in both brown rots and white rots must be important and that study of the brown rots may lead to a better understanding of fungal degradation mechanisms in general.

Biology of Brown Rot Fungi and Physical Properties of the Degraded Wood

The fungi that cause brown rot are in the Basidiomycotina subdivision. Decay by these fungi in the natural environment is typically initiated via the deposition of spores or mycelial fragments that are carried to wood surfaces by the wind, water, or by insect or animal vectors; or because of mycelial growth into wood in direct soil contact. As with all wood inhabiting fungi, germination of spores and/or the initiation of fungal growth into the wood will not occur until appropriate conditions of moisture and temperature are met. In general, most types of brown rot decay can be initiated when: 1) The wood moisture content is above the fiber saturation point, but the cell lumen void space is not saturated, and 2) The temperature is between 10 and 45° C. Optimal temperatures for growth and degradative activity varies with fungal species (*2*).

The brown rot fungi initially invade wood by passing through the cell lumens and colonizing ray cells and axial parenchyma where stored carbohydrate is accessible as a ready energy source for the fungus. Once established in the parenchyma, the fungal mycelia penetrate through pit membranes to access tracheid lumens where they can further proliferate (*16*). These fungi develop the capacity to penetrate the wood cell walls via bore hole production, but for some brown rot fungi penetration via pit membranes is the predominant mode of passage (*16*). In other species there is no preference for pit penetration as opposed to bore hole production (*17*). Formation of bore holes is initiated via the production of smaller hyphae (bore hyphae) with enlargement of the bore hole occurring as the decay process proceeds. The mechanism for bore hole production is not known but is thought to include a combination of mechanical, chemical, and enzymatic processes.

As decay progresses, the hyphae of brown-rot fungi become established and proliferate in the lumen of wood cells (*17-19*) The hyphae secrete a glucan layer which coats the wood cell walls, and this 'hyphal sheath' helps to bind the hyphae to the S_3 layer of the wood cell wall (*20*). While the S_3 layer may remain relatively undegraded until late in decay (*20,21*), the S_2 layer of the wood cell wall is intensely degraded. Degradation of the S_2 layer is not localized near the hyphae of fungi, which indicates that the degradation reagents produced by brown rot fungi are capable of diffusing through the S_3 into the S_2 layer of the wood cell wall. The S_2 layer of wood cell walls undergoing brown rot attack is depolymerized preferentially to other layers. This layer has a reduced lignin content and lignin density compared to the S_1 or S_3 layers (*22*), which may help explain the preferential and early depolymerization of the cellulose there. In advanced stages of decay, an intact residual S_3 layer is often observed within the cell wall while at the same time much of the S_2 layer has been severely degraded and metabolized (*20,21*).

Brown rot degradation is considered to be more problematic in structural wood products than white rot because of the way these fungi attack the wood cell wall and promote rapid strength losses. Although selective delignification of the cell wall has been observed in some white rot fungi (*5*), white rot attack of the wood cell wall typically occurs via thinning from the lumen outward. This is thought to occur because enzymes secreted by the fungus can act only on exposed wood layer surfaces. They are confined to initiating reactions only at the wall layer surface because spatial considerations and diffusion do not permit access into interior regions of the cell wall. Since all known enzymes exceed the micropore size of the intact structure of the wood matrix, in the absence of a diffusible low-molecular weight degradation system, enzymatic action is at least initially confined to thinning or erosion of the wood cell wall starting at the S_3 layer surface. Considering the strength properties of a wood cell, assuming a cylindrically shaped model approximates cells such as tracheids or fibers, the interior thinning of a cylindrical wall has only a limited effect on strength. (The bending strength properties of a hollow pipe vs. a solid rod of the same size are similar). This explains why white rot fungi cause only a gradual loss of wood strength as decay progresses. Conversely, brown rot fungi employ a low molecular weight decay system and rapidly depolymerize the cellulosic fraction of the cell wall. This rapid decrease in cellulose DP throughout the wood cell wall is perhaps the key defining feature of brown rot attack at the chemical level. Although the brown rot fungi metabolize only limited amounts of the cellulose breakdown products, at least in the initial stages of degradation, the extensive depolymerization of the cellulose in the wood cell wall dramatically reduces the strength of the wood.

Strength losses caused by brown rot fungi have been reported to be as much as 70% of modulus of elasticity (MOE) and modulus of rupture MOR (*18*) in early, or incipient, decay stages. In other work (*23*) with *G. trabeum* brown rotted wood, loss of mass from the wood was undetectable until 40% loss in strength (MOR) had occurred. As discussed above, extensive cellulose depolymerization with limited glucose metabolism has been well documented as the cause of this strength loss in early stages of degradation. However, Winandy and Morrell have shown that hemicellulose losses in early stages of brown rot degradation also correlate with wood strength loss (*24*).

The Chemistry of Holocellulose and Lignin Degradation by Brown Rot

Holocellulose degradation

Brown rot fungi have the capacity to depolymerize and metabolize the holocellulose fraction (*25*) of wood, but they also chemically modify (*1,26*) and remove (*27,28*) the lignin fraction. The carbohydrate fraction is depolymerized to soluble oligosaccharides or monosaccharides. These diffuse though the watery fungal glucan matrix (hyphal sheath) to the wood cell lumen where fungal extracellular enzymes may further act on some of the complex sugars to produce simple sugars. The fungus can absorb the resulting sugars through the fungal cell membrane, and the sugars are then metabolized. This metabolism by the fungi results in mass loss of the wood which can be readily quantified. Mass loss values are commonly used as a measure of the extent of decay in samples tested in academic and industrial laboratories (*29,30*). However, mass loss is not a sensitive predictor of strength loss for brown rot in early decay stages (*23*), and in later stages of decay can be misleading because it fails to account for wood mass conversion into fungal hyphae mass (*31,32*).

Brown rot fungi have been reported to be incapable of degrading pure cellulose in the absence of other wood components (*33-35*). However in other work, brown rot fungi are commonly grown on media containing only cellulose or glucose alone (*36-38*). It has been suggested therefore that either hemicellulose or perhaps lignin is critical to the mechanisms involved in brown rot degradation. Some researchers have suggested a mechanism whereby modified phenolic groups derived from lignin breakdown products may participate in Fenton based reactions to promote brown rot oxidative activity (*1,38*).

Early workers studying brown rot fungi assumed that enzymes were responsible for the degradative action, and attention focussed on these metabolites. Cellulolytic activity was initially described using terminology of C_1 or C_x (*39*) and this was later refined to refer to the action of general classes of exoglucanases and endoglucanases, respectively. As analytical methods were further refined and applied to the brown rot fungi more specific functionalities were defined and newly isolated enzymes were catalogued in brown rot fungi

(*40*). Brown rot fungi have been reported to produce several endo ß-1,4-glucanases and ß-glucosidases, but typically are described as lacking cellobiohydrolase (CBH) and other exoglucanases. However, one brown rot species *Coniophora puteana*, has been reported to produce two types of cellobiohydrolases (*41,42*). Brown rot fungi also typically lack cellobiose dehydrogenase (CDH) enzyme; but CDH has been isolated from the brown rot fungus *C. puteana* (*42*), and other brown rot fungi have recently been reported to posses cellobiohydrolase activity (*43*). It is now recognized that some fungi produce cellobiohydrolases and endoglucanases that have mixed functions. Glucanases have been isolated that act in a classical endo-manner of cleaving cellulose to shorter chains, but also having exo-activity in decreasing chain length through endwise hydrolysis (*44*). Other glucanases with broad specificity in degrading xylans and mannans have also been isolated (*45*). The brown rot fungal enzyme systems have not been studied adequately to provide extensive information on mixed functionality. However, endoglucanases isolated from *G. trabeum* have been reported to be active on galactomannan and glucomannan (*44*) and also on xylan (*46*).

The enzymes produced by brown rot fungi are thought to act in concert with each other as well as with non-enzymatic systems (see section below) after the initial stages of cell wall degradation allow access to sites within the wood matrix as decay progresses. Hemicellulose degradation generally occurs in early stages of cell wall degradation (*24*) as these branched and relatively low molecular weight polysaccharides are more accessible to chemical attack than the partially crystalline cellulose regions that the hemicellulose complexes with and surrounds (*47*).

Lignin Degradation

The view that lignin is slowly metabolized over time by most brown rot fungi has not been well recognized in the past (*39,48,49*), but substantial evidence now supports this view (*16,21,27,28,50*), and in some cases the oxidative removal of lignin by brown rot fungi has been reported to be as great as 25% (*27*). Lignin degrading enzymes are generally considered not to be present in the brown rot fungi. However in artificial media laccase production has been observed in the brown rot fungi *Gloeophyllum trabeum* and *Postia placenta (51)*. Laccase gene-specific sequences were also reported to be present in *G. trabeum* (*51*). Isolated reports of manganese peroxidase and lignin peroxidase in *Polyporous* spp. and other brown rot species have also been noted (*52,53*). Lignin loss or metabolism by brown rot fungi has been reported, particularly in later stages of degradation, and the middle lamella of wood (composed predominantly of lignin) has been observed to undergo attack by brown rot fungi (*21*). Brown rot fungi also regularly penetrate the wood cell wall via the production of bore holes which pass through both cell wall and middle lamellae regions, removing lignin in the process. The literature on this subject therefore suggests that low molecular weight lignin degrading agents and potentially even lignin degrading enzymes (*26*) may be present in some brown rot fungi, and these are capable of at least localized activity. A diffusible, non-enzymatic system that oxidizes lignin must also exist to account for the broader oxidative chemical changes that lignin is subjected to during brown rot decay, as discussed below.

The residual lignin in wood degraded by brown rot fungi is dealkylated (*26,54*), demethoxylated and demethylated (*1*), with some oxidation of the alkyl side chain. The aromatic ring of lignin is not attacked in brown rot degradation. Because the pore size in the native wood cell wall structure will not permit penetration of compounds the size of known cellulolytic enzymes (*55-57*), and the three dimensional matrix of lignin also prevents ready access by enzymes into interior lignin, the nature of the chemical attack on lignin suggests that a low molecular weight agent participates in the chemical modifications which occur. Barr and Aust (*58*) have suggested that hydroxyl radicals may produce a rapid depolymerization and repolymerization of lignin. This may be one way that the observed modifications of lignin structure occur in the brown rots. Non-enzymatic systems for production of hydroxyl radicals in the brown rot fungi are reviewed later in this chapter.

Importance of pH, Metals, and Low Molecular Metabolites

Oxalate and the reduction of pH

The pH of sound wood usually varies between 3 and 6 (*59-61*). Following initial colonization of the wood by hyphae however, the brown rot fungi begin to lower the pH of their wood environment (*62*). Both brown- and white rot fungi will decrease the pH of wood significantly (*63*); however, the brown rot fungi, in general, reduce the pH to lower levels than do the white rots. Highley found that the cellulolytic activity of brown rot fungi, as well as the activity of brown rot filtrate, was inhibited in high pH environments (*64*). High pH was not as inhibitory to white rot activity. Green and others (*25*) used a relatively small pH probe, approximately 1mm diameter, to obtain localized pH values of wood undergoing decay by the brown rot fungus *P. placenta*. During the first week of incubation the wood pH was reduced to 2.5 and 1.6. Since average pH values of ground, extracted wood undergoing decay by this fungus (*63*) have been reported to average 3.7, the lower localized values suggest that some brown rot fungi may be able to maintain extremely low pH microsite environments.

Most brown rot fungi appear to initiate the reduction of pH in their environment, at least in part, through the production of oxalate (*62,65*). Maintenance of pH, and micro-pH gradients (*42*) or micro-pH differentials (*1,38,66,67*) may occur as oxalate crystals are solubilized and the fungus regulates the pH via oxalic acid formation. The reduction of pH by oxalate is important to the fungus for the function of extracellular enzymes and has been discussed as a key factor in several hypotheses related to low molecular weight degradation schemes as outlined in a following section of this chapter. However, oxalate has also been hypothesized to function in a number of other ways in the fungi (*62,68*).

Oxalate is produced in large amounts by many brown rot (*25,65,69-72*) and white rot fungi (*65,68,72,73*). However only limited work has been done in the brown rot fungi to determine the biochemical pathways for its production (*74*). Within the literature, reports vary on levels of accumulation of oxalate, differ in distinguishing between production and accumulation of oxalate, and there are disagreements on production by certain species (*75,76*). The oxalate produced by decay fungi is predominantly present as calcium oxalate, and in the white and brown rot decay fungi that have been studied a dihydrate form of calcium oxalate crystal is usually formed. This form of oxalate can be slowly solubilized in dilute acid solution. Caution should therefore be exercised when assaying for oxalate in fungi because both soluble and insoluble forms (oxalic acid and calcium oxalate crystals) may not be detected with some assay methods. In addition, fungi that produce oxalate may do so only on appropriate media and under certain environmental and pH conditions. Some fungi accumulate oxalate over time while others degrade crystalline forms to more soluble forms, so the timing of oxalate sampling is also important.

Connolly and others (*72*) point out that the brown rot fungus *G. trabeum* initially produces oxalate crystals, but that these crystals are corroded or solubilized over time. Oxalate crystal corrosion was not observed in crystals produced by white rot mycelia (*72*), and instead oxalate crystal formation continued throughout the period of growth. The appearance of oxalate crystals on the surface of newly formed brown rot hyphae and the dissolution of these crystals as the hyphae age suggests that brown rot fungi may use the crystalline and acid forms of oxalate to regulate the pH of their environment allowing the brown rot fungi to produce the lower pH environments typically observed in brown rot degradation as compared to white rot. Concepts of brown rot fungi accumulating oxalate in growth media while white rot fungi do not (*77*), may reflect more the pH of the media used and other environmental or temporal factors (*62*). Species and strain differences, rather than the class of fungus, may dictate factors of oxalate accumulation and production (*25,62,72*). The hyper-production of oxalate by both brown and white rot fungi grown on wood in high pH environments (*78*) further supports this argument. Hyde and Wood (*42*), and other authors, have proposed a scheme in brown rots which requires the presence of a pH gradient, potentially supported by production of oxalate (See sections below on Fenton mechanisms for this discussion).

Low pH environments are often deleterious to living cells and protective systems are employed by some cells to provide protection from acids. The mechanism brown rot fungi employ to withstand their acidic environment is not known. However previous reports of glucan (fungal or hyphal sheath) production which enveloped and then exposed oxalate crystals produced by wood degrading fungi (*79*) suggest the possibility that the extracellular glucan matrix secreted by wood degrading fungi may not only help to maintain the pH of the fungal environment, but help protect the organism in acidic environments as well.

Metals

The ability to sequester, mobilize, and/or accumulate metals is important for virtually all microorganisms. Metals are needed in trace (or higher) amounts by all organisms for metabolic functions. Alternately, some metals may be sequestered and precipitated if present in concentrations in localized environments that would be deleterious to biological activities of the fungi. Cation uptake in Douglas-fir decayed by the brown rot *P. placenta* was examined by Shortle (*80*) who found that magnesium and calcium were actively transported into wood during the decay process. In spruce wood decayed by the same fungus, researchers found increased uptake of calcium, magnesium, iron and aluminum (*81*). The brown rot *G. trabeum* was able to translocate cations both into and out of wood blocks (*82*). Interestingly, considering the development of Fenton-based hypotheses to explain brown rot degradation processes (see later sections of this chapter), the amount of transition metals accumulated by brown rot fungi in wood are relatively low (*60*). This may be a mechanism that brown rot fungi employ to control the levels of these potentially reactive metals. When these metals are used in cell wall oxidation schemes (*36-38,83*), they may be cycled through oxidative reactions to better control transition metal-based reactions in the fungal-wood environment as well as to maximize their utilization in this environment where transition metals are frequently unavailable because of complexation with oxygen other compounds.

The emerging concept of non-enzymatic degradation

Cowling (*49*) first recognized that the known enzymes of the time were too large to penetrate into the interior structure of the wood cell wall. Therefore in brown rot degradation processes, which involved rapid depolymerization of the cellulosic fraction within the wall as opposed to a gradual thinning all components of the cell structure, a smaller compound was likely involved. Initial hypotheses focussed on the possible existence of a very small mass enzyme (*49*). However, when no such enzyme could be identified attention turned to examination of acids and components generally termed as low-molecular weight systems.

Research in the 1960's and 1970's by Halliwell (*84*), Koenigs (*85*) and others proposed that Fenton chemistry may be involved in brown rot degradation of wood. Halliwell, in work with cellulose as well as wood powders, was the first to show that cellulose could be degraded by Fenton reagents. Koenigs (*85*) showed that the action of Fenton chemistry oxidatively degraded the cellulosic fraction of the wood and, at least in softwoods, did not appear to oxidize the lignin fraction. Since ferrous iron is required in Fenton reactions and this oxidation state of iron is absent in oxygenated environments where decay processes would occur (*86,87*), a mechanism to reduce iron was required. Work by Schmidt and others (*88*) suggested that oxalate could potentially function as this reducing agent; however, later work revealed that this type of reaction with oxalate was unlikely to occur in the absence of light. It is well established that brown rot decay processes are not dependant on light (*87,89-90*).

Brown rot degradation mechanisms have been proposed which involve oxalate, separate from its role in reducing pH, as a primary agent in the acid hydrolysis of cellulose and hemicellulose (*25*). Cellulose is well known to be resistant to the action of weak, non-oxidizing acids (*91*), but hemicellulose with its non-crystalline structure may be more susceptible to attack. Brown rot fungi have been reported to preferentially deplete mannan from the hemicellulose fraction in early stages of degradation (*24,39*). However, when wood is treated with high concentrations of oxalic acid, depletion of mannan is not observed (*25*), and the loss of other sugars except arabinose (a relatively minor pentose sugar which occurs as a side chain in some hemicelluloses) is also minimal. This suggests that the action of oxalate on wood does not mimic the action of brown rot fungi. However, the slow action of oxalate on the hemicellulose matrix may help to open up the wood structure to permit later penetration of fungal metabolites.

Fenton-based free radical mechanisms

Based on the earlier work conducted with Fenton based systems for hydroxyl radical production, several research groups have developed hypotheses explaining the function of low molecular weight metabolites, metals, and radicals that may initiate degradation of the wood cell wall in brown rot degradation. Several of these hypotheses are reviewed in the following sub-sections. Potential concerns and areas of convergence and divergence in these hypothesis are discussed within each subsection:

Glycopeptide Degradation

Glycopeptide activity in white rot, brown rot, and soft-rot fungi is reviewed in depth in a later chapter (see Enoki, Tanaka, and Itakura) and is reviewed here only briefly, with focus on the brown rot fungi.

Low-molecular weight compounds, termed 'glycopeptides', were first isolated from brown rot fungi in the late 1980's (*92,93*). These compounds initially were described as having a 1.5kD-5kD mass, containing a small percentage of iron, with approximately 22% neutral carbohydrate and 12%

protein (*94*). A new molecular mass of 7.2kD-12kD for the glycopeptide from the brown rot *Tyromyces palustris* has recently been reported (*95*). The glycopeptide possesses the capacity to bind ferric iron, reduce it, and then bind the reduced iron. In the presence of hydrogen peroxide the glycopeptide generates one-electron oxidation potential as determined by 2-keto-4-thiomethylbutyric acid oxidation to carbon dioxide. The glycopeptide also possesses the ability to oxidize NADH in the presence of oxygen and thereby produces hydrogen peroxide so that Fenton type reactions can occur.

The structure of the glycopeptide has not yet been elucidated but a heme prosthetic group functioning as an active site has been ruled out because of the lack of absorbance from 240 to 700 nm. Reaction with dimethylsulfoxide indicates that the one electron oxidation activity comes from the production of hydroxyl radicals. Enoki, Tanaka and others from this group have proposed that the glycopeptide produced by brown rot fungi diffuses into the wood cell wall to initiate one-electron oxidation reactions via the reaction of the bound ferrous iron and hydrogen peroxide (*83*). These reactions promote the oxidative degradation of both lignin and cellulosic components.

One potential issue with the brown rot glycoprotein hypothesis is that the molecular size of pores in the intact structure of intact wood cell walls has been reported to exclude compounds with kD mass greater than 6,000 (*21,55,96*). If the glycopeptide is at the upper range of 7.2-12kD mass as reported, then it may be excluded from the wood cell wall; however, the authors discuss the diffusion of a deglycosylated 'effector' form of the glycopeptide capable of penetrating the wood matrix (*95*). It is also conceivable that the shape of the glycopeptide is elongated and this may allow cell wall penetration. Alternately, the glycopeptide may be responsible for the generation of longer-lived radicals such as superoxide (*95*) which are capable of penetrating the wall microvoid spaces.

Cellobiose Dehydrogenase Iron Reduction – Autoxidation

Hyde, Wood and co-workers (*42,89,97*) proposed a model where cellobiose dehydrogenase (CDH), produced by the brown rot fungus *Coniophora puteana*, would initially bind ferric iron in the low pH environment immediately surrounding the fungal hyphae. Critical to this hypothesis is the development of a pH gradient by the fungus where, as shown in an example model by the authors, the pH at the surface of the hyphae is at 2.5 and increases at a distance from the hyphae to pH 4.5. CDH is a multifaceted enzyme capable of carrying out the extraction of electrons from cellobiose as well as the higher cellodextrans. The enzyme also catalyses the reduction of quinones and will bind and reduce ferric iron.

In the presence of oxalate, which the fungus employs to generate and maintain the low pH environment surrounding the fungal hyphae, CDH is capable of binding and reducing iron. The reduced iron is then released from the enzyme where, in the low pH environment, it is stable to autoxidation. It has been estimated (*42*) that the reduced valence state of iron may have a lifetime of an hour or more at the pH's surrounding the fungal hyphae before autoxidation will occur. However, as the free ferrous iron diffuses away from the fungal hyphae, a higher pH environment is encountered and rapid autoxidation can then occur with this reaction also resulting in the production of superoxide and hydrogen peroxide (*98*). The peroxide is able to diffuse through membranes or into the wood cell wall where it is able to react with any reduced (*94*) iron in the environment in a Fenton reaction to produce hydroxyl radicals.

CDH is widespread in the fungi and has been proposed to function in the white rot fungi in the generation of hydroxyl radicals (*99*). In addition to its potential importance in Fenton chemistry initiation, it also has been postulated to inhibit lignin repolymerization in white rot degradation (*100*). CDH production in brown rots had previously been observed only in *C. puteana* (*42,101*) and this mechanism was therefore considered unlikely to play a role in other brown rot fungal degradation schemes (*42*). However, in the presence of cellulose, cellobiose dehydrogenase (CDH) and quinone reducing activity was recently found in *P. chrysosporium*, *G. trabeum* and the non-decay fungus *Trichoderma viride (43)*. The overall role CDH and reductase enzymes may play in non-enzymatic biodegradative systems has still not been clearly elucidated and should be further explored.

A concern with the CDH iron reduction-autoxidation hypothesis pointed out by the authors of the concept (*42*) is that the iron-oxalate reaction is very slow, even at pH 2.5, and therefore, even if the enzyme is found in other brown rot fungi, iron reduction may be too limited to be effective. Another issue is that autoxidation of iron to produce hydrogen peroxide at a specific pH may be problematic because iron cannot be readily maintained in two different oxidation states in the same environment. If the pH is high enough for ferrous iron to autoxidize and form hydrogen peroxide, then this oxidant will diffuse and react with any ferrous iron in the immediate vicinity. It is unlikely given this scenario that significant amounts of ferrous iron would then be available to diffuse deep into the wood cell wall. Given the extremely short half-life of hydroxyl radicals, Fenton reactions would be required to occur immediately adjacent to sites where they are proposed to react (*67,102,103*). All components of Fenton reactions must therefore be present within the S_2 layer of the wood cell wall to initiate wood oxidation of cellulose and lignin in this layer.

Chelator-Mediated Fenton System (CMFS)

Work in this area was initiated in the late 1980's with the discovery of phenolate or catecholate derivative compounds produced by wood degrading fungi (*13,104*). These compounds are able to bind and reduce transition metals, particularly ferric iron, and are able to carry out multiple reductions in a non-stochiometric (greater than 1:1), 'pseudo-catalytic' manner. Initially these chelators were termed siderophores because the laboratory assays used to characterize their iron-binding capabilities were taken from the bacterial and fungal siderophore literature (*105-108*). Some of the siderophores isolated from bacteria, non-wood-degrading fungi, and other microorganisms are undoubtedly similar to chelators produced by wood degrading fungi (*14,109*). However, unlike the proposed function of brown rot chelators in mediating Fenton chemistry reactions, the function of siderophores in other microorganisms has been described in terms of iron sequestration and transport to allow the organism to carry out critical internal metabolic processes. Criteria for acceptance as a microbial siderophore also require that the producing organism bear a specific receptor site on the membrane surface, and to date these receptors have not been isolated from wood-degrading fungi. Given the ambiguity in the microbial literature on the function of fungal chelators, and because several compounds have been isolated from *G. trabeum* and their structures identified (*36-38,83*), these compounds are now generally described simply as 'low molecular weight fungal chelators'. Although two catechol-quinone pairs have been characterized as to structure in the Gt chelator fraction (*37,83*), because most of the structures in this active low molecular weight fraction still have incompletely assigned functional groups, the general term 'Gt chelator' has been used to define the low molecular weight fraction that contains the active metal binding phenolate/catecholate compounds (*38,66*).

Studies on the function of Gt chelator (*14,38,66,67,110,111*) have indicated that these compounds are produced by the fungus and can be immunolocalized in the S_2 layer of wood cell walls undergoing active degradation (*13,112*). These fungal biochelators also mediate the production of hydroxyl radicals within the wood cell wall through what has been termed a chelator-mediated Fenton system (CMFS). In this system iron is reduced and then repeatedly 're-reduced' (*112*), exceeding a 1:1 ratio for reduction of iron by catechol. This mechanism has not been well described in the free radical literature, nor in the medical literature, where much of the published work on metal sequestration and oxygen radical production is found. However, previous work (*113,114*) also suggested a greater than 1:1, non-stochiometric, reduction of iron by caffeic acid and by catechols (*115*). More recent work has also shown that some catechols reduce iron via oxidation and complete or partial mineralization of the phenolic ring to CO_2 (*116*). Other work on this pseudo-catalytic mechanism using model compounds simulating the action of brown rot degradation showed similar oxidative degradation of 2,3-dihydroxybenzoic acid with higher amounts of iron produced than would be expected from a 1:1 stochiometric reduction (*66*). This suggests that other orthodihydroxy phenolic groups or metal binding/reducing compounds produced by brown rot fungi may function in a similar manner.

Treatment of wood with Gt chelator in CMFS reactions resulted in a reduction in cellulose crystallinity similar to that produced by brown rot fungi (*12,38*) as revealed by X-ray diffraction. A reduction in the molecular weight of crystalline cellulose was also observed when Avicell was treated using Gt chelator in CMFS (*67*). Work also showed that oxalic acid could be used by the fungus, solubilizing iron to maintain its availability at the low pH values typical of those produced at microsites by brown rot fungi (*25*). When the pH was raised to levels typical of the wood cell wall however, the Gt chelator phenolates pulled iron from both oxalate (and from cellulose components) allowing CMFS to proceed. This research (*66*) suggests that a pH differential must exist between the fungal hyphal environment and the wood cell wall. The pH differential may be manifested as a relatively abrupt pH transition at the interface of the wood cell lumen (or hyphal sheath) and S_3 layer (*38,67*) consistent with the presence of an intact S_3 cell wall layer throughout the decay process in some brown rots. The concept of a pH differential may fit previously reviewed observations of oxalate production and cell wall buffering better than the concept of a gradual pH transition within the lumen void space as suggested by Hyde and Wood (*42*).

One issue related to the CMFS is that previous workers, in research related to the study of white rot degradation mechanisms (*117*), were unable to identify demethylated products from model lignin compounds following treatment with Fenton reagent. However, unidentified compounds were the major product from Fenton degradation of the model compounds, accounting for 70% of the initial model compounds. Most of these compounds were polar phenolics. More complete work, comparing products from the reactions of Fenton chemistry on wood and lignin to the products from brown rot wood degradation by fungi, is warranted (*1*).

Quinone Redox Cycling

Identification of some of the active compounds involved in the CMFS by the brown rot fungus *G. trabeum* has allowed alternative hypotheses focussed on the action of these compounds and the generation of free radicals to be developed. Paszczyski and others (*37*) first identified 4,5-dimethoxycatechol (DMC) (*37*) and 2,5-dimethoxyhydroquinone (DMH) as important components of CMFS and suggested that redox cycling mediated by quinone reductases may be involved. Kerem and others (*36*) also found the DMH compound as well as its oxidized benzoquinone form (2,5-DMBQ) was produced by *G. trabeum*. In later work this group also found the oxidized benzoquinone form of DMC (4,5-DMBQ) (*83*). Other low molecular weight compounds are also produced by this fungus using different media (*38*) and it is likely that under different culture conditions, or on different natural substrates, that other compounds may be produced (*37*). The finding of two catechol/quinone pairs, and the strong reductive capability of fungal cultures toward quinones, suggested that enzyme-mediated oxidative reactions could be occurring which involve these catechol-quinone couples (*36,37*).

2,5-DMBQ was found to cleave a polyethylene glycol (PEG) model compound in the presence of ferric iron, but only in the presence of *G. trabeum* mycelium (*36*). This suggested that the reducing capacity of the fungus is important in producing DMH, the hydroquinone form capable of binding and reducing metal, and this reduction was demonstrated (*36*). With DMH present, ferric iron was reduced and the ferrous iron was then able to react with hydrogen peroxide, also produced by the fungus, in Fenton reactions. Although production of hydroxyl radicals was not demonstrated in this work, they were presumed to be responsible for the attack on PEG.

More recent work (*83*) has furthered the concept that the benzoquinones 2,5-DMBQ and 4,5-DMBQ can be redox cycled using extracellular quinone reductases produced by the fungus. The enzyme provides a source of electrons to reduce these compounds to their hydroquinone states (DMH and DMC, respectively) that can bind and reduce iron. This mechanism provides an alternative explanation to other work already reviewed (*66,67*) where the metal binding chelators would be oxidized to produce iron reducing potential.

For either mechanism to work however, the reactive components must be capable of diffusing into the wood cell wall. This is because: 1) Hydroxyl radicals must be generated in very close proximity to their reaction site to be effective (*102,103,118*), and 2) Free ferric iron, a very limited metal in oxygenated environments (*87,119*), is readily bound to the wood cell wall (*67*) and after moving into the wall will not readily diffuse back into the wood cell lumen void. Because, as reviewed earlier in this chapter, enzymes cannot penetrate the intact matrix of the wood cell wall, it is unlikely that a quinone reductase enzyme-based system is actively promoting Fenton reactions within the wood cell wall (*1,66*). It is possible that enzymes may be involved in producing longer-lived radical species which are capable of diffusing into the wood cell wall to effect oxidative damage, or to generate low molecular weight compounds (mediators) with reducing potential (see chapter by Li, K. in this text). However, these mechanisms have not yet been found to be active in the brown rot fungi.

Quinone reductase enzymes are important for reducing quinones in the immediate environment of the fungal hyphae because these compounds may be toxic to the fungus and can mediate undesirable reactions. These enzymes may also be important from the standpoint of Fenton-based chemistry because they can reduce quinones produced by the fungus to hydroquinone or catechol type compounds which can bind and reduce iron. Once these reduced compounds diffuse deep within the S_2 layer of the wood cell wall to mediate Fenton reactions there, it is unlikely that the oxidized quinonic forms would diffuse back repeatedly to the wood lumen to be reduced by the enzyme. Enzymatic redox-cycling of these compounds as it has been described (*36*), would be limited. Instead the quinone reductases may simply carry out the important function of reducing quinones in the wood lumen region, limiting toxicity to the fungus and producing metal binding capacity as described earlier in this chapter (*67,111*). Another possible role for the membrane bound quinone redox system is pH regulation. The redox system in brown rots is thought to be linked to a proton pump (*120*) which may function, as an alternate to oxalate, to lower the pH in the area adjacent to the hyphae.

Fenton Chemistry and Metal Binding Compounds as Related to Differences Between Decay and Non-Decay Fungi

Brown and white rot fungi have both been shown to produce biochelators, although the pattern of production and properties of the biochelators produced appears to differ (*43,121*). In studies which examined both decay and non-decay fungi, all species tested produced iron-binding compounds. This is to be expected since all microorganisms studied have been shown to produce some form of siderophore or low molecular weight iron binding compound (*122*). Among the decay fungi that have been examined, the brown rot fungi displayed greater amounts of iron-reducing activity than the white rot fungi (*43*), but some non-decay mold species were also found to possess significant iron binding capability in the low molecular weight fraction. Both decay and non-decay fungi produced hydroxamic acid type iron chelators (*123*). Within the decay fungi however, hydroxamic type chelator production was greater in the white rot isolates examined, whereas more phenolate (or catecholate) chelators were produced by the brown rot isolates studied (*43*). The compounds hypothesized to be active in Fenton reactions in the brown rots, as reviewed previously in this chapter, would all fall into the phenolate class of chelator as described in classical microbiology terminology (*122*).

Even though the white rot and non-decay fungi also produce some level of low molecular weight iron binding compounds, it is important to recognize that adaptive changes in the manner which metabolites are used is the classic evolutionary stategy employed in biology for evolutionary development. Developmental change that allows metabolites to be used for new functions is one mechanism that species employ to branch out to exploit new niches in the environment. This is the essence of evolutionary principle and may have allowed some of the higher fungi (basidiomycetes in particular) to use metal binding metabolites in different ways than microorganisms that are unable to degrade wood. The manner in which brown rot fungi manipulate the pH of their environment, the availability of metals, production of hydrogen peroxide, and many other metabolite and microenvironmental differences are all different than those observed in white rot or non-decay fungi. Just as some *Trichoderma* mold species produce copious amounts of cellulase enzyme, but do not degrade cellulose when they inhabit wood, most fungi do not degrade wood simply by producing low molecular weight metal binding compounds. The absence or presence of a key component, or the repression or expression of a single gene, may be all that separates degradative ability or mechanisms for lignocellulose degradation. This would also help to explain the presence of gene sequences for lacasse in some brown rot fungi that do not normally produce lacasse, as reviewed earlier in this chapter (*51*).

The addition of a catechol derivative to wood in the absence of soluble iron or appropriate pH to allow iron reduction, will not result in the appropriate valence state of iron to initiate Fenton reactions. Similarly, even if appropriate conditions exist to promote iron reduction, hydroxyl radicals will not be produced unless a system for generating hydrogen peroxide in proximity to the reduced iron exists. Further, even if conditions exist to generate Fenton reactions, these reactions must occur very close to the site of oxidative action on lignocellulose for oxidative degradation of wood to occur. Site-directed hydroxyl radical production mediated by the binding of Fenton reactants to cellulose, just as has been discussed in site-directed hydroxyl radical damage to DNA, is also required (*67,118*). Lignin, which surrounds or encrusts the cellulose where iron may be bound may be depolymerized, and rapidly repolymerized (*58*), when Fenton reactions occur at these sites. Discussion (*1,78*) on the possible participation of lignin degradation products in mediated Fenton reactions is an area which may prove fruitful to our further understanding of brown rot degradation processes.

Conclusions

Our view of brown rot decay has changed considerably over the last 50 years. It is well established that the brown rot fungi rapidly depolymerize holocellulose in the wood cell wall and degrade hemicellulose components early in the decay process. Lignin is also oxidatively demethylated and its propyl side chains are oxidized. Our earlier concept that enzymes were the primary agents involved in the degradative process, has now changed to an evolving understanding that low molecular weight metabolites must be involved as precursors and/or co-agents with enzymatic degradation. The view that brown rots do not remove or oxidize lignin has also changed. Scant evidence exists for the production of lignin degrading enzymes by the brown rot fungi. However, non-enzymatic systems have clearly been shown to oxidize lignin as well as cellulose in these fungi, and evidence of lignin metabolism, although much slower than that seen in the white rot fungi, is clear.

The brown rot fungi, in general, reduce the pH of their microenvironment to a greater extent than do the white rot fungi. This is thought to favor activity by some non-enzymatic systems hypothesized to be active in the bronw rots, as well as cellulolytic enzyme activity. The brown rot fungi appear to have the ability to regulate the pH of their fungal microenvironments within the wood. Non enzymatic systems currently being examined include work with low molecular weight glycopeptides, cellobiose dehydrogenase enzyme systems, phenolate chelators, and quinones. Further work must be done on all systems to examine convergent and divergent concepts on mechanisms.

Acknowledgements

I would like to thank the many graduate students and colleagues I have worked with over the years who have all educated me on the ways of fungi and fungal degradation of wood, and therefore have helped in the writing of this chapter. I also acknowledge the support of the UMaine WUR program. This is publication number 2545 of the Maine Agricultural and Forest Experiment Station.

References

1. Filley, T. R.; Cody, G. D.; Goodell, B.; Jellison, J.; Noser, C.; Ostrofsky, A. *Org. Geochem.* **2002**. *33* (2) 111-124.
2. *Wood Microbiology: Decay and its prevention*; Zabel, R. A.; Morrell, J. J.; Academic Press, Inc.: San Diego, CA, 1992; pp 476.
3. Thomas, G. P.; Thomas, R. W. *Can. J. Bot.* **1954,** *32*, 630-653.
4. *Common tree diseases of British Columbia*; Allen, E. D.; Morrison, D.; Wallis, G.; Canadian Forest Service: Victoria, BC, 1996; pp 178.
5. Blanchette, R. A.; Burnes, T. A.; Eerdmans, M. M.; Akhtar, M. *Holzforschung* **1992,** *46*, 106-115.
6. Akhtar, M.; Attridge, M. C.; Myers, G.; Blanchette, R. A. *Holzforshung* **1993,** *47*, 36-40.
7. Paszczynski, A.; Crawford, R. L. *Biotechnol. Prog.* **1995,** *11*, 368-379.
8. Hammel, K. E. In *Metal Ions in Biological Systems;* Sigel, A.; Marcel Dekker, Inc.: New York, 1992; pp. 41-60.
9. Lamar, R. T. *International Research Group on Wood Preservation .Cannes conference proceedings* **1995** *95-50040.*
10. Fahr, K.; Wetzstein, H.-G.; Grey, R.; Schlosser, D. *FEMS Microbiol. Lett.* **1999,** *175*, 127-132.
11. Wetzstein, H.-G.; Stadler, M.; Tichy, H.-V.; Dalhoff, A.; Karl, W. *Appl. Environ. Microb.* **1999,** *65*, 1556-1563.
12. Jellison, J.; Chandhoke, V.; Goodell, B.; Fekete, F.; Hayashi, N.; Ishihara, M.; Yamamoto, K. *International Research Group on Wood Preservation* **1991,** 90-1479.
13. Jellison, J.; Chandhoke, V.; Goodell, B.; Fekete, F. A. *Appl. Microbiol. Biotechnol.* **1991,** *35*, 805-809.
14. Goodell, B.; Jellison, J. In *Forest Products Biotechnology;* Palfreyman, J.; Taylor and Francis Publishers: London, 1998; pp. 235-250.
15. Chandhoke, V.; Goodell, B.; Jellison, J.; Fekete, F. A. *FEMS Microbiol. Lett.* **1992,** *90*, 263-266.
16. Daniel, G. *FEMS Microbiol. Rev.* **1994,** *13*, 199-233.
17. Liese, W. *Ann. Rev. Phytopath.* **1970,** *8*, 231-257.
18. Wilcox, W. W. *Wood Fiber Sci.* **1978,** *9*, 252-257.
19. Kuo, M.-L.; Stokke, D. D.; McNabb, H. S. *Wood and Fiber Science* **1998,** *20*, 405-414.
20. Illman, B.; Highley, T. L. In *Biodeterioration Research: Vol 2;* Llewellyn, G. C.; Perseus Books, Basic Books: NY, 1989.
21. Kim, Y. S.; Goodell, B.; Jellison, J. *Holzforschung* **1991,** *45*, 389-393.
22. Sachs, I. B.; Clark, I. T.; Pew, J. C. *J. Polym. Sci.* **1963,** *C-2*, 203-212.

23. Curling, S.; Clausen, C. A.; Winandy, J. E. *International Research Group on Wood Preservation* **2001,** 01-20219.
24. Winandy, J.; Morrell, J. *Wood and Fiber Science* **1993,** *25*, 278-288.
25. Green, F., III; Larsen, M. J.; Winandy, J. E.; Highley, T. L. *Mater. und Organis.* **1991,** *26*.
26. Jin, L.; Schultz, T. P.; Nicholas, D. D. *Holzforschung* **1990,** *44*, 133-138.
27. Irbe, I.; Andersone, I.; Andersons, B.; Chirkova, J. *Inter. Biodeter. Biodeg.* **2001,** *47*, 37-45.
28. Machuca, A.; Ferraz, A. *Enzyme Microb. Tech.* **2001,** *29*, 386-391.
29. Anon. In *ASTM standards on materials and environmental microbiology. Second edition.;* Agents, A. S. E. o. A.; ASTM: Philadelphia, PA, 1993; pp. 110-113.
30. Anon. In *AWPA Standards;* Anon.; AWPA. American Wood-Preservers' Association: Granbury, TX, 2001.
31. Jones, H. L.; Worrall, J. J. *Mycologia* **1995,** *87*, 459-466.
32. Jones, H. L.; Worrall, J. J. *Phytopathology* **1993,** *83*, 1364.
33. Highley, T. L. *Wood Fiber Sci.* **1973,** *5*, 50-58.
34. Nilsson, T. *Mater. und Organis.* **1974,** *9*, 173-198.
35. Enoki, A.; Takahashi, M.; Tanaka, H.; Fuse, G. *Mokuzai Gakkaishi* **1985,** *31*, 397-408.
36. Kerem, Z.; Jensen, K. A.; Hammel, K. E. *FEBS Lett.* **1999,** *446*, 49-54.
37. Paszczynski, A.; Crawford, R.; Funk, D.; Goodell, B. *Appl. Environ. Microb..* **1999,** *65*, 674-679.
38. Goodell, B.; Jellison, J.; Liu, J.; Daniel, G.; Paszczynski, A.; Fekete, F.; Krishnamurthy, S.; Jun, L.; Xu, G. *J. Biotechnol.* **1997,** *53*, 133-162.
39. Kirk, T. K.; Highley, T. L. *Phytopathology* **1973,** *63*, 1338-1342.
40. Kleman-Leyer, K. M.; Kirk, T. K. *Appl. Environ. Microb.* **1994,** *60*, 2839-2845.
41. Schmidhalter, D. R.; Canevescini, G. *Arch. Biochem. Biophys* **1993,** *300*, 551–558.
42. Hyde, S. M.; Wood, P. M. *Microbiology* **1997,** *143*, 259-266.
43. Qi, W.; Jellison, J. *International Research Group on Wood Preservation* **2000,** *00-10364.*
44. Mansfield, S. D.; Saddler, J. N.; Gübitz, G. M. *Enzyme Microb. Technol.* **1998,** *23*, 133-140.
45. Kvesitadze, E. G.; Lomitashvili, T. B.; Khutsishvili, M. P.; Lamed, R.; Bayer, E. A. *Appl. Biochem. Biotechnol.* **1995,** *50*, 137–143.
46. Herr, D.; Baumer, F.; Dellweg, H. *Arch. Microbiol.* **1978,** *117*, 287–292.
47. Jeffries, T. W. In *Biochemistry of Microbial Degradation;* Ratledge, C.; Kluwer Academic Publishers: The Netherlands, 1994; pp. 233-277.
48. Danninger, E. V.; Messner, K.; Röhr, M.; Stachelberger, H. *Mater. und Organis.* **1984,** *19*, 287-299.
49. Cowling, E. B. **1961**. USDA Technical Bulletin No. 1258, pp 79
50. Highley, T. L.; Dashek, W. V. In *Forest Products Biotechnology;* Palfreyman, J. W.; Taylor and Francis: London, UK, 1998; pp. 15-36.

51. D'Souza, T. M.; Boominathan , K.; Reddy, C. A. *Appl. Environ. Microb.* **1996,** *62*, 3739-3744.
52. Szklarz, G. D.; Antibus, R. K.; Sinsabaugh, R. L.; Linkins, A. E. *Mycologia* **1989,** *81*, 234-240.
53. Dey, S.; Miti, T. K.; Bhattacharyya, B. C. *J. Ferment. Bioeng.* **1991,** *72*, 402-404.
54. Kirk, T. K.; Farrel, R. L. *Annu. Rev. Microbiol.* **1987,** *41*, 465-505.
55. Flournoy, D. S.; Paul, J. A.; Kirk, T. K.; Highley, T. L. *Holzforschung* **1993,** *47*, 297-301.
56. Kleman-Leyer, K.; Agosin, E.; Conner, A. H.; Kirk, T. K. *Appl. Environ. Microb.* **1992,** *58*, 1266-1270.
57. Stone, J. E.; Scallan, A. M. *Tappi J.* **1967,** *50*, 496-501.
58. Barr, D. P.; S. D. Aust, S. D. *Environ. Sci. Technol.* **1994,** *28*, 78A-87A.
59. Gray, V. R. *J. I. Wood Sci.* **1958,** *1*, 58-64.
60. Jellison, J.; Connolly, J. H.; Goodell, B.; Doyle, B.; Illman, B.; Fekete, F.; Ostrofsky, A. *Int.Biodeter Biodegr.* **1997,** *39*, 165-179.
61. *Fungal decomposition of wood: its biology and ecology.*; Rayner, A. D. M.; Boddy, L.; John Wiley and Sons: NY, 1988; pp 578.
62. Connolly, J. H.; Jellison, J. *International Research Group on Wood Preservation* **1994,** *94-10075.*
63. Jellison, J.; Smith, K. C.; Shortle, W. T. *International Research Group on Wood Preservation* **1992,** 1552-92.
64. Highley, T. L. *Mater. und Organis.* **1976,** *11*, 161-170.
65. Espejo, E.; Agosin, E. *Appl. Environ. Microb.* **1991,** *57*, 1980-1986.
66. Goodell, B.; Qian, Y.; Jellison, J.; Richard, M.; Qi, W. *Biotechnology in the Pulp and Paper Industries* **2002**.
67. Xu, G.; Goodell, B. *J. Biotechnol.* **2001,** *87*, 43-57.
68. Dutton, M. V.; Evans, C. S.; Atkey, P. T.; Wood, D. A. *Appl. Microbiol. Biotechnol.* **1993,** *39*, 5-10.
69. Green III, F.; Larsen, M. J.; Hackney, J. M.; Clausen, C. A.; Highley, T. L. *5th International Conference on Biotechnology in the Pulp and Paper Industry* **1992**.
70. Green III, F.; Clausen, C. A. *International Research Group on Wood Preservation* **2001,** *01-10388.*
71. Bech-Anderson, J. *International Research Group on Wood Preservation* **1987,** *1330-87.*
72. Connolly, J. H.; Arnott, H. J.; Jellison, J. *Scanning Microscopy* **1996,** *10*, 385-400.
73. Connolly, J. H.; Jelllison, J. *Can. J. Bot.* **1995,** *73*, 927-963.
74. Munir, E.; Yoon, J. J.; Tokimatsu, T.; T.Hattori; Shimada, M. *Proc. Natl. Acad. Sci. USA.* **2001,** *98*, 11126-11130.
75. Jordan, C. R.; Dashek, W. V.; Highley, T. L. *Holzforschung* **1996,** *50*, 312-318.
76. Jordan, C. R.; Dashek, W. V.; Highley, T. L. *International Research Group on Wood Preservation* **1994,** *94-10063.*

77. Munir, E.; Hattori, T.; Shimada, M. *International Research Group on Wood Preservation* **2001,** *01-10394.*
78. Goodell, B.; Daniel, G.; Liu, J.; Mott, L.; Frank, R. *Forest Prod. J.* **1997,** *47*, 75-80.
79. Arnott, H. J.; Webb, M. A. *Scanning Electron. Microsc.* **1983,** *IV*, 1747-1758.
80. Shortle, W. T. *Wood Science* **1982,** *15*, 29-32.
81. Ostrofsky, A.; Jellison, J.; Smith, K. C.; Shortle, W. T. *Can. J. For. Res.* **1997,** *27*, 567-571.
82. Connolly, J. H.; Jellison, J. *Int. Biodeter. Biodegr.* **1997,** *39*, 181-188.
83. Jensen Jr., K. A.; Houtman, C. J.; Ryan, Z. C.; Hammel, K. E. *Appl. Environ. Microbiol.* **2001,** *67*, 2705-27011.
84. Halliwell, G. *Biochem. J.* **1965,** *95*, 35-40.
85. Koenigs, J. W. *Wood Fiber Sci.* **1974,** *6*, 66-80.
86. Sulzberger, B.; Laubscher, H. *Mar. Chem.* **1995,** *50*, 103-115.
87. Wu, J.; Luther III, G. W. *Mar. Chem.* **1995,** *50*, 159-177.
88. Schmidt, C. J.; Whitten, B. K.; Nicholas, D. D. *Proc. Am. Wood-Preserv. Assoc.* **1981,** *77*, 157-164.
89. Hyde, S. M.; Wood, P. M. *International Research Group on Wood Preservation* **1995,** 95-15010.
90. Zepp, R. G.; Faust, B. C.; Hoigné, J. *Environ. Sci. Technol.* **1992,** *26*, 313-319.
91. Green, D. W.; Winandy, J. E.; Kretschmann, D. E. In *Wood handbook - wood as an engineering material;* Anon.; USDA, Forest Service, Forest Products Laboratory: Madison, WI, 1999; pp. 463.
92. Enoki, A.; Tanaka, H.; Fuse, G. *Wood Sci Technol.* **1989,** *23*, 1-12.
93. Tanaka, H.; Fuse, G.; Enoki, A. *Mokuzai Gakkaishi* **1991,** *37*, 986-988.
94. Hirano, T.; Tanaka, H.; Enoki, A. *Mokuzai Gakkaishi* **1995,** *41*, 334-341.
95. Enoki, A.; Tanaka, H.; Itakura. In *This ACS book;* American Chemical Society, 2003.
96. Srebotnik, E.; Messner, K. *Holzforschung* **1991,** *45*, 95-101.
97. Wood, P. M. *FEMS Microbiol. Rev.* **1994,** *13*, 313-320.
98. Miller, D. M.; Buettner, G. R.; Aust, S. D. *Free Radical Biol. Med.* **1990,** *8*, 95-108.
99. Cameron, M. D.; Aust, S. D. *Arch. Biochem Biophys.* **1999,** *367*, 115-121.
100. Temp, U.; Eggert, C. *Appl. Environ. Microbiol.* **1999,** *65*, 389-395.
101. Ander, P.; Eriksson, K.-E. *Prog. Ind. Microbiol,* **1978,** *14*, 1-58.
102. Dreosti, I. E. In *Trace Elements, Micronutrients, and Free Radicals;* Dreosti, I. E.; Humana Press, Inc.: Clifton, NJ, 1991; pp. 149-170.
103. Cadenas, E. In *Oxidative Stress and Antioxidant Defenses in Biology;* Ahmad, S.; Chapman and Hall, International Thompson Publishing Co.: NY, 1995; pp. 1-61.
104. Fekete, F. A.; Chandhoke, V.; Jellison, J. *Appl. Environ. Microbiol.* **1989,** *55*, 2720-2722.

105. Hofte, M. In *Iron Chelation in Plants and Soil Microorganisms;* Hemming, B.; Academic Press, 1993; pp. 3-26.
106. Raymond, K. N.; Carrano, C. J. *Accounts Chem. Res.* **1979,** *12*, 183-190.
107. Hider, R. C. In *Iron, Siderophores and Plant Diseases;* Swinburne, T. R.; Plenum Press, 1985; pp. 49-60.
108. Neilands, J. B.; Konopka, K.; Schwyn, B.; Coy, M.; Francis, R. T.; Paw, P. W.; Bagg, A. In *Iron Transport in Microbes, Plants and Animals;* Winkelmann, G., van der Helm,_D. and Neilands, J. B; VCH Publishers: NY, 1987; pp. 3-33.
109. Lopez-Goni, I.; Moriyon, I.; Neilands, J. B. *Infect. Immun.* **1992,** *60.*
110. Lu, J.; Goodell, B.; Liu, J.; Enoki, A.; Jellison, J.; Tanaka, H.; Fekete, F. *International Research Group on Wood Preservation* **1994,** 94-7010.
111. Goodell, B.; Jellison, J.; Qian, Y.; Daniel, G. *TAPPI Biol. Sci. Symp.* **1997,** 463-471.
112. Goodell, B.; Daniel, G.; Jellison, J.; Nilsson, T. *International Research Group on Wood Preservation* **1988,** *88-1361.*
113. Iwahashi, H.; Morishita, H.; Ishii, T.; Sugata, R.; Kido, R. *J. Biochem.* **1989,** *105*, 429-434.
114. Deiana, S.; Gessa, C.; Marchetti, M.; Usai, M. *Soil Sci. Soc. Am. J.* **1995,** *59*, 1301-1307.
115. Deiana, S.; Gessa, C.; Manunza, B.; Marchetti, M.; Usai, M. *Plant Soil* **1992,** *145*, 287-294.
116. Pracht, J.; Boenigk, J.; Isen-Schröter, M.; Keppler, F.; Schöler, H. F. *Chemosphere* **2001,** *44*, 613-619.
117. Kirk, T. K.; Mozuch, M. D.; Tien, M. *Biochem. J.* **1985,** *226*, 455-460.
118. Halliwell, B.; Gutteridge, J. M. C. In *Method Enzymol.* **1990,** 1-85.
119. Hartwig, R. C.; Loepper, R. H. In *Iron chelation in plants and soil microorganisms;* Hemming, B. C.; Academic Press: NY, 1993; pp. 465-482.
120. Qi, W., Ph.D. thesis, University of Maine,Orono, ME, 2001.
121. Srinivasan, U.; Highley, T. L.; Bruce, A. *Int. Biodeter. Biodegr.* **1995,** *9*, 226-230.
122. *Metal Ions in Fungi*; Winkelmann, G.; Winge, D. R.; Marcel Dekker, Inc.: NY, 1994; pp 507.
123. Helm, D. v. d.; Winkelmann, G. In *Metal Ions in Fungi;* Winge, D. R.; Marcel Dekker, Inc: NY, 1994; pp. 39-98.

Chapter 7

Assessment of Fungal Wood Decay by Lignin Analysis Using Tetramethylammonium Hydroxide (TMAH) and ^{13}C-Labeled TMAH Thermochemolysis

Timothy R. Filley

Department of Earth and Atmospheric Sciences, Purdue University, 1397 Civil Engineering Building, West Lafayette, IN 47907

Thermochemolysis using tetramethylammonium hydroxide (TMAH) and ^{13}C-labeled TMAH is an analytical tool recently applied to the study of white- and brown-rot decay of wood. It has several advantages over more conventional molecular pyrolytic and chemolytic techniques in that lignin side-chain oxidation and demethylation can be assessed simultaneously, while using only tens of micrograms of material. In samples analyzed thus far clear distinction in the lignin decay patterns between white- and brown-rotted wood residue has provided insightful clues into the mechanism of fungal decay at the earliest stages of fungal activity. This chapter reviews the application of this technique to lignin decay experiments from recent literature.

Significant effort has been expended to detect, monitor, and mitigate the degradation of wood by decay fungi particularly in regard to the protection of wood in service and the study of wooden archeological artifacts where large economic and historical issues are of importance. Early detection of decay is critical to avoid structural damage, as strength loss can precede visible damage.

Additionally, early detection is necessary in inoculation experiments where the decay mechanisms of wood-rot fungi as well as the effectiveness of anti-fungal agents are evaluated. The study of wood decay is also of agricultural and ecological importance because vascular plant tissue represents the largest pool of actively cycling organic carbon and the rates and type of microbial conversion of its constituents exert an important control on soil chemistry (*1,2*). Because woody tissue is a complex composite material made up of a variety of crystalline and amorphous biopolymers, it does not easily lend itself to rapid analysis of its chemical and physical structure, particularly if sample quantities are small (~ 1mg).

Lignin, a biopolymer composed of phenylpropanoid monomers, accounts for approximately 30% of wood mass and represents the second most abundant biopolymer constituent in vascular plant tissue as a whole (*3*). As lignin is unique to vascular plants and varies in monomer chemistry and structure based upon plant type (*4*), its structural components in soil and water can serve as indicators of both the presence of vascular tissue and, broadly, the taxonomic source (*5-7*). Additionally, the chemistry of lignin monomers provides clues to the extent and type of decomposition as wood decay fungi impart "characteristic" chemical signatures in degraded lignin (*8-12*). Thus, a common approach to track and study the chemistry of wood decay has been to analyze the abundance and structure of lignin during decay..

The wide variety of polymeric structures and its infusion within a predominantly polysaccharide matrix creates an analytical barrier to analysis of lignin in both its native and degraded states. Often multiple techniques, such as solid-state ^{13}C-NMR, FTIR, and molecular studies such as analytical pyrolysis, alkaline CuO oxidation, and tetramethylammonium hydroxide (TMAH) thermochemolysis, are used to infer changing lignin chemistry during microbial alteration of samples (*11-16*). Molecular characterizations typically utilize either analytical pyrolysis (*17*) or chemolysis, such as chemical decomposition using TMAH thermochemolysis (*10,11*) or alkaline CuO oxidation (*9*) to break the lignin macromolecule into monomer constituents. Lignin monomers generated by molecular techniques are generally analyzed by gas chromatographic (GC), mass spectroscopic (MS), or coupled GC/MS techniques. Sample size, sample matrix, or cost, however, frequently do not always permit a combined approach and necessitate using microanalytical molecular tools.

TMAH thermochemolysis is a relatively new technique for the analysis of lignin structure and decay (*6,10,18-21,24*). This chemolytic method has been used in recent years as a tool to assay the composition and concentration of other biopolymeric and polar plant components in a variety of sample matrices (*22-24*). With regard to the analysis of lignin, TMAH thermochemolysis is an analytical procedure that combines the base catalyzed decomposition of specific structural bonds in lignin with simultaneous derivatization of phenols, alcohols and acids with a methyl group (*6,24,25*). The technique minimizes decarboxylation of compounds and yields derivatized polar moieties that are amenable to gas chromatography. This is an advantage over traditional analytical pyrolysis which excludes highly polar compounds from analysis. Additionally, ortho di-hydroxyaromatic compounds, common in hydrolyzable tannins and brown-rotted lignin, are readily analyzed by TMAH thermochemolysis, in contrast to the analytical alkaline CuO oxidation which cleaves these types of bonds. A recent advancement in thermochemolysis involves the use of ^{13}C-labeled TMAH (*25*), which permits assessment of the degree of aromatic hydroxyl content in the released lignin monomers and a compound profile identical to convential TMAH thermochemolysis. This latter capability has proven quite useful in the analysis of brown-rot alteration of wood (*10,16*).

This chapter will review the chemistry and application of both TMAH and ^{13}C-labeled TMAH thermochemolysis to the study of fungal decomposition of lignin. Following a brief introduction to the native chemistry of lignin and the mechanisms by which TMAH thermochemolysis decomposes it, recent field and laboratory decomposition studies will be reviewed to demonstrate the applicability of this method to detect, at the very early stages of decay, fungal degradation of woody and herbaceous tissue.

The Chemistry of Lignin

Lignin is a complex biopolymer made up of cross-linked phenyl propanoid monomers of cinnamyl alcohol derivatives. These monomers vary in the extent of ring methoxylation at the 3 and 5 position with botanical source, wood type, and even position within the wood cell wall (*26,27*). The relative distribution of guaiacyl (3-methoxy, 4-hydroxy) and syringyl (3,5-dimethoxy, 4-hydroxy)phenols vary considerably between angiosperm and gymnosperm woody tissue. For example, conifer lignin is based upon guaiacyl units whereas angiosperm wood contains both guaiacyl and syringyl monomers. A characteristic of non-woody tissues is that they contain significant *p*-hydroxyphenols in addition to syringyl and guaiacyl units. In addition to the cinnamyl alcohol derivatives which have alcohol groups at the γ-carbon, herbaceous plants may contain a significant proportion of ester bound phenolics with a –COOH group at the γ-carbon. It has been suggested that these acidic phenolic monomers serve as crosslinking agents between the lignin and carbohydrate components (*28*).

The chemical structure of lignin is highly variable owing to the random polymerization of phenoxy radicals of the constituent monomers by the action of peroxidases (*3,26*). Models for softwood and hardwood lignins propose multiple linkage-types between monomers. These include aliphatic C-C, aromatic C-C, aromatic ether, aliphatic ether, and ester bonds (*26,29*). The lignin monomers are linked predominantly at the β-O-4 positions (Figure 1); however, the relative abundance of the β-O-4 linkage among lignins from different plant sources is variable (*26*). In wheat straw it has been estimated as much as 74% percent of lignin monomers are linked by β-O-4 bonds (*30*). As will be shown below, it is the predominance of the β-O-4 linkage in lignin that permits the application of TMAH thermochemolysis to the study of lignin chemistry.

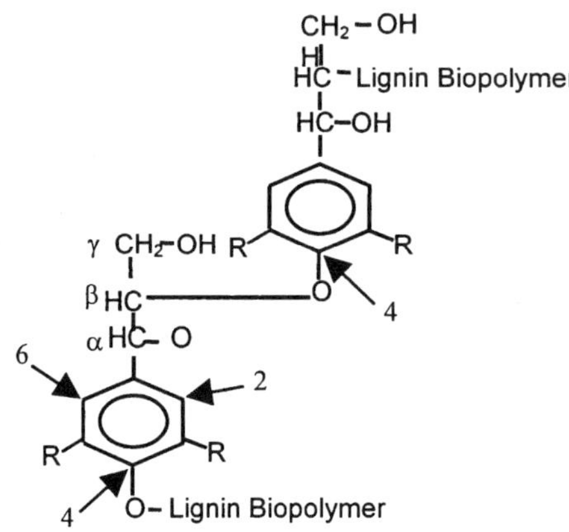

Figure 1. Chemical structure of the β-0-4 bond within lignin.

Thermochemolysis of Lignin by TMAH

The decomposition of lignin by TMAH thermochemolysis primarily results from breaking β-O-4 bonds that have hydroxyl functional groups on the adjacent α or γ carbons. The main biopolymer-cleaving reaction is thought to proceed through an intramolecular oxirane cyclization via the α or γ hydroxyl. (*25*). A series of base catalyzed (e.g. reverse aldol, hydride shift, and Cannizaro) reactions results in additional depolymerization of the lignin and C-C bond breakage within the glyceryl side-chain (*25*). This is similar to the chemistry of soda pulping (*31*).

The methylammonium salt serves as a alkylating agent above about 150°C for acidic groups on the lignin, e.g, aromatic and aliphatic acids, phenols, and aliphatic alcohols, yielding a characteristic suite of methyl ethers and esters (Figure 2) (*6,25*). Derivatization of the monomers also helps to minimize

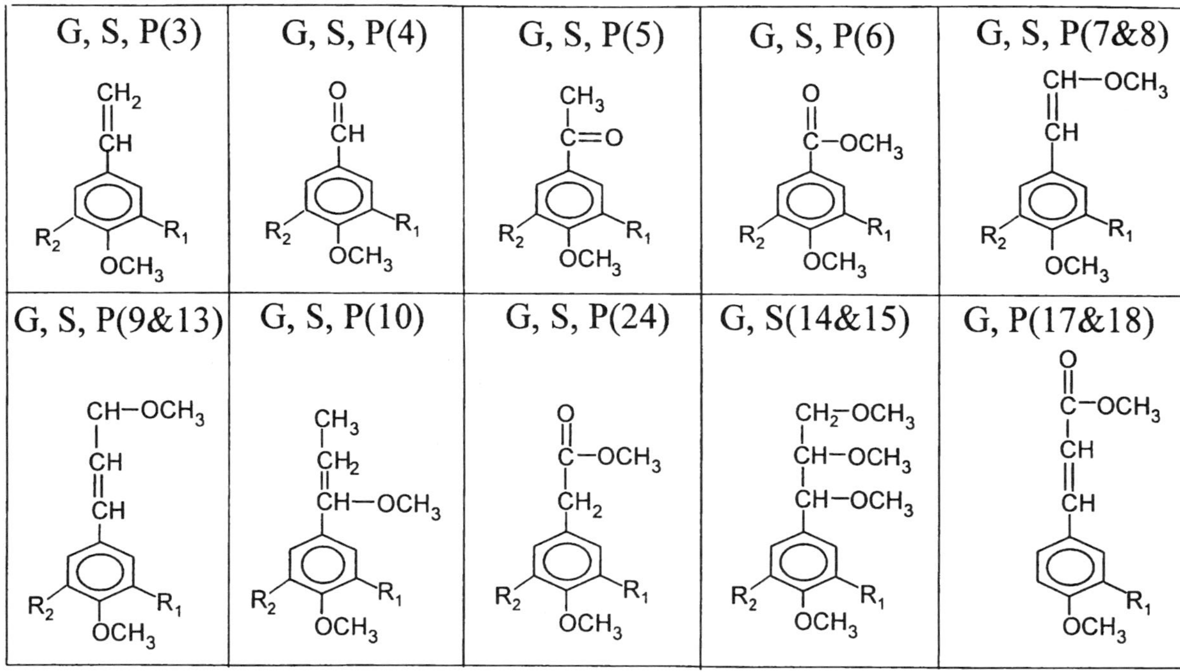

G (R_1=H; R_2=OCH_3); S (R_1= R_2=OCH_3); P (R_1= R_2=H)

Figure 2. Chemical structures for the dominant TMAH monomers released upon thermochemolysis of lignin. G, S, and P represent guaiacyl, syringyl, and p-hydroxyphenyl groups.

further base-catalyzed decomposition. An additional distinguishing characteristic of TMAH thermochemolysis of lignin is the release of the four enantiomers of 1-(3,4-dimethoxyphenyl)-1,2,3-trimethoxy propane or 1-(3,4,5-trimethoxyphenyl)-1,2,3-trimethoxy propane (*19*). When analyzed by column chromatography, two peaks consisting of the diastereomeric pairs result (see structures G(S)14 and G(S)15 in Figure 2). These compounds retain the full glyceryl side-chain from lignin and indicate structurally intact β-O-4 bonds. No other chemolytic or pyrolytic techniques form these compounds.

The Use of ^{13}C-TMAH in Thermochemolysis

When ^{13}C-labelled TMAH is used in the thermochemolysis reaction, all of the methyl groups added to the lignin phenols carry an extra mass unit due to the stable isotope label. The exact number of labeled methyl groups added to each compound can be determined by structural mass spectrometry (*25*). Figure 3 illustrates this point with the TMAH thermochemolysis of a bound vanillic acid residue on a lignin backbone. The molecular weight of the permethylated product exhibits an increase of one mass unit from 198 to 199 when the starting material is the 3,4-dihydroxy functionality (R = H) rather than the 3-methoxy, 4-hydroxy (R = CH_3) arrangement found in undegraded lignin. ^{13}C-TMAH thermochemolysis maintains the same chemical properties as standard TMAH but adds this additional dimension of information concerning the yield of hydroxylated phenyl groups in the degraded residue. Demethylation of lignin during fungal degradation is determined by subtracting the ^{13}C content of monomers in the degraded samples by the ^{13}C content of the respective monomers in the control (fresh) samples (*10,16*). Positive deviations from the ^{13}C content in the control are attributed to demethylation by the fungi. A detailed description of the synthesis of ^{13}C-TMAH, the mechanism by which it depolymerizes lignin, and the calculation used to determine the number of methyl groups added is provided elsewhere (*25*).

Method of Application

The thermochemolysis reactions are typically performed in two modes: as an in-line reaction in a pyrolyzer with direct transfer of products to a GC injector (*7,32*) or in an off-line mode where TMAH is reacted with the wood in an ampoule and processed separately prior to analysis (*32*). Each method has benefits that relate to sample size and ease of quantification. These are briefly discussed below.

In a typical off-line analysis of wood, approximately 15 mg of TMAH (added dry or as a solution in methanol or water) is weighed into glass ampoules with approximately 1 mg of wood residue. Samples that contain only small amounts of lignin will require a larger amount of wood sample (5-10 mg). The ampoules are then evacuated on a vacuum line, flame sealed and then heated, typically at 250 °C for 30 minutes. The methylated products are then extracted from the ampoules, taken up in an organic solvent, spiked with a chromatographic standard, and analyzed by GC. When running analyses using ^{13}C-labeled TMAH it is important to use water not methanol to make the ^{13}C-TMAH solution that is added to the reactor as methanol will participate in the reaction and its natural abundance carbon isotope signature will confuse the isotope ratios used to determine extent of demethylation. Alternatively, addition of dry TMAH/^{13}C-TMAH (without a carrier solution of water or methanol) to a sample has also been shown to yield lignin monomers from wood (*16*).

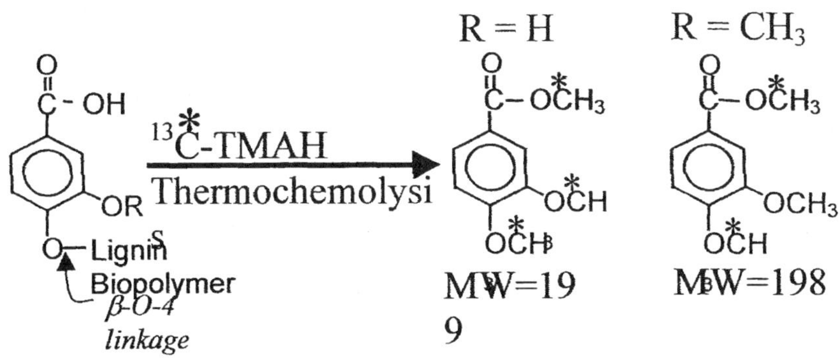

Figure 3. Comparison of the molecular weight difference between G6 produced after ^{13}C-TMAH thermochemolysis of a lignin monomer with and without a methoxyl carbon at position 3. G6, derived from methylation of a fragment with a 3,4-dihydroxy functionality, has a molecular weight 1 mass unit higher than if the fragment has a methoxyl group in the 3 position.

Generally, in-line methods for thermochemolysis of lignin employ a commercial pyrolyzer as a reactor that is interfaced directly to the injection port of a GC. A variety of pyrolysis units, such as Curie point, quartz tube with Pt wire resistive heating and isothermal drop Pt bucket, have been adapted for this purpose. Some of these units are available with autosamplers that permit the unattended processing of large numbers of samples. The amount needed for analysis depends upon the lignin quantity, but 50 to 100 μg of wood residue is usually sufficient and still requires a large injector split ratio (~100/1) for chromatographic analysis. No solvent extraction steps are required and the amount of TMAH used is generally 1/20th of that used in the off-line protocol. The ability to add standards depends upon the design of the apparatus with the greatest ease being that of the drop bucket system where the Pt cups can be pre-charged. Most in-line thermochemolysis has been designed to obtain qualitative distributions of compounds.

Comparing off-line and in-line methods, the former requires a relatively greater investment of time, space (a hood for the extraction procedure), solvents, and sample. Benefits include the ability to use larger quantities of sample in the ampoules, which is important when dealing with highly degraded samples containing only trace quantities of lignin. The latter procedure, however, has limitations in its ability to be a quantitative procedure and requires the added purchase of a pyrolyzer while requiring very little time investment for attainment of the data. Because so little organic matter is used in the in-line procedure the homogeneity of the sample is a principle concern for reproducibility.

^{13}C-TMAH/TMAH Thermochemolysis of Undegraded Lignin

Figure 4 compares the GC/MS profiles for *Pinus taeda* (southern pine) (a) and *Betula papyrifera* (birch) wood (b) reacted with ^{13}C-TMAH under thermochemolysis conditions and is provided as an example for distinguishing plant source. The woods were analyzed in-line using a Shimadzu Pyr4a pyrolyzer interfaced to a GC17A containing a J&W DB-5M fused capillary column (30m x 0.32 mm i.d. x 0.50 μm film thickness) and temperature programmed from 35 to 300°C at 6°C/min with a flow of 1.5 mL/min. Detection was by a QP5050A quadrupole mass spectrometer. The pyrolyzer, a Pt drop-bucket design, was run isothermally at 350 °C with split flow at 50/1. Approximately 50 μg of dry wood was used in each analysis. The peak notations correspond to assignments in Figure 3.

Because the relative abundance of p-hydroxy, guaiacyl, and syringyl lignin monomers (Figure 2) vary among different plant types, their distribution can be used to distinguish sources when comparing thermochemolysis product profiles (*6*). A clear distinction between the two traces in Figure 4 is evident when comparing the composition and distribution of peaks. As expected, pine (a) is characterized almost exclusively by guaiacyl derivatives while the birch wood (b) is composed of both syringyl and guaiacyl analogues.

TMAH thermochemolysis of undegraded lignin is characterized by relatively high amounts of the reduced G(S)4, G(S)7, G(S)8, G(S)14, G(S)15

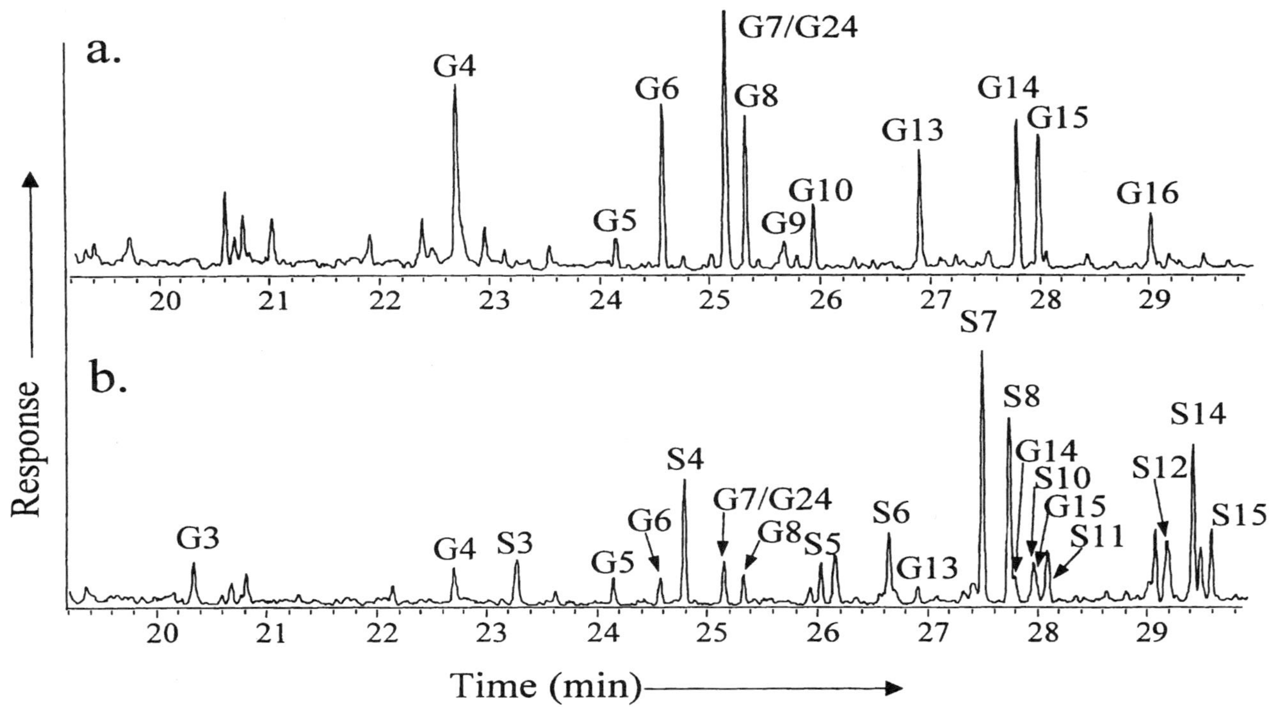

Figure 4. Partial chromatograms for GC/MS analysis of the products from the ^{13}C-TMAH thermochemolysis of fresh southern pine (a) and fresh birch (b) wood. Notations above peaks refer to compounds depicted in Figure 3.

with respect to the oxidized G(S)6. However, a fair amount of variability exists between plant species which reflects the complex nature of lignin among plants (*6,10,11,19,21*). It has been proposed that G4 forms primarily as the result of thermochemolysis of the α or γ hydroxy β-O-4 linkages and as such, the relative yield of G4 should be indicative of the degree of alteration of those β-O-4 linkages (*25*). Present in significant yield in the thermochemolysis products of undegraded lignin are G(S)14 and G(S)15 units, dependent upon the lignin type. These compounds contain the fully methylated glycerol side-chain and are near to the structure of the complete lignin monomer (*19*). These compounds are also thought to be indicative of undegraded β-O-4 linkages with adjacent hydroxyl groups in the sample and are frequently used to asses the abundance of this dominant linkage in fresh and degraded samples (*10,11,22,34*).

Carbon normalized yields for TMAH monomers from undegraded lignin vary considerably depending upon the plant source, but typically range from 3.0 to 6.0 mg monomer/100 mg of organic carbon (*10,11,21*). Significant differences in the relative yields of the individual compounds in the undegraded woods have also been documented and most likely reflect differences in the nature of the linkages of the lignin macromolecule among the samples or in the association of lignin within the lignocellulosic network of the wood cells.

^{13}C-TMAH thermochemolysis on a limited set of fresh woods (*Picea rubens, P. glauca, Abies balsamea, Pinus taeda, Betula papyrifera*) yields lignin monomers with an average 3,4-dihydroxy and 3,4,5-trihydroxy (as for the case of *B. papyrifera*) content of about 5.5% (*10,16,35*). The range in values among monomers from a particular wood, however, is from 2 to 16% with G6 and S6 exhibiting the highest hydroxyl content. In fact, the range of all of the monomers, excluding G6 and S6, are typically 2-6%. These values represent the background level of the nonmethoxylated lignin structure in lignin. Applying a mass balance it can be determined that approximately 11% of the measured G6 and S6 is not methoxylated and may represent either breakdown products from extractives, phenolic monomers used to bridge lignin and carbohydrates or proteins, or potentially an enhanced base-catalyzed demethylation of methoxyls on electron-withdrawing aromatic acids during the TMAH thermochemolysis (*25*).

It is important to reiterate that TMAH thermochemolysis cleaves specific bonds within lignin and the yields upon reaction represent the lability of such bonds during the reaction. For example, C-C bonds and ether bonds without adjacent hydroxyl groups have minimal reactivity and those moieties bound as such are not represented in the released monomers. This is an important consideration when comparing the chemistry of lignin from different plants sources and investigating the relative decomposition rates of the syringyl, guaiacyl, and *p*-hydroxy phenol components in a lignin, as the proportion of the dominant linkages such as β-O-4, β-5, and β-β, for example, will vary both with taxonomic affinity and degree of degradation.

^{13}C-TMAH/TMAH Thermochemolysis of Decayed Lignin

The fungal alteration of vascular tissue proceeds to a greater or lesser degree of lignin modification for brown-, white-, and soft-rot decay (*27*). At the most basic chemical level, fungal alteration of lignin results in demethylation or demethoxylation, side-chain oxidation, and ring cleavage. White-rot degradation of isolated lignin and model lignin compounds generally involves extensive side-chain and ring oxidation (*28,36,37*) where ring cleavage is preceded by demethylation (*27,36,37*). Brown-rot lignin decomposition is generally characterized by side-chain oxidation and extensive ring demethylation as well as partial depolymerization (*38,39,40*). Thus, depending upon the type of fungal decay the remaining wood residues may contain lignin that is more enriched in oxidized phenolic fragments, as in white-rot and soft-rot decay, or high molecular weight, demethylated phenolics with oxidized sidechains, as in brown-rot decay. This difference between these decay types is also reflected in the yield and chemistry of the lignin monomers.

TMAH and ^{13}C-TMAH thermochemolysis has been recently used to track lignin decomposition by decay (mostly white- and brown-rot) fungi in field-scale and laboratory-scale fungal decomposition experiments and to infer changes to the lignin structure and decay mechanisms (*10-12,15,16,18,24*). The chemistry of the degraded lignin is monitored by quantifying changes in the concentration of TMAH monomers, as influenced by side-chain oxidation and recondensation reactions, and their ^{13}C-enrichement upon methylation by ^{13}C-labeled methyl groups, as influenced by demethylation. In fact, TMAH thermochemolysis of white- and brown-rotted residues suggests that this method can be used to distinguish the two decay pathways (*10*). In the following section recent reports of the use of ^{13}C-TMAH and TMAH thermochemolysis to track fungal decay are reviewed to demonstrate the applicability of thermochemolysis methods to monitor fungal decomposition of lignin.

Changes in Lignin Monomer Composition With Fungal Decay

Figure 5 compares the partial GC/MS chromatograms of sapwood of fresh southern pine (also depicted in Figure 4) versus samples degraded by an 8-week modified soil block decay test (*41*) by the brown-rot fungus *Gloeophylum trabeum* or the white-rot fungus *Phlebia tremellosa*. Three parameters typically used to relate the degree of lignin oxidation are noted below and in Figure 5.

1. Λ, which is the sum of the concentration of lignin monomers in mg/100 mg organic carbon content.

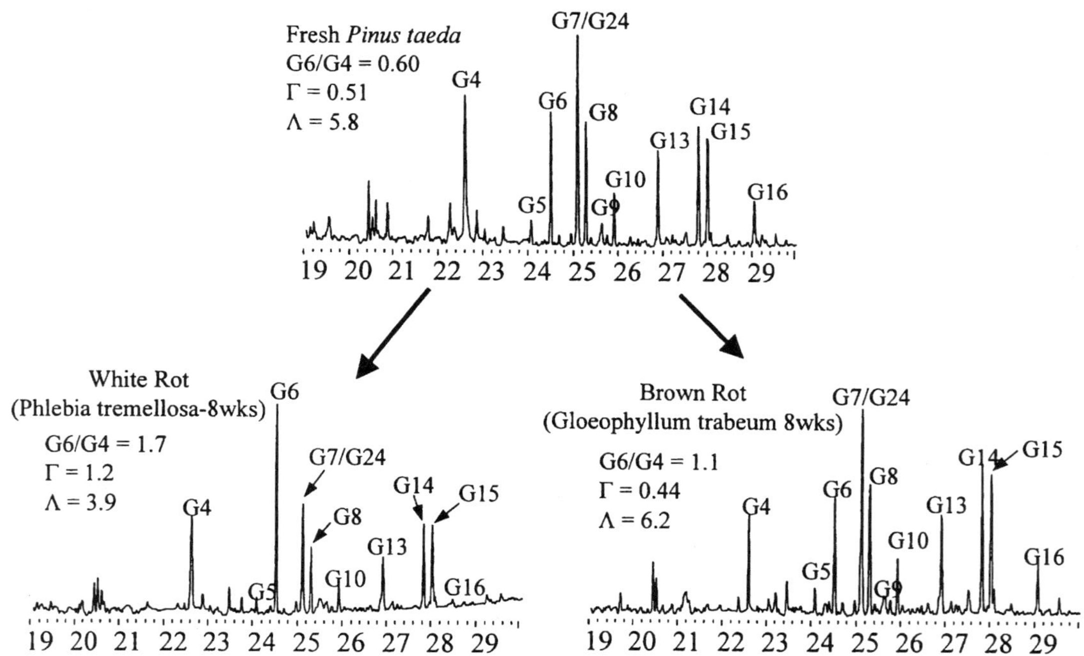

Figure 5. Partial chromatograms for GC/MS analysis of the products from the ^{13}*C-TMAH thermochemolysis of fresh southern pine, southern pine degraded 8 weeks by the white-rot fungus Phlebia tremellosa, and 8 weeks by the brown-rot fungus Gloeophyllum trabeum. Notations above peaks refer to compounds depicted in Figure 3. G6/G4, Γ, and Λ parameters noted on the figure are described in the text.*

2. Γ, which is the concentration of G6 divided by the sum of G14 and G15
3. Ad/Al (or G6/G4), which is the concentration of G6 divided by G4.

The compositional differences between the chromatograms are illustrative of the molecular changes typically seen for these two types of wood decay by TMAH thermochemolysis. The overall composition of compounds among the chromatograms is nearly identical but the absolute concentrations and relative abundances vary considerably. The Λ values decrease in the white-rot decay samples and show a small increase in the brown-rot residue. This trend would be anticipated as this white-rot fungus selectively oxidizes lignin while enriching polysaccharides, while the brown-rot fungus enriches lignin at the expense of cellulose. Decay of wheat lignin by *Pleurotus ostreatus* lowered the Λ values with respect to the starting wheat straw. (*11,12*). Similar trends for lignin phenolic compounds after white- and brown-rot decay have been documented using CuO on inoculated woods (*8,9*). Brown-rot decay does not consistently lead to an increase in lignin yield as one would expect given its concentration upon removal of cellulose. Recent investigations of brown-rot (*Fomitopsis pinicola*) using three gymnosperm woods showed decreased yields upon TMAH thermochemolysis (*10*). A likely explanation for a decreasing Λ in brown-rotted residues, even though the overall lignin content is increasing, may have to do with radical condensation of lignin phenols.

The G6/G4 ratio is a common proxy for lignin side-chain oxidation in degraded wood (*24*). Both the brown-rot and white-rot degraded woods in Figure 5 exhibit increases in this ratio after degradation. Molecular studies using both TMAH thermochemolysis (*10-12,24*), alkaline CuO oxidation (*8*), and alkaline nitrobenzene oxidation (*42*) of white-rot degraded lignin demonstrate a reduction in the overall yield of monomers and progression in the molecular distribution toward oxidized side-chains (e.g. G6 and S6 or vanillic acid and syringic acid). The high yield of G6 obtained from white-rotted wood is most likely derived from vanillic acid subunits as white-rot degradation is known to increase aromatic carboxyl content during the enzyme mediated (e.g. laccase and lignin peroxidase) oxidation of lignin (*13,37*).

In a comparison of three softwoods degraded by either *Phlebia tremellosa* or *Armillaria sp.*, it was found that each exhibited significant increases in G6 with respect to all other components upon analysis with ^{13}C-labeled TMAH (*10*). In particular, the monomers with the full glyceryl side-chain (G14 and G15) and G4 each showed a significant decrease in concentration with decay, indicating oxidative cleavage of the Cα–Cβ bond on the side-chain. Therefore, side-chain oxidation results in an increase in both G6/G4 and Γ. Similarly, a comparison of the intensities of the TMAH thermochemolysis products from wheat straw degraded by *Pleurotus ostreatus* showed a decrease in G4, S4, G14,

S14 and G15, S15 relative to other products. Consequently, Γ, G6/G4, and S6/S4 all showed significant increases over 63 days of decay (*11*). In Figure 5, while there is an increase in G6/G4 in both degradation experiments, Γ shows an increase only in the white-rot system and a small decrease in the brown-rotted wood. A decrease in Γ with a concomitant increase in G6 may indicate that lignin is being depolymerized to some extent without substantial side-chain oxidation but does not necessarily indicate that no side chain cleavage is occurring in the brown rotted wood.

Currently, there is only one published TMAH thermochemolysis analysis of soft-rot wood decay (*24*). The study involved archeological samples from the "King Midas Tomb" that exhibited extensive soft-rot decay. Those findings indicate chemical properties of the soft-rot similar to both white- and brown-rotted wood as the Λ values increased with decay, probably due to a preferential loss of cellulose, while the G6/G4 ratio also increased suggesting substantial side chain oxidation.

Lignin Demethylation Examined by ^{13}C-TMAH Thermochemolysis

Demethylation of lignin is a process common to both white- and brown-rot fungi (*13,39*). A characteristic of brown-rot, however, is that the remaining residue contains highly demethylated, high molecular weight lignin fragments (*38,40,43*), while lignin remaining after white-rot exhibits a low degree of demethylated lignin, as ring cleavage (and metabolism of the breakdown products) proceeds rapidly after demethylation (*13,40,44,45*).

^{13}C-TMAH thermochemolysis is well suited to track lignin demethylation in wood residues (*10,16*). The woods represented in Figure 5, when analyzed for demethylation among the lignin monomers, exhibit chemistry that would be considered consistent with the particular type of decay. Figure 6 shows histograms of the % dihydroxyl content of the dominant monomers released after ^{13}C-TMAH thermochemolysis. The control wood shows background dihydroxyl levels of approximately 4% for all of the monomers except G6, which is approximately 15%. As previously stated, the high level of hydroxyl content for G6 is normally observed. A comparison of the two histograms reveals the general relationship between white- and brown-rot residues in that each of the monomers from the brown-rotted wood show a significant increase in dihydroxyl content whereas the white-rot residues show little or no increase. Similar relationships were demonstrated in both field and laboratory inoculation studies of softwoods (*10*). Some brown-rotted field samples have demethylation values in TMAH monomers as high as 25% (*35*).

The assessment of lignin demethylation or aromatic hydroxyl content by alkylammonium hydroxide thermochemolysis was first performed using tetrabutylammonium hydroxide (TBAH) (19,46,47). TBAH thermochemolysis effectively butylates phenols leaving methoxyl groups in tact thereby permitting

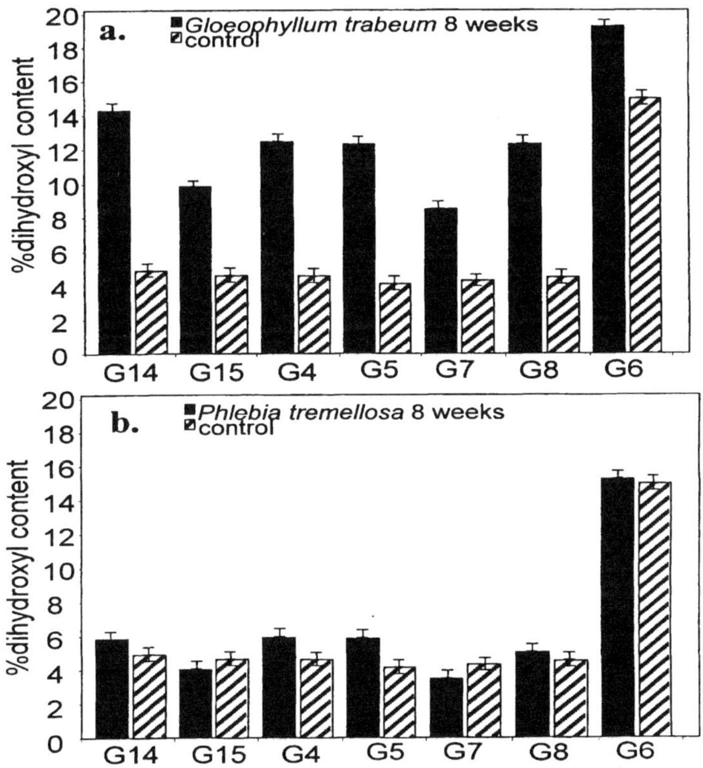

Figure. 6. Graph of the % dihydroxyl content for the seven major TMAH lignin monomers released after ^{13}C-TMAH thermochemolysis of fresh southern pine and southern pine degraded 8 weeks by the brown-rot fungus Gloeophyllum trabeum (a) or 8 weeks by the white-rot fungus Phlebia tremellosa (b).

an accounting of hydroxyl content. Thus, TMAH and TBAH thermochemolysis can been employed as complimentary techniques. TBAH has an advantage over ^{13}C-TMAH in that it is commercially available while ^{13}C-labeled TMAH requires synthesis (*25*). The product distribution obtained from TBAH thermochemolysis does not however resemble the compound distribution obtained by TMAH as TBAH does not butylate aliphatic alcohols. For example, no compounds with the full glycerol side chain are produced (e.g. G(S)14 and G(S)15) by the butylation process. TMAH and ^{13}C-labeled TMAH thermochemolysis products are identical such that labeled reagent can be used in place of regular TMAH. These added benefits of ^{13}C-TMAH can only be realized when analysis is performed by GC/MS.

Long-term inoculation studies of red spruce sapwood by *Postia placenta* or *Gloeophyllum trabeum* have demonstrated a close correlation between polysaccharide loss and demethylation, determined by ^{13}C-TMAH thermochemolysis (*16*). The authors speculated that lignin demethylation may aid brown-rot fungi, as the dihydroxy lignin fragments may participate in quinone/hydroquinone–mediated Fenton chemistry (*46*) to produce hydroxyl radicals for cellulose decomposition. Figure 7 replots previously published data (*16*) to show the relationship between average demethylation of the lignin monomers and polysaccharide content as determined by solid-state ^{13}C-NMR over 32 weeks of decay. Lignin demethylation and loss of polysaccharides, the form of carbon metabolized by brown-rot fungi, occur simultaneously. The monomers exhibit a progressive increase in demethylation while the wood loses polysaccharides over the first 16 weeks of degradation. A point to note in Figure

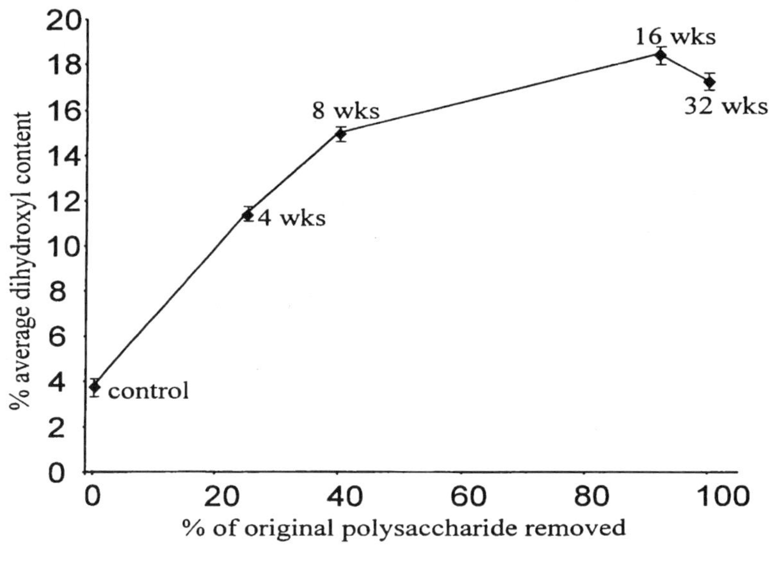

Figure 7. Plot of average % dihydroxyl content of lignin monomers released from ^{13} C-TMAH thermochemolysis of brown-rot (red spruce inoculated with Gloeophyllum trabeum) residue and polysaccharide content of the same residue as determined by solid-state ^{13}C-NMR. Values plotted from Filley et al. (2002) (Figures 6 and 8). The inoculation duration is noted on each data point.

7 is the decrease in demethylation from 16 to 32 weeks. Both *Postia placenta* and *Gloeophyllum trabeum* degraded wood exhibit this phenomenon. By 32 weeks there is no longer any polysaccharide carbon identifiable in the residue; therefore, it was speculated that either the fungi metabolize the demethylated lignin or a radical condensation of the phenolic residues makes them inaccessible to the TMAH thermochemolysis technique.

Relationship Between Demethylation and Side-chain Oxidation

^{13}C-TMAH thermochemolysis analyses of white- and brown-rotted wood residues demonstrates that these distinctive decay types have characteristic residues. Each of the brown-rotted woods exhibited a high level of demethylation and a low level of acidic functionalities (e.g. G6), while white-rot

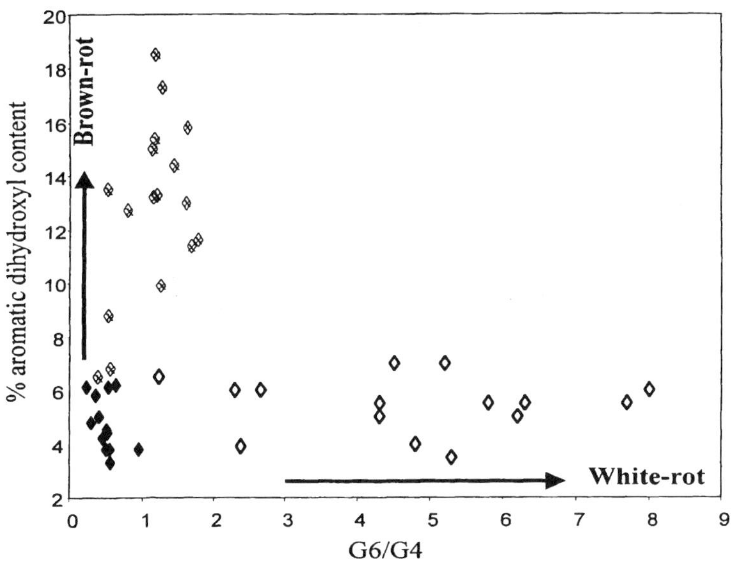

Figure 8. Cross plot of G6/G4 and % aromatic dihydroxyl content for 32 softwood samples from fresh and associated brown- and white-rotted woods. Open diamonds indicate decay by white-rot fungi (Phlebia tremellosa, Armillaria sp., Phlebia gigantia). Cross-filled diamonds indicate decay by brown-rot fungi (Fomitopsis pinicola, Gloeophyllum trabeum, Postia placenta). Filled diamonds indicate control woods. The data is a compilation from Filley et al. (10,16, 24), comprised of inoculation studies greater than 4 weeks duration or degraded wood collected from field samples. Arrows indicate direction of chemical evolution of lignin during progressive decay.

residues showed low levels of demethylation and a significant amount of oxidized aromatic units. These relationships could possibly be utilized to rapidly differentiate between the decay types. Figure 8 shows the results from 32 ^{13}C-TMAH thermochemolysis analyses of white- and brown-rot fungal inoculation experiments, each from a separate modified soil block experiment (*41*) of 4 weeks or longer duration, or decayed woods collected from field samples (*10,16,35*). The G6/G4 parameter is plotted with the average % dihydroxy content of the major monomers released upon ^{13}C-TMAH thermochemolysis. Three points are immediately apparent:

1) Undegraded control wood (pine, balsam fir, red spruce) plot all had a dihydroxy content of 3% to 6% and with a G6/G4 of less than 1.
2) Most (> 90%) of the white-rot (*Phlebia tremellosa, Armillaria sp., Phlebiopsis gigantea*) residues have a dihydroxy content of less than 6% and a range in G6/G4 of 1 to 8.
3) The average G6/G4 for the brown-rot (*Fomtopsis pinicola, Gloeophyllum trabeum, Postia placenta*) degraded lignin is 1.2 with a wide range in dihydroxyl content (6.5 to 19) among TMAH monomers.

These relationships are consistent with known white- and brown-rot activity and should be representative of lignin in decayed wood.

Conclusions

In summary, as analytical tools ^{13}C-TMAH and TMAH thermochemolysis have a number of benefits that make them chemically appropriate for the detection of early wood decay. Using ^{13}C-TMAH thermochemolysis one can simultaneously assess side-chain oxidation (G6/G4), the prevalence of G14(S14) and G15(S15) which can be used in Γ value determination, and demethylation determination, all in one analysis. This is a unique feature among molecular chemolytic and pyrolytic techniques and provides multidimensional information about the extent and type of decay. In addition, as with other molecular analytical techniques TMAH thermochemolysis can be used to assess taxonomic affinity by relating the distribution of lignin monomers containing *p*-hydroxylphenol, guaiacyl, or syringyl chemical structures to known bulk lignin composition among taxa. When using a pyrolyzer interfaced to a GC or GC/MS this molecular information can be obtained with very small samples amounts, typically on the order of 0.10 mg dry wood, which permits noninvasive sampling of lignocellulosic samples. From a limited data set, it appears that ^{13}C-TMAH thermochemolysis may also be an effective means to rapidly discriminate between white- and brown-rot decay when relating demethylation and side-chain oxidation. Because only a minimal sample amount is required this capability may be most useful at the very earliest stages of decay, prior to extensive visible damage, where early detection can dictate appropriate treatment.

Acknowledgements

The author would like to acknowledge helpful comments regarding this manuscript from Juan Carlos Del Rio, Rose Filley, and three anonymous reviewers. Support from USDA grant 2001-35103-09935 is also acknowledged. The author would like to thank Robert Blanchette, Wally Shortle, Barry Goodell, and Jody Jellison for many of the laboratory and field samples analyzed in this manuscript.

References

1. Hedges, J.I.; Oades, J.M. *Org. Geoch.* **1997**, *27*, 319-361.
2. Schlesinger, W.H., *Biogeochemistry: An Analysis of Global Change*; Acad. Press; San Diego, CA 1997, 558 p.
3. Sjostrom, E., *Wood Chemistry: Fund. & Appl.*, 2nd ed., 1993, New York, Academic Press.
4. Higuchi, T., In *Lignin Biodegradation: Microbiology, Chem., and Potential Appl.*, Vol. 1, 2nd ed. (eds. T. Kirk, T. Higuchi and H. Chang), 1981 (pp 1-19). CRC Press, Boca Raton, FL.
5. Ertel, J. R.; Hedges J.I. *Geochim. Cosmochim. Acta* **1984**, *48*, 2065-2074.
6. Challinor, J.M. *J. Anal. App. Pyrol.* **1995**, *35*, 93-107.
7. Nierop K.G.J. *J. Anal. App. Pyrol.* **2001**, *61*, 111-132.
8. Hedges, J.I.; Blanchette, R.A.; Weliky, K.; Devol, A.H. *Geochim. Cosmochim. Acta* **1988**, *52*, 2717-2726.
9. Goñi, M.A.; Nelson, B.; Blanchette, R.A.; Hedges, J. *Geochim. Cosmochim. Acta* **1993**, *57*, 3985-4002.
10. Filley, T.R.; Hatcher, P.G.; Shortle, W. *Org. Geochem.* **2000**, *31*, 181-198.
11. Vane, C.H.; Martin, S.C.; Snape, C.E.; Abbott, G.D. *J. Agric. Food Chem.* **2001**, *49*, 2709-2716.
12. Vane, C.H.; Abbott, G.D.; Head, I.M. *J. Anal. App. Pyrol.* **2001**, *60*, 69-78.
13. Robert, D.; Chen, C.L. *Holzforschung* **1989**, *43*, 323-332.
14. Agosin, E.; Jarpa, S.; Rojas, E.; Espejo, E. *Enzyme Microbial Tech.* **1989**, *11*, 511-517.
15. Chen, Y.; Chefetz, B.; Rosario, R.; van Heemst, J.D.H.; Romaine, P.C.; Hatcher, P.G. *Compost Sci. Util.* **2000**, *8*, 347-359.

16. Filley, T.R.; Cody, G.C.; Goodell, B.; Jellison, J.; Noser, C.; Ostrofsky, A. *Org. Geochem.* **2002**, *33*, 111-124.
17. Klap, V.A.; Boon, J.J.; Marten, A.H.; Soelen, J.V. *Org. Geochem.* **1998**, *28*, 707-727.
18. Hatcher, P.G.; Minard, R.D. *Org. Geochem.* **1995**, *23*, 991-994.
19. Martin, F.; del Rio, J.C.; Gonzalez-Vila, F.J.; Verdejo, T. *J. Anal. App. Pyrol.* **1995**, *35*, 1-13.
20. Hatcher, P.G.; Minard, R.D. *Org. Geochem.* **1996**, *24*, 593-600.
21. Kuroda, K.; Ozawa T.; Ueno, T. *J. Agric. Food Chem.* **2001**, *49*, 1840-1847.
22. McKinney, D.E.; Bortiatynski J.M.; Carson, D.M.; Clifford, D.J.; De Leeuw, J.W.; Hatcher, P.G. *Org. Geochem.* **1996**, *24*, 641-650.
23. Del Rio, J.C.; Hatcher, P.G. *Org. Geochem.* **1998**, *29*, 1441-1451.
24. Del Rio, J.C.; McKinney, D.E.; Knicker, H.; Nanny, M.A.; Minard, R.D.; Hatcher, P.G. *J. Chrom. A* **1998**, *823*, 433-448.
25. Filley, T.R.; Minard, R.D.; Hatcher, P.G. *Org. Geochem.* **1999**, *30*, 607-621.
26. Adler, E. *Wood Sci. Technol.* **1977**, *11*, 169-218.
27. Eriksson, K.-E.L.; Blanchette, R.A.; Ander, P. *Microbial and Enzymatic Degradation of Wood and Wood Components;* Springer-Verlag; New York, **1990**, 397 pp.
28. Crestini, C.; Argyropoulos, D.S. *J. Agric. Food Chem.* **1997**, *45*, 1212-1219.
29. Nimz, H.H. *Angew. Chem. Int. Ed.* **1974**, 13, 313-321.
30. Terashima, N.; Atalla, R.H.; Vanderhart, D.L. *Phytochemistry* **1997**, 46, 863-870.
31. Gierer, J. *Holzforschung* **1982**, *36*, 43-51.
32. Clifford, D.J.; Carson, D.M.; McKinney, D.E.; Bortiatynski, J.M.; Hatcher, P.G. *Org. Geochem.* **1995**, *23*, 169-175.
33. McKinney, D.E.; Carson, D.M.; Clifford, D.J.; Minard, R.D.; Hatcher, P.G. *J. Anal. App. Pyrol.* **1995**, *34*, 41-46.
34. Hatcher, P.G.; Minard, R.D. *Org. Geochem.* **1995**, *23*, 991-994.
35. Filley, T.R. unpublished results
36. Umezawa, T.; Higuchi, T. *Fed. European Biochem. Soc. Letts.* **1987**, *218*, 255-260
37. Srebotnik, E.; Jensen, K.A.; Jensen, A.; Kawai, S.; Hammel, K. *Appl. Environ. Microbiol.* **1997**, *63*, 4435-4440.
38. Enoki, A.; Tanaka, H.; Fuse, G. *Holzforschung* **1988**, *42*, 85-93.
39. Ander, P.; Stoytschev, I.; Eriksson, K-E. *Cellulose Chem. Tech.* **1988**, *22*, 255-266.
40. Jin, L.; Schultz, T.P.; Nicholas, D.D. *Holzforschung* **1990**, *44*, 133-138.

41. *Annual Book of ASTM Standards*; D 1413-760; American Society for Testing and Materials: Philadelphia, PA. **1994**, section 4, vol. 04.10 P. 218-224 American Society for Testing and Materials

42. Iiyama, K., Stone, B. A., Macauley, B. *Appl. Environ. Microbiol.* **1994**, *60*, 1538-1546.

43. Kirk, T.K. In: *Microbial Degradation of Organic Compounds* (ed. D. Gibson); Marcel Dekker, New York, 1984.

44. Ander, P.; Eriksson, K-E . *Appl. Microbiol. Biotech.* **1985**, *21*, 96-102.

45. Ander, P.;Eriksson, M.E.R.;Eriksson, K-E. *Physiol. Plant.* **1985**, *65*, 317-321.

46. Del Rio, J.C. and Hatcher, P.G. In: *Colloidal Materials in the Environment;* Gaffney, J. and Clark, S. eds.; ACS Symposium Series No. 651; ACS Washington, DC, **1996**, pp. 78-95

47. Gonzalez-Vila, F.J.; Del Rio, J.C.; Martin, F.; Verdejo, T. *J. Chrom.* **1996**, *A 750*, 155-160

48. Paszczynski, A.; Crawford, R.; Funk, D.; Goodell, B. *Appl. Environ. Microbiol.* **1999**, *65*, 674-679.

Chapter 8

Physical and Chemical Characteristics of Glycopeptide from Wood Decay Fungi

Akio Enoki, Hiromi Tanaka, and Shuji Itakura

Department of Agricultural Chemistry, Faculty of Agriculture, Kinki University, Nara, Japan

A low-molecular-weight (7,200~12,000) substance was isolated from cultures of brown-rot basidiomycete *Tyromyces palustris*. It contained about 61% protein, 24% neutral carbohydrate, and 0.12% Fe(II) by weight. The glycopeptide containing Fe(II) alone reduced O_2 to ·OH and catalyzed redox reactions between an electron donor, such as NADH, and O_2 to produce H_2O_2 and to reduce H_2O_2 to ·OH. The glycopeptide reduced Fe(III) to Fe(II) (1.7 μM/mg). The glycopeptide was a glycosylated peptide and had about 0.83 μM glycosylamine/mg.

The Role of Hydroxyl Radicals in Brown-rot, White-rot, and Soft-rot Degradation of Wood

The one-electron oxidation activity in intact cultures of wood-decay fungi is generally related with the rate of weight loss during wood degradation in those cultures. In brown-rot, most of the oxidation activity is caused by hydroxyl radical, while in white- and soft-rots, most of the activity is due to a combination of phenol oxidase and • OH. Most hydroxyl radicals produced in cultures of some white- and brown-rot fungi are caused in the redox reaction between O_2 and one-electron donors catalized by extracellular low-molecular-weight glycopeptides. Therefore, hydroxyl radicals in the redox reaction mediated by the glycopeptides are important in wood degradation by wood decay fungi. The role of the hydroxyl radical in wood degradation by wood decay fungi, and the physical and chemical characteristics of the glycopeptide are discussed.

Brown-rots

Brown-rot fungi can degrade crystalline cellulose in wood, even though they lack exo-1,4-glucanse activity (1,2). Although these fungi can also degrade lignin, they preferentially metabolize the cellulose and hemicellulose, leaving an amorphous, brown, crumbly residue that is rich in lignin (3,4). Despite this, brown-rot fungi degrade and metabolize crystalline cellulose only when their ligninolytic systems are active. Furthermore, the cellulolytic systems of brown-rot fungi are always active when these organisms are degrading lignin (4). This indicates that brown-rot fungi may possess a unique wood-component-degrading system that is capable of fully degrading cellulose as well as modifying lignin. Hydroxyl radicals (• OH) have been suggested as being involved in the degradation of wood by brown-rot fungi, since they can depolymerise cellulose (5) and attacks the aromatic rings in lignin, causing a variety of reactions including hydroxylation and ring-opening (6) without delignifying wood significantly (7). There is increasing experimental evidence to suggest that •OH is involved in the wood-component-degrading system of brown-rots. For example, some brown-rot fungi can degrade three 1,2-diarylethane lignin-related model compounds, yielding the same products in relatively large yields as those rising from the degradation of the lignin model compounds by hydroxyl radicals (8). High concentration of hydroxyl radicals are also detected in cultures of brown-rot fungi in large yields (9). Brown-rot fungi cause rapid depolymerization of cellulose in wood before losses in total wood substance are detected. The pattern of progressive change in the mean degree of polymerization of the holocellulose in wood caused by Fenton's reagent is similar to that caused by brown-rot fungi (1,10). The rates of •OH-production in cultures of brown-rot fungi are directly proportional to the degradation rates of wood, and of crystalline cellulose and lignin substrate model compounds when these are added to fungal cultures (11,12). In the early stages of degradation of wood by brown-rot fungi, the S_2 layer of wood cell wall is degraded extensively. The S_3 layer adjacent to the cell lumen is less affected when it is attacked by fungal hyphae from the lumens (13,14). During these early stages of degradation, fungal enzymes such as cellulase and peroxidase are too large to penetrate wood cell walls (15,16). This further suggests that brown-rot fungi have a unique non enzymatic degradative system possibly involving $^{\bullet}OH$ that degrades wood cell walls at some distance from hyphae in the lumen. Pretreatment of wood with $\cdot OH$ results in a marked increase in the amount of hydrolysis of the cellulose in the wood by endoglucanases which degrade little of the cellulose if the wood is not pretreated with $\cdot OH$ (5). When wood is treated with $\cdot OH$, the hemicellulose is mainly lost and the contents of the lignin and cellulose increase (5). This suggests that in brown-rot lacking exo-1,4-glucanase activity, endo-1,4-glucanse play an important role since brown-rot fungi preferentially metabolize the cellulose in wood.

Extracellular low-molecular-weight glycopeptides that catalyze a redox reaction between O_2 and electron donors to produce the hydroxyl radicals have been isolated from wood degrading cultures of the brown-rot fungi, *Gloeophyllum trabeum* and *Tyromyces palustris* (Figure 1) (17,18). The extracelluar substance from *T. palustris* reduces Fe(III) to Fe(II) and strongly adsorbes Fe(II) (18). Most of the hydroxyl radical yield from cultures of *T. palustris* are produced in the redox reaction between O_2 and certain electron donors catalyzed by the low-molecular-weight glycopeptide (19). During early stages of degradation, the glycopeptide is localized in the fungal cytoplasm and fungal cell wall, and in the extracellular slime sheath surrounding the fungal cell wall. The cell wall remains almost intact as long as the fungal hyphae remains in the lumen (20). Subsequently the glycopeptide is found throughout the wood cell wall, suggesting that it diffuses through the S_3 layer into the S_2 layer and the middle lamella.

White-rots

White-rot fungi deplete all components of the wood cell wall during decay, but some species cause selective removal of the lignin in wood. In both the decay types, the lignin in wood cell walls being decayed is completely depleted. Some white-rot fungi can degrade the lignin in the wood preferentially to cellulose. Only in white-rot cultures with high levels of phenol oxidase activity (that is, at least one phenol-oxidizing enzyme such as LiP, MnP, or laccase) can significant degradation of wood with preferential degradation of the lignin occur (21-23). However, the phenol oxidase-activity in cultures of the white-rot fungi is not necessarily correlated with the rate of wood degradation by the

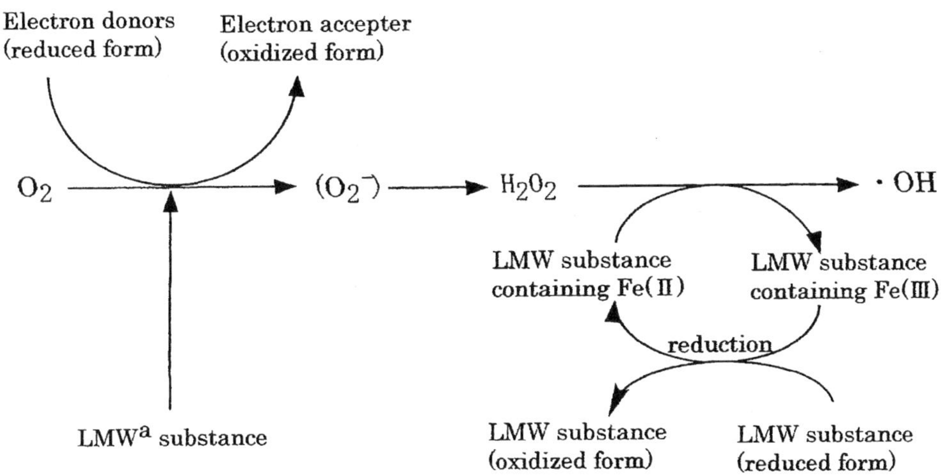

Fig . 1. Proposed mechanism for generation of activated oxygen species by the extracellular one-electron oxidizing substance produced by wood decay fungi.
[a] Low-molecular-weight

fungi in the cultures (23-25). In contrast, the generation of $^{\cdot}OH$ in cultures of white-rot fungi is proportional to the rate of wood degradation (23,25,26). Although neither MnP or laccase are able to modify or degrade the non-phenolic lignins, both can degrade the non-phenolic portions of the polymer that are linked by β-O-4-alkyl aryl ether bonds, provided that some terminal phenols are present (27). Thus phenoloxidases such as LiP, MnP or laccase can degrade the lignin attacked by $\cdot OH$ since hydroxyl radicals attack the aromatic rings in lignin, causing aromatic hydroxylations.

During early stages of wood degradation by white-rot fungi, only low-molecular-weight agents are able to diffuse into the wood cell wall (28-30). During wood decay, white-rot fungi secrete a low-molecular-weight glycopeptide that catalyzes a redox reaction between O_2 and electron donors to produce $^{\cdot}OH$; furthermore the glycopeptide reduces Fe(III) to Fe(II) and strongly binds Fe(II) (Figure 1) (23,24,31). Hydroxyl radical in combination with phenol oxidase may play a role in lignin degradation by white-rot fungi (23,25,26). Most of the hydroxyl radicals produced in wood-degrading cultures of the white-rot fungi, *Phanerochaete crysosporium and Trametes versicolor* are produced in the redox reaction between certain electron donors and O_2 catalyzed by the low-molecular-weight glycopeptide (23,25).

Soft-rots

The typical microstuctural features of soft rot decay are cavity formation in the S_2 layers of wood cell walls (Type 1), and erosion of the S_3 layers adjacent to cell lumens (Type 2). In both of these decay types, all the woods chemical components in an area proximal to the hypha are simultaneously degraded and eliminated, that is, simultaneous removal of all the components at one location of the cell wall results from soft-rot. This means that soft-rot fungi as well as white-rot fungi completely deplete the lignin in wood cell walls being decayed. Soft-rot and white-rot fungi generally produce the full cellulolytic enzyme complement (endo-1,4-glucanase, exo-1,4- β -glucanase, and 1,4- β -glucosidase) and can hydrolyze highly-crystalline cellulose substrates. But a complete cellulase sysytem alone is insufficient for degrading the cellulose in wood cell walls because the lignin covering the cellulose prevents their contact (32). Therefore in soft-rot as well as white-rot, a ligninolytic system must be produced.

The one-electron oxidation activity in cultures of soft-rot fungi is related to the rate of weight loss during wood degradation in the cultures; most of the one-electron oxidation activity results from phenol oxidase and $^{\cdot}OH$ action; although soft-rot fungi can degrade wood and the lignin in the wood only in cultures with significant levels of phenol oxidase activity, the phenol oxidase activity in cultures of soft-rot fungi is not correlated with the rate of wood degradation in the cultures. (32). This suggests that in soft-rot, phenol oxidase activity is essential for lignin degradation, but that $^{\cdot}OH$ may be involved since $^{\cdot}OH$ does not delignify wood significantly, and works in combination with phenol oxidase, to play an important role in lignin degradation by white-rot fungi(25,26). Low-molecular-weight glycopeptides that catalyze a redox reaction to produce $^{\cdot}OH$ between O_2 and electron donors have been isolated from wood-degrading cultures of the soft-rot fungi *Chaetomiun globosum* and *Xylaria polymorpha* (33).

Thus, hydroxyl radicals produced in the redox reaction between O_2 and electron donors, mediated by extracellular low-molecular-weight gycopeptides seem to be involved in wood decay by brown-rot, white-rot and soft-rot fungi.

Glycopeptide Background

An extracellular substance that showed one-electron oxidation activity was isolated from wood-degrading cultures of the brown-rot fungus *T. palusutris*. The substance was partially purified by acetone- and ammonium sulfate-precipitations, and gel filtration chromatography on Sephadex G-50 and G-25, and DEAE Affi-Gel Blue gel chromatography according to the procedures previously described (31). The content of protein, neutral carbohydrates, and Fe(II) in the substance are shown in Table 1. The protein content assayed by the method of Lowry et al. (34) is 61%. Carbohydrate content, as measured by the phenol-sulfuric acid method with glucose as the standard (35) was 24%.

Table 1. Content of Protein, Neutral Carbohydrate and Fe(II) in the Glycopeptide Isolated from Wood Degrading Cultures of *T. palustris*, and the Molecular Weight of the Glycopeptide.

Protein	*Neutral carbohydrate*	*Fe(II)*	*Molecular weight*
61%	24%	0.12%	12,000~7,200

The Fe(II) content measured with ferrozine (18) was 0.12%. The molecular weight determined by size-exclusion chromatography and gel-filtration on Sephadex G-25, was in a range of 1,000-5,000 D. But Tricine-SDS-PAGE showed that the glycopeptide consisted of at least three bands of 12.0, 10.5, and 7.2 kDa. The amino acid composition analyzed with an amino acid analyzer is shown in Table 2. The glycopeptide has an abundance of amino acids which have side chain amino groups.

Table 2. Amino Acid composition of the Glycopeptide.

Amino acid	*%*	*Amino acid*	*%*
asparagine	---	tryptophan	---
aspartic acid	13.0	threonine	2.7
alanine	4.1	valine	1.9
arginine	0.7	histidine	1.1
isoleucine	0.9	phenylalanine	0.8
glycine	11.6	proline	1.9
glutamine	---	methionine	0.1
glutamic acid	18.7	lysine	0.6
cystine	0.7	leucine	1.7
serine	4.0	NH_3	35.2
tyrosine	0.2		

The carbohydrate composition determined by the method of Clamp et al. (36) is shown in Table 3. The galactose content is very high (74%), whereas the levels of N-acetyl-D-glucosamine observed were very limited. The production of $^{\bullet}OH$ in the reduction of O_2 with the glycopeptide under various conditions is shown in Table 4. The generation of $^{\bullet}OH$ was measured with dimethyl sulfoxide using the method of Fukui et al. (37). The assay for $^{\bullet}OH$, based on the conversion of dimethyl sulfoxide to methane sulfinic acid, and is specific for $^{\bullet}OH$ (19).

Table 3. Carbohydrate Composition of the Glycopeptide (Sugar %).

Fucose	*Xylose*	*Mannose*	*Galactose*	*Glucose*	*N-acetyl-glucosamine*
5	**5**	**5**	**74**	**11**	**0**

The results shown in Table 4 suggest that the glycopeptide containing Fe(II) alone reduces O_2 to $^{\bullet}OH$ via O_2^- and H_2O_2, and catalyzes the oxidation of electron donors, such as NADH, in the presence of O_2 to produce $^{\bullet}OH$.

Table 4. Effects of O_2 and N_2 Atmospheres, and of H_2O_2, electron Donors, Superoxide Dismutase, Catalase and a Hydroxyl Radical Scavenger on the Generation of Hydroxyl Radicals by an Extracellular One-Electron Oxidation Substance Isolated from *T. palustris*.

Additions to reaction mixtures[a)]	*Atmospheres*	*Hydroxyl radical generation*	
		x 10^{-8}mol/2h	*Relative values(%)*
None	100%O_2	0.6	6
$3x10^{-6}$mol H_2O_2	100%N_2	4.0	44
$3x10^{-6}$mol NADH	100%O_2	9.2	100
$3x10^{-6}$mol NADH,0.3mg SOD[b)]	100%O_2	3.6	39
$3x10^{-6}$mol NADH,0.3mg cata[c)]	100%O_2	0.9	10
$3x10^{-6}$mol ascorbic acid	100%O_2	28.2	307
$3x10^{-6}$mols NADH and DMNA [d)]	100%O_2	3.3	36

[a)] All reaction mixtures contained 3.0mg of the extacellular substance,$3x10^{-4}$mol dimethyl sulphoxide , and acetate buffer (40mM, pH 4.5) to total 2ml in a 33ml test tube. Test tube were purged with 100% O_2 or N_2. Each value represents the average of a triplicate experiment.

[b)] Superoxide dismutase,2,670unit/mg,

[c)] Catalase,42,000unit/mg.

[d)] N,N-dimethyl-4-nitrosoaniline.

The ferric reducing ability of the glycopeptide is shown in Figure 2 which indicates that 1 mg of the glycopeptide reduces 1.7 μM of Fe(III) to Fe(II). that is, a single mole of the glycopeptide reduces about 20 moles of Fe(III) to Fe(II), provided the molecular weight is 12,000. The Fe(III)- and Cu(II)-reducing ability of the glycopeptide, and the content of aldehyde and carbonyl groups, excluding the peptide linkages in the glycopeptide, are shown in Table 5. The

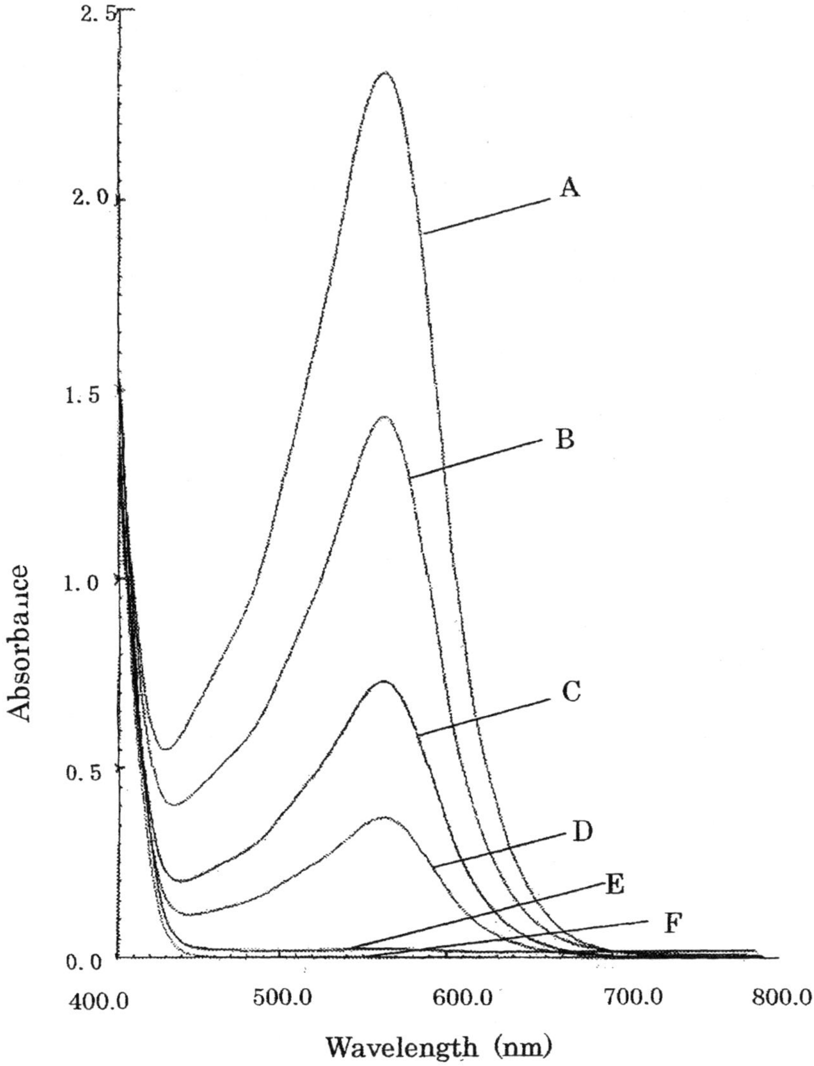

Fig. 2. Absorption spectra of solution of Fe(II) and ferrozine, Fe(III) and ferrozine, the glycopeptide and ferrozine, and the glycopeptide, Fe(III) and ferrozine. A solution (500 μ l) of Fe(III) (20ppm) was added to solutions of the glycopetide (0.066mg, 0.033mg or 0.0165mg) and the solutions were permitted to stand for 24 hours at 28℃. Ferrozine was added to the solutions, and the solutions of Fe(II)(10×10^{-6}g), Fe(III)(10×10^{-6}g) or the glycopeptide. Then the reaction mixtures were allowed to stand for 2 days and the absorption spectra were measured. All reaction mixtures contained 0.4% ferrozine to total 1.5ml. A: 6.7 ppm Fe(II), B: 0.066 mg of the glycopeptide and 6.7 ppm Fe(III), C: 0.033 mg of the glycopeptide and 6.7 ppm Fe(III), D: 0.0165mg of the glycopetide and 6.7 ppm Fe(II), E: 0.066mg of the glycopeptide, F: 6.7 ppm Fe(III).

amount of carbonyl groups in the glycopeptide, excepting those of the peptide itself, measured with 2,4-dinitrophenylhydrazine is 0.8 μmol carbonyl groups/mg of glucopeptide, that is, about 9 mol/mol provided the molecular weight is 12,000. The aldehyde group in the glycopeptide was determined with thiobituric acid or methone (dimethyldihydroresorcin) which stoichiometrically condenses with aldehyde groups with the loss of water. The aldehyde group is less than 2.8×10^{-9} mol/mg or 0.04 mol/mol provided the molecular weight is 12,000. The Cu(II)-reducing ability of the glycopeptide measured by the method of Somogi-Nelson (38) is 1.6 μmol/mg or 19 mol/mol provided the molecular weight is 12,000. This method is used for the measurement of aldehyde and α -hydroxyketone or endiol groups. These results mentioned above indicate that the glycopeptide is a glycosylated peptide and has about 10 mol of glycosylamine/mol provided the molecular weight is 12,000.

Table 5. Fe(III)- and Cu(II) Reducing Ability of the Glycopeptide,and Contents of the Carbonyl Group (Except that of Peptide Linkage in the Glycopeptide).

	μmol/mg	*mol/1.2x10⁴g*
Fe(III)-reducing ability	1.7	20
Cu(II)-reducing ability	1.6	19
Carbonyl group-content	0.8	9
Aldehyde group-content	0.0	0

The side chain amino groups of proteins will condense nonenzymatically with the aldehyde groups of a carbohydrate to give a glycosylamine (Schiff base) which undergoes nucleophile-catalyzed rearragements to 1-amino -1-deoxy-D-fructose derivatives (an Amadori compound). Amadori compounds reduce O_2 to H_2O_2 and Fe(III) to Fe(II) (39). Thus glycosylated peptides containing iron ions can reduce O_2 to $^{\bullet}OH$ as shown in Figure 3. Extracellular glycopeptides have been isolated from wood-degrading cultures of the white-rot fungi, *P. chrysosporium*, *Irpex lacteus* and *Trametes versicolor*, and the brown-rot fungus *G. trabeum*. The glycopeptides are essentially identical to the glycopeptide of *T. palustris* with regard to their physical and chemical characteristics.

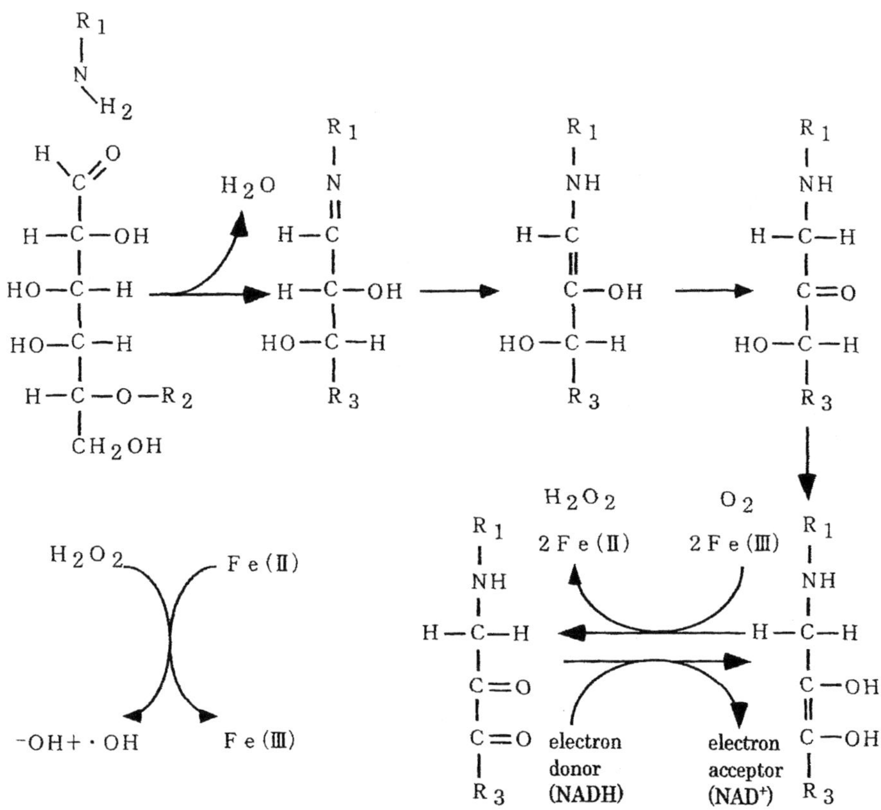

Fig 3. Proposed mechanism for the generation of · OH and the reduction of Fe(III) to Fe(II) by the glycated peptide of *T.palustris* .

Glycopeptide Proposed Mechanism

Based on the results discussed above, we propose the following mechanism for wood degradation by fungi . In wood decay by brown-rot fungi, during the initial fungal attack on the wood cell wall, low-molecular-weight glycosylated peptides are secreted by fungal hyphae into the wood cell lumen. This effector is able to reduce the Fe(III) in wood to Fe(II), chelate the Fe(II), and penetrate or diffuse into the wood cell wall. Together with the bound Fe(II), the effector catalyzes the oxidation of electron donors present in the presence of O_2 within the wood cell wall to produce $^{\bullet}OH$ *via* H_2O_2. The $^{\bullet}OH$ produced in the wood cell wall attacks chemical constituents of the wall and causes the depolymerization of both crystalline and noncrystalline cellulose. Modification of the lignin also occurs and the fungus then transforms the cell wall layer by cutting channels through the S_3 layer for enzyme-diffusion. After that, enzymes such as endoglucanases can penetrate the cell wall to act on the cellulose and hemicellulose. The proposed mechnism is consistent with the rapid increase of the solubility of wood in various solvents, and the rapid initial decrease in average degree of depolymerization (DP) of the holocellulose in wood during decay by brown-rot fungi (1).

In wood decay by white-rot and soft-rot fungi, hydroxyl radicals produced in wood cell walls by low-molecular weight glycosylated peptides modify lignin, resulting in new phenolic, benzyl radical, and cation radical substructures that are susceptible to attack by phenol oxidases. The modified part of lignin is completely depleted by phenol oxidases such as LiP, MnP or laccase. The hydroxyl radicals also depolymerize polysaccharides. The depolymerized and exposed polysaccharides are efficiently hydrolyzed and soon depleted by a complete cellulase system and hemicellulase. Some white-rot fungi lack exo-1,4- β -glucanase (40, 4) and such species seem to decompose the lignin component in preference to the cellulosic component (41,4). The gradual decrease in average DP of the holocellulose in wood, and the very small change of the solubility of wood in various solvents during decay by white-rot fungi (1) can be interpreted in this proposed mechanism for white-rot.

References

1. Cowling, E. B. *US Dep. Agric. Tech. Bull.* 1961, *No. 1258*, 1-75.
2. Kirk, T. K.; Highley, T. L. *Phytopathology* 1973, *63*, 1355-1342.
3. Highley, T. L. *Mater. Org.* 1987, *22*, 39-45.
4. Enoki, A.; Tanaka, H.; Fuse, G. *Holzforschung* 1988, *42*, 85-93.

5. Rätto, M.; Ritschkoff, A. C.; Viikari, L. *Appl. Microbiol. Biotechnol* 1997, *48*, 53-57.
6. Gierer, J.; Yang, E.; Reitberger, T. *Holzforschung* 1992, *46*, 495-504.
7. Chirat, C.; Lachenal, D. *Holzforschung* 1997, *51*, 147-154.
8. Espejo, E.; Agosin, E.; Vicuña, R. *Arch. Microbiol.* 1990, *154*, 370-374.
9. Bacha, S.; Gierer, J.; Reitberger, T.; Nilsson, T. *Holzforschung* 1992, *46*, 61-67.
10. Koenigs, J. W. *Wood Fiber*, 1974, *6*, 66-80.
11. Itakura, S.; Hirano, T.; Tanaka, H.; Enoki, A. Res. Group Wood Preserv. IRG Secretariat, Stockholm, 1994; Doc. No. IRG/WP/94-10062.
12. Enoki, A.; Tanaka, H.; Fuse, G. *Wood Sci. Technol.* 1989, *23*, 1-12.
13. Highley, T. L.; Murmanis, L. *Holzforschung* 1985, *39*, 73-78.
14. Kuo, M.; Stokke, D. D.; MacNabb, H. S. Jr. *Wood Fiber Sci.* 1988, *20*, 405-414.
15. Flournoy, D. S.; Kirk, T. K.; Highley, T. L. *Holzforschung* 1991, *45*, 383-388.
16. Srebotnik, E.; Messner, K. *Holzforschung* 1991, *45*, 95-101.
17. Enoki, A.; Hirano, T.; Tanaka, H. *Mater. Org.* 1992, *27*, 247-261.
18. Hirano, T.; Tanaka, H.; Enoki, A. *Mokuzai Gakkaishi* 1995, *41*, 334-341.
19. Hirano, T.; Tanaka, H.; Enoki, A. *Holzforschung* 1997, *51*, 389-395.
20. Hirano, T.; Enoki, A.; Tanaka, H. *J. Wood Sci.* 2000, *46*, 45-51.
21. Tanaka, H.; Enoki, A.; Fuse, G. *Mokuzai Gakkaishi* 1985, *31*, 935-945.
22. Tanaka, H.; Enoki, A.; Fuse, G. *Mokuzai Gakkaishi* 1986, *32*, 125-135.
23. Tanaka, H.; Itakura, S.; Enoki, A. *Holzforschung* 1999, 53, 21-28.
24. Tanaka, H.; Hirano, T.; Enoki, A. *Mokuzai Gakkaishi* 1993, *39*, 493-499.
25. Tanaka, H.; Itakura, S.; Enoki, A. *J. Biotechnol.* 1999, 75, 57-70.
26. Tanaka, H.; Itakura, S.; Enoki, A. *Mater. Org.* 1999/2000, *33*, 91-105.
27. Martinez-Inigo, M. J.; Kurek, B. *Holzforschung* 1997, *51*, 543-548.
28. Srebotnik, E.; Messner, K.; Foisner, R. Penetrability of white-rot degraded pine wood by the lignin peroxidase of *Phanerochaete chrysosporium. Appl. Environ. Microbial.* 1988, *54*, 2608-2614.
29. Blanchette, R. A.; Kruger, E. W.; Haight, J. E.; Akhtar, M.; Akin, D. E. *J. Biotechnol.* 1997, *53*, 203-213.
30. Goodell, B.; Yamamoto, K.; Jellison, J.; Nakamura, M.; Fujii, T.; Takabe, K.; Hayashi, N. *Holzforschung* 1998, *52*, 345-350.
31. Tanaka, H.; Itakura, S.; Hirano, T.; Enoki, A. *Holzforschung* 1996, *50*, 541-548.
32. Tanaka, H.; Itakura, S.; Enoki, A. *Holzforschung* 2000, *54*, 463-468.
33. Enoki, A.; Fuse, G.; Tanaka, H. Inter. Res. Group Wood Preserv. IRG Secretariat, Stockholm, 1991; Doc. No. IRG/WP/1516.
34. Lowry, O. H.; Rosebrough, N.; Farr, A. L.; Randall, R. J. *J. Biol. Chem.* 1951, *193*, 265-275.

35. Dubois, M.; Gilles, K. A.; Hamilton, J. K.; Rebers, P. A.; Smith, F. *Anal. Chem.* 1956, *28*, 350-356.
36. Clamp, J. R.; Bhatti, T.; Chambers, R. E. "Glycoproteins", Gottschalk, A. Ed. Elsevier, Amsterdam, 1972; p. 300-321.
37. Fukui, S.; Hanasaki, Y.; Ogawa, S. *J. Chromatogr.* 1993, *630*, 187-193.
38. Somogyi, M. *J. Biol. Chem.* 1951, *195*, 19-23.
39. Oak, J. H.; Nakagawa, K.; Miyazawa, T. *FEBS Lett.* 2000, *481*, 26-30.
40. Sethuraman, A.; Akin, D. E.; Eriksson, Karl-Erik, L.*Biotechnol. Appl. Biochem.* 1998, *27*, 37-47.
41. Srebonik, E. ; Messner, K. *Appl. Environ. Microbiol.* 1994, *60*, 1383-1386.

Chapter 9

Role of Metals in Wood Biodegradation

Jaime Rodriguez[1], André Ferraz[2], and Maricilda P. de Mello[3]

[1]Renewable Resources Laboratory, Universidad de Concepción, Casilla 160–C, Concepción, Chile
[2]Departamento de Biotecnologia, Faculdade de Engenharia Química de Lorena, Caixa Postal 116, Lorena, SP, Brazil
[3]Centro de Biologia Molecular e Engenharia Genética-Universidade Estadual de Campinas, Caixa Postal 6010, Campinas, SP, Brazil

Wood degrading fungi produce several metal-containing enzymes that participate in wood biodegradation. In addition, they produce iron-reducing compounds that take part in the process as well. Ferric ions present in woods and soils are essential for living organisms that, in general, produce high affinity iron-chelating compounds to scavenge the metal necessary for their metabolism. The importance of low molecular mass compounds produced by fungi in wood biodegradation has been progressively confirmed. Increasing evidence has been reported that both brown- and white-rot fungi use the Fenton reaction to decay wood. In this chapter, the role of metals in wood biodegradation and the results of using fungal iron-reducing mimetic compounds are presented. Furthermore, the application of a bio-inspired fungal degradation system as an alternative to current methods of waste-stream effluent treatment is discussed.

Organisms live in equilibrium with the atmosphere and lithosphere, taking from it necessary substances for growth. To survive under different environmental conditions, microorganisms have developed a number of strategies to: a) take up essential but scarce elements in environments; b) protect themselves against toxic heavy metal concentrations; and c) utilize metals in the biodegradation of organic substances.

Living systems involve complex mechanisms. In general, microorganisms use considerable energy to create and maintain concentration gradients for inorganic ions between the membrane and separate organelles inside the cells. Similarly, an efficient biological mechanism exists to accumulate both silica and Fe(III) ions, each practically insoluble at pH 7, and thus make them bioavailable for structural or biochemical purposes (*1*).

Several metal-containing enzymes are necessary to life. In wood-rot fungi, for instance, iron-containing (lignin and manganese peroxidases, and cellobiose dehydrogenase) and copper-containing enzymes (laccases) participate in lignin degradation. The enzyme's structures and mechanisms of action, as well as the regulation of their expression, will be discussed later. However, the extracellular enzymes produced by wood-degrading fungi are too large to penetrate the intact wood cell wall (*2*, *3*) to initiate the delignification process. Therefore there must be cooperative mechanisms, which first involve low molecular mass compounds (LMMCs), such as metal chelators (*4*) or enzyme mediators, capable of preceding enzymes into the microvoids of lignocellulose structure and opening up the cell wall.

From the first report about iron binding compounds produced by wood-decaying basidiomycetes by Fekete *et al.* (*5*), several researchers have presented evidence of their participation in wood decay processes. Jellison *et al.* (*6*) reported the production of low molecular mass iron-binding compounds by the brown-rot fungus *Gloeophyllum trabeum*. Fenton type mechanisms involving the participation of iron chelators were proposed to explain wood biodegradation. Today it is well known that brown-rot fungi use such a mechanism, but a similar process in white-rot fungi is not yet clear. To carry out the Fenton reaction, white- and brown-rot fungi have developed different systems to reduce Fe(III), among them dihydroxybenzenes, low molecular mass glycopeptides (*7*) and the enzyme cellobiose dehydrogenase (CDH). In this chapter they will be called iron reducing compounds (IRCs) as a general designation.

This chapter does not intend to be all-inclusive, as different mechanisms used by wood degrading fungi and their complexity cannot be discussed in detail within a single space-limited chapter. Special emphasis is dedicated to the use of synthetic IRCs as mimics of fungal compounds involved in wood degradation.

Metals in Wood

Mineral elements enter living plants predominantly through plant root systems, therefore in a sense plants act as the "miners" of the earth's crust (*8*). In this task, they frequently are helped by mycorrhizae fungi. Certain elements present in wood are essential for the tree growth, whereas others are not necessary. The wood inhabiting fungi use metals present in wood to degrade it and for their growth. In addition, metals present in forest soil can also be a source of ions for fungi, enhancing fungal capacity to degrade the wood. The role of cations in wood biodegradation by brown-rot fungi was previously reviewed by Jellison *et al.* (*9*), and special emphasis was devoted to iron, manganese, zinc, copper, calcium, magnesium, potassium, and hydrogen ion.

Inorganic compounds comprise 0.1 to 0.5 %, oxide basis, of total wood components in temperate zones and up to 3-4 % in tropical woods. The main inorganic components found in wood ash are potassium, calcium, magnesium, sodium, iron, silica, sulfate, phosphate, chloride and carbonate (*10, 11*).

Saka and Mimori (*12*) analyzed the distribution of inorganic constituents from the pith to the outer ring of Japanese birch wood by SEM-EDXA. The secondary walls of wood fibers, vessels, and ray parenchyma cells usually contained detectable concentrations of only three elements, sulfur, chloride and calcium. In contrast, almost all other elements detected (Na, Mg, Al, Si, P, S, Cl, K, Ca, Fe and Zn) were found to be localized and concentrated in the ray parenchyma cells and the pit membranes between vessels and ray parenchyma cells. The distribution of some of these elements (Ca, Mg, S, K and Zn) seemed to be correlated to the presence of pectic materials. The total level of iron, manganese and other transition metals are usually significantly lower than found in soil. These metals would be expected to be either bound to components in the wood cell wall, or alternatively as insoluble oxy(hydr)oxide forms (*13*). Also, some metals can be transported via fungal hyphae from soil to wood. Further research is still needed to refine the relationship between minerals in wood and its biodegradation.

An attempt to correlate the minerals in wood and the selective delignification produced in "palo-podrido" (a term employed to describe selectively biodelignified logs found in evergreen temperate rain forests of South Chile) was carried out by Agosin *et al.* (*14*). Elemental analysis of sound wood showed a variety of concentrations for P, K, Ca, Mg, Al, Fe, Na, Mn, and Zn among the different species. However, correlation between metal content and selective delignification was not observed, although it is important to highlight that copper content was not evaluated in this research.

The contribution of manganese to wood degradation was first reported by Hartig already in 1878 (*15*) as black deposits in white-rotten trees and after that, by Boyce in 1961(*16*). Blanchette (*17*) demonstrated that the black deposits consist mainly of MnO_2 and attributed the deposits especially to white-pocket rot. Martinez *et al.* (*18*) and Barrasa *et al.* (*19*) found them also in the Chilean "palo podrido". Recently, Schmidt *et al.* (*20*) were able to produce the black manganese deposits also in laboratory wood samples by the action of the white-rot fungus *Physisporinus vitreus* which they had isolated from cooling-tower woods. Inductively coupled plasma emission analysis (ICP) of *Pinus sylvestris* sapwood samples revealed 46-107 ppm Mn for the controls, 55-88 ppm Mn for the non-discolored wood parts of the sample and an increase of up to 518 ppm Mn in the black parts. TEM/EDXA showed this manganese to occur in the S_2-layer of the woody cell wall directly beneath the fungal hypha and especially in the hyphal extracellular slime layer, both indicating the correlation to the manganese-dependent peroxidase (MnP).

While metal ions contribute to wood degradation mechanisms, they are also required in the maintenance of metabolism in living organisms. This is especially true for iron. Although it is the fourth most abundant element, by mass, in the crust on earth, the availability of ferric iron (in an oxidizing environment) is limited by the extreme insolubility of $Fe(OH)_3$. In general, microorganisms produce high affinity iron-chelating compounds to scavenge metals required for their metabolic needs.

Metal-Containing Enzymes

Several metallo-proteins are involved in wood biodegradation, especially in lignin degradation. These enzymes comprise iron- and copper-containing proteins. The first iron-containing proteins described in white-rot fungi were the peroxidases viz., lignin peroxidase (LiP) in 1983 (*21, 22*) and manganese-dependent peroxidase (MnP) in 1984 (*23*). More recently, some "versatile" peroxidases that oxidize Mn(II), but also oxidize phenolic and non-phenolic lignin structures independent of manganese-mediated reactions, have been reported in *Pleurotus* and *Bjerkandera* species (*24-26*). Some flavoheme-proteins, namely cellobiose-dehydrogenase (CDH), have also been reported and, apparently, function to catalyze both hydrolytic and ligninolytic reactions in wood-decay fungi (*27,28*). In addition, copper-containing enzymes viz., laccases, have a well-documented record of being produced by white-rot fungi.

Peroxidases are found not only in white-rot fungi but also in bacteria, plants, and animals, and have molecular masses ranging from 35 to 100 kDa. In peroxidases, the heme-prosthetic group has an iron in the +3 oxidation state. The iron coordinated to the four pyrrole nitrogens of the heme-group and to nitrogen of an axial histidine (*29*).

LiPs are the fungal peroxidases with the highest redox potential. These enzymes undergoes a two-electron oxidation by hydrogen peroxide as the electron acceptor, leading to the formation of the Fe(IV)/cation-radical intermediate, usually referred as compound I (C_I). This intermediate is reduced in two one-electron oxidation steps, regenerating the native enzyme, passing through the one-electron deficient form named compound II (C_{II}). The one-electron reduction pathways occur via enzymatic oxidation of non-phenolic lignin structures, giving rise to lignin radical-cations that undergo further degradation reactions (*30, 31*). MnP are similar to LiPs but possess lower redox potentials and can use Mn(II) as an electron donor for the catalytic cycle. The MnPs, are dependent on Mn(II) oxidation to convert the C_{II} compound to the native enzyme form (*32*). It is believed that the Mn(III) ion generated in the catalytic cycle is stabilized by chelating compounds, such as oxalic acid produced by the fungi. The chelated Mn(III) diffuses into the wood cell walls and oxidize phenolic lignin structures. More recently, some data has shown that chelated Mn(III) can also initiate the peroxidation of unsaturated fatty acids (UFAs), giving peroxyl radicals and UFA-hydroperoxides that could oxidize non-phenolic lignin structures (*33, 34*). The "versatile" peroxidases are similar to MnPs, but can oxidize Mn(II) as well as phenolic and non-phenolic lignin structures independently of a Mn(III)-mediated reaction (*24, 26*).

Laccases are copper-containing glycoproteins. The copper content of laccases varies between two and four atoms per enzyme molecule (*35*). Laccases are coordinated to Cu(II), which is reduced during the oxidation of a broad range of substrates such as phenols and aromatic amines. Subsequently, Cu(I) is re-oxidized to the divalent stage by oxygen in acidic media yielding water as a final product. Many white-rot fungi produce laccases. Even though the redox potentials of known laccases are too low to directly oxidize non-phenolic structures of lignin, they may be important in lignin degradation since some laccase-producing fungi degrade lignin very efficiently. At least one of these, *Pycnoporus cinnabarinus*, degrades lignin yet it produces only laccase. To overcome the redox potential barrier, it was believed that *P. cinnabarinus* produces the metabolite 3-hydroxyantranilic acid that mediates the oxidation of non-phenolic substrates by laccase as a laccase-mediator (*36*). However, recent data have shown that a 3-hydroxyantranilate-less mutant of this fungus is still able to degrade lignin without producing this mediator. Moreover, a laccase-less mutant of *P. cinnabarinus* lost its ability to degrade lignin, thus, additional research efforts are clearly necessary to understand the role of laccases in wood biodegradation (*37*).

The flavoheme-enzyme CDH is also produced by many wood-decaying fungi (*27, 28*). This enzyme oxidizes the reducing end of cellodextrins, coupling this reaction to single-electron acceptors such as quinones, molecular oxygen and Fe(III). CDH plays an important role in wood biodegradation by reducing Fe(III) to Fe(II), which serves as a reductant of both dioxygen to form superoxide radicals in the presence of oxalate, and hydrogen peroxide to form hydroxyl radicals.

Metal Regulation of Enzyme Gene Expressions

Different environmental factors may influence the expression of various ligninolitic enzymes. Also, depending on the fungal species, there are several LiP genes that encode different isoenzymes (*38*). The transcription of some LiP genes is regulated, particularly by nitrogen, carbon and sulfur availability (*33*). Likewise, many fungi produce more than one manganese peroxidase enzyme, and these are encoded by different gene copies (*40*). The expression of those genes is modulated by nutritional conditions as well as by manganese concentration (*41, 42*), but the mechanism by which the metal exerts its regulation is presently unknown. Metal-responsive cis elements (MRE) have been found in the upstream region of manganese peroxidase genes from different fungal species (*41, 42*). This may indicate that other metals are involved in MnP gene regulation. MREs, which do not respond to manganese (*43*), control the expression of genes and allow organisms to maintain tight control over metabollically appropriate concentrations of metal ions. While these metals, including zinc, copper, and iron, are maintained at the required levels by the fungal system, other toxic metals are efficiently removed. The expression of MRE regulated genes in eukaryotes is strongly induced by potentially toxic amounts of zinc and cadmium; and also by strong oxidizing agents, such as hydrogen peroxide, which is known to regulate MnP genes at the level of gene transcription (*44, 45*).

There are two allelic variants coded by one gene for CDH in *Phanerochaete chrysosporium* (*46*). CDH is produced in the presence of cellulose, but the possibility that other unknown substrates or environmental conditions trigger the production of CDH cannot be ruled out. In yeast, CDH may act as an external reductant in a high-affinity iron-uptake mechanism which involves Fe(III) reduction by a transmembrane reductase with overlapping roles in Fe(III) and Cu(II) reduction (*47*).

It has been reported that several fungi have more than one laccase-encoding gene. Differential regulation of ligninolytic-enzyme-encoding genes in response to culture conditions has been documented previously (*48-50*). Palmieri *et al.* (*51*) described the effect of copper on the production of specific laccase isoenzyme patterns and regulation at the level of gene transcription. This work indicates that laccase genes appear to be differentially regulated by the metal. Moreover, the induction of each of the four laccase isozyme genes in

Pleurotus sajor-caju by carbon, nitrogen, copper, manganese and the two aromatic compounds, 2,5-xylidine and ferulic acid, occurred at the level of gene transcription (*52*).

In a recent paper, we described the expression of iron-regulated proteins in both *P. chrysosporium* and *Lentinula edodes* (*53*). Several polypeptides with molecular mass (MM) ranging from 41 to 45 kDa and pIs varying from 4.3 to 5.00 were induced by iron starvation. Since these MM and pI ranges are characteristic of lignin and manganese-peroxidases, it is possible that some of their isoforms are also iron-responsive. Examining the current literature it can be concluded that, for most wood-decay fungi, metals present in the natural environment are not only important as an integral component of enzymes, but are also essential elements in the regulation of gene transcription for several enzymes important in wood degradation.

Role of Enzymes in Wood Decay

There are three main types of wood decay in nature, brown-, soft- and white-rot. In brown-rot the polysaccharides are primarily removed and lignin is degraded (transformed) only to a limited extent. Soft-rot involves a slow rate of polysaccharides degradation and a limited degradation of lignin. White-rot can be divided into two additional sub-types. One involves the decay of all wood components simultaneously. In this case, wood decay occurs by formation of erosion troughs and by progressive thinning of wood cell walls. This type of decay is consistent with a model in which several polymer-degrading enzymes act on the exposed surfaces of the wood cell walls producing a progressive erosion of wood components from the lumen to the middle lamella. The second type of white-rot account for a relatively small number of fungi that are selective for lignin degradation. In this case, removal of lignin and polyoses from the wood cell walls occurs without progressive thinning of the wall. For this type of white-rot, it has been demonstrated that even after long biodegradation periods when significant amounts of lignin and polyoses have been removed, the wood cell walls remain inaccessible to enzymes of molecular masses around 40 kDa. This indicates therefore that selective white-rot processes cannot be caused by the direct action of the enzymes on wood cell walls (*54*).

In a recent study evaluating the levels of enzymes produced during wood decay by white- and brown-rot fungi, Machuca and Ferraz (*55*) demonstrated that these fungi produced several extracellular enzymes when wood chips underwent solid-state fermentation. The profiles of enzymatic activities produced during wood decay varied among the different fungi studied, but some general features were noted. All these fungi produce polysaccharide hydrolytic activities, but brown-rot fungi produced higher levels of cellulases and xylanases than did white-rot fungi. On the other hand, phenoloxidases were only found in white-rot fungal extracts. In general, high enzymatic activity did not necessarily result in large amount of enzyme-specific wood component removal. Nevertheless, lignin removal (either by enzymatic or a hypothetical non-enzymatic mechanism) was confirmed as a key process for wood decay. The extent of polysaccharide degradation was closely dependent on the extent of lignin removal. For example, the highest polyose losses were observed in cultures of the white-rot *Trametes versicolor*, *Pycnoporus coccineus* and *Phebia tremellosa*. However, these fungi were not the best producers of xylanases. The xylanase activities detected in the cultures of these fungi at advanced stages of decay (after 90 days of biodegradation) were almost 10 times lower than the activities detected in cultures of the brown-rot fungi. Interestingly, *T. versicolor*, *P. coccineus* and *P. tremellosa* were also the best lignin degraders, which suggested that lignin removal resulted in increased wood cell wall permeability, facilitating the diffusion and action of hydrolytic enzymes. In the brown-rot fungi, with their limited (or absent) capacity to remove lignin from the wood cell walls, the capacity for polysaccharides degradation was not proportional to the high level of hydrolytic activity produced. These data also confirm that polysaccharide degradation by brown-rot fungi depends, at least for a portion of cell wall fragmentation, on the action of LMMCs that diffuse into the wood cell wall even in the presence of lignified material.

Iron Reducing Compounds

The important role of LMMCs in wood biodegradation processes has been repeatedly confirmed through experimental validation (*4, 56*). Among LMMCs, those with metal-chelating characteristics, especially iron-chelating compounds, have been well reviewed (*4*). In a first approach to evaluate the role of *G. trabeum* iron chelators in brown-rot decay, Goodell *et al.* (4) reported that cathecol-type compounds were involved in a Fenton reaction employed by the fungus in the wood degradation process. Two years later, Paszczynski *et al.* (*57*) isolated two ferric chelators, 4,5-dimethoxycatechol (DMC) and 2,5-dimethoxyhydroquinone (DMH), from stationary cultures of the same fungus. They considered several possibilities for the role of DMC and DMH such as oxygen-reducing agents and redox-cycling molecules, which could act as electron transport carriers in Fenton reactions. Further evidence for an extracellular hydroquinone-driven Fenton reaction in the biodegradative mechanism of the brown-rot *G. trabeum* was reported by Kerem *et al.* (*58*) and Jensen *et al.* (*59*). Several papers demonstrated that those LMMCs were iron-chelating compounds, but not necessarily of the siderophore-type as initially thought. Actually, these compounds promote Fe(III) reduction with consequent H_2O_2 production. The production of Fe(II) and H_2O_2 promote an extracellular Fenton reaction, which in turn provide the fungus with the capability to degrade a great variety of organic compounds. The production of these IRCs by white- (*T. versicolor*, *Perenniporia medulla-panis* previously known as *Poria medulla-panis*) and brown-rot (*G. trabeum*, *Wolfiporia cocos*) fungi was recently evaluated by Milagres *et al.* (*60*). *G. trabeum* showed the highest Fe(III) reduction ability as determined by ferrozine assay.

The occurrence of IRCs associated with "palo-podrido" (a good example of extensive and selective wood biodelignification) has been recently evaluated, and was tentatively related to the lignin biodegradation process (*61*). Water extract of field-collected samples of "palo-podrido" contained iron-chelating compounds based on a chrome azurol S (CAS) positive test (*62*). To recover LMMCs with iron-chelating capacity, decayed wood samples were extracted with ethyl acetate. After solvent evaporation, residual solids were suspended in water and filtered. Filtrates also reacted positively with CAS, which indicated the presence of iron chelators. Based on a negative Czacky test (*63*) and a positive Arnow test (*64*), chelators were classified as catechol compounds. The filtrates containing iron chelators were analyzed by GC/MS after methylation. The structures of the most abundant compounds are shown in Figure 1. Almost all compounds possessed hydroxyl and carboxylic acid groups, which were methylated in the derivatization procedure. The presence of carboxylic acids and phenolic compounds in the extracts explains the iron chelating capacity revealed by CAS positive reaction. Compound **6** was the major component found.

The origin of the compounds detected in "palo-podrido" is not clear. It is still uncertain whether these compounds are products from lignin or aromatic extractive biodegradation, or metabolites synthesized by the fungal species involved in "palo podrido" formation. Several low molecular mass aromatic carboxyllic acids have been found in decayed wood samples (*65*). Vanillic acid was the major compound extracted from decayed spruce, a softwood. Syringic and vanillic acids together were major compounds found in decayed birch, a hardwood, whereas 4,5-dihydroxyphthalic acid was a minor component. Tri-, tetra- and penta-aromatic carboxyllic acids (compounds **7** to **9**) have not been reported in decayed wood as products of lignin degradation, and they seem to be fungal metabolites.

As discussed before, IRCs have been postulated to play an essential role in wood cell wall degradation in brown-rot fungi. However, that mechanism is not

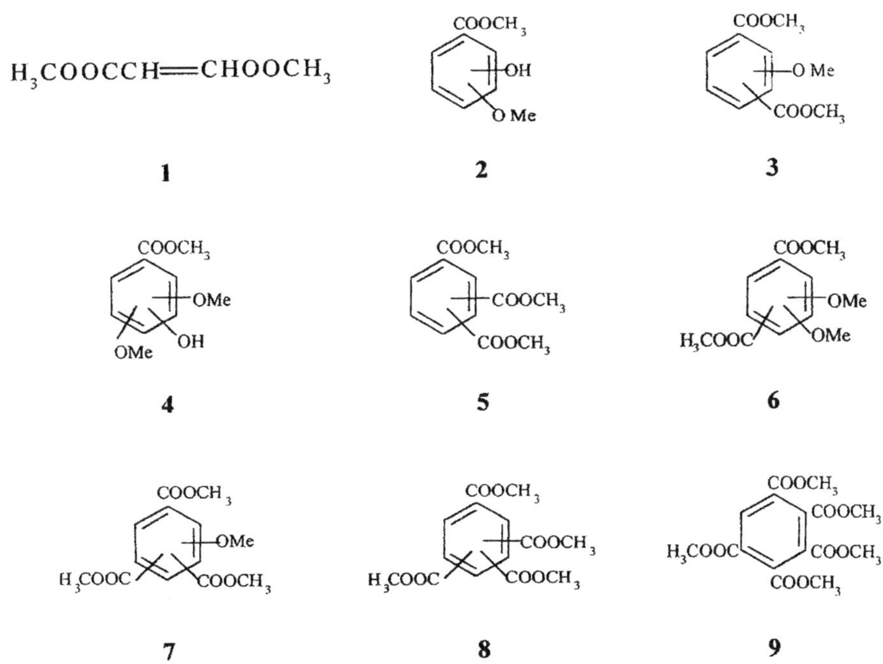

Figure 1. Proposed structures for the LMM iron chelators found in "palo podrido" extracts after methylation

compatible with "palo-podrido" formation, since the activated oxygen species produced are highly reactive and unselective therefore it could not explain selective lignin degradation. To investigate this, preliminary work was conducted to determine whether the aromatic hydroxyacids found in "palo-podrido" could reduce iron simultaneously with oxidation of an external electron donor (instead of self-oxidation of the chelator) to form activated oxygen species Thus, filtrates from "palo-podrido" samples were assayed for iron reduction and concurrent oxidation of *o*-dianisidine. Figure 2 shows that significant iron reduction occurred. Approximately 3.1% of the initial Fe(III) concentration (500 micro-M) was reduced after the first minute of reaction and after 10 minutes, 9.5% of the initial Fe(III) was reduced. Reduction activity, expressed on the basis of extracted dry mass of "palo-podrido", was 12.5 micro-moles of Fe(III) reduced/min/kg of dry wood. The capacity for Fe(III) reduction in the filtrate was ascribed to the presence of catechol derivatives (*57, 58, 66*) such as compound **6**, which was present in high concentrations in the filtrate. Measuring the oxidation rate of freshly prepared *o*-dianisidine at different ferric iron concentrations, as shown in Figure 3a, provided a determination of the oxidative capacity of the filtrate. The rate of *o*-dianisidine oxidation was low and dependent on iron concentration (Figure 3b). At the maximal iron concentration assayed, the estimated activity was 0.9 micro-moles of oxidized *o*-dianisidine per minute per kg of extracted dry wood. A comparison of the reduction activity of the iron chelators (12.5 micro-moles of reduced iron/min) with the oxidation rate of *o*-dianisidine (0.9 micro-moles of oxidized *o*-dianisidine/min) suggested that auto-oxidation of the iron chelators was preferred to oxidation of external electron donors, such as *o*-dianisidine or lignin derivatives. These results demonstrate that if the IRCs found in "palo-podrido" are involved in the biodegradation process mediated by Fe(III) reduction, direct oxidation of phenolic substrates should not occur to a significant extent.

The mechanism hypothesized to function in brown-rot decay mechanisms, where Fe(II) reacts with H_2O_2 and generates an OH radical, is unspecific and should unselectively degrade all cell wall components. This would not be compatible with the selective lignin removal pattern observed in "palo-podrido". These contradictory observations suggest that in white-rot decay, the chelators (if such is the case) may act through a different mechanism. Perhaps the production of activated species mediated by IRCs might govern just the initial stages of wood biodegradation. Such as the initial opening up of the cell wall proceed lignin removal could continue through the action of lignin specific oxidative enzymes. This hypothesis is supported by the high concentrations of IRCs found only in palo-podrido samples that were not completely collapsed. In samples showing very advanced decay stages (collapsed), IRCs activity was

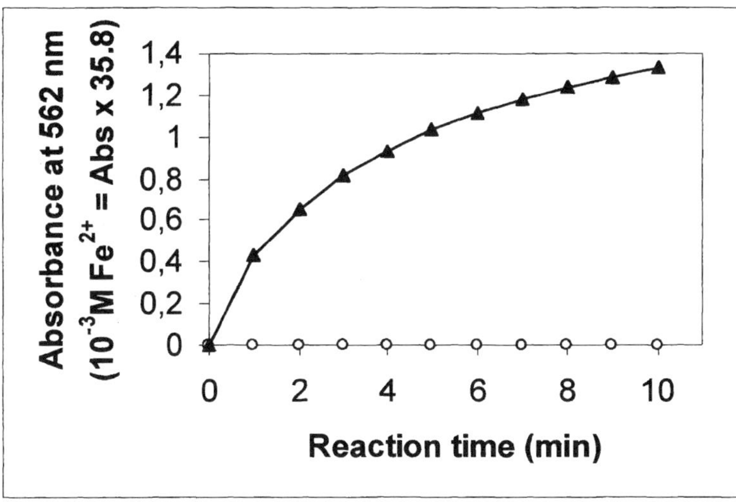

Figure 2. Iron reduction promoted by filtrates recovered from "palo podrido" samples. Absorption at 562 nm corresponds to the formation of ferrozine-Fe^{2+} complex. (-▲-) "palo-podrido" filtrate; and (-o-) control without "palo-podrido" filtrate.

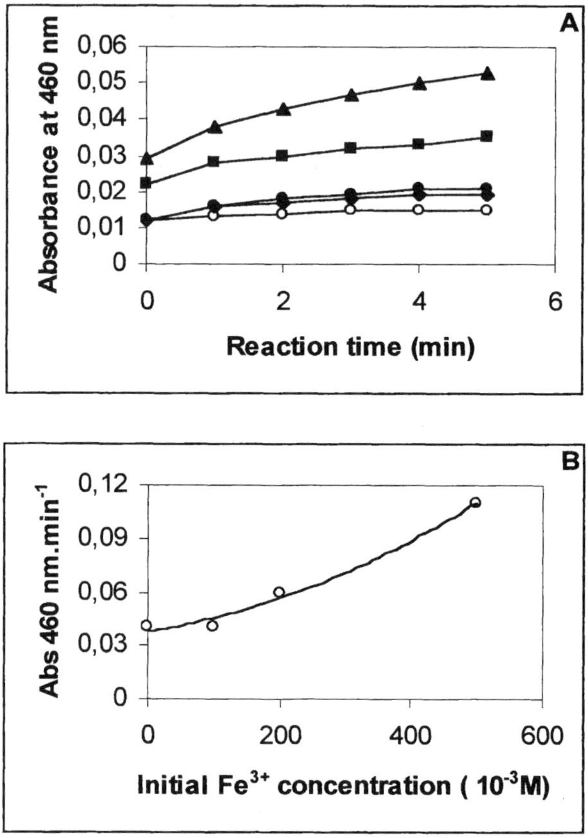

Figure 3. (A) o-Dianisidine oxidation at different ferric iron concentrations mediated by "palo-podrido" filtrate. Initial concentration of ferric iron: (-▲-) 500 μM, (-■-) 200 μM, (-●-) 100 μM, (-◆-) no added external iron and, (-o-) control reaction without filtrate. (B) Rate of o-Dianisidine oxidation by "palo podrido" filtrate as a function of the initial ferric iron concentrations (-o-). All reactions contained 8 mM oxalate to prevent o-dianisidine oxidation by water-solvated ferric iron.

negligible. Further research into the involvement of these iron-reducing compounds in "palo-podrido" formation is needed to better understand the biochemical mechanisms of this unique natural wood decay process.

Finally, it is important to consider not only OH radical production by wood degrading fungi, but also the different activated species formed in the presence of a chelator during Fenton reactions. According to Winterbourn (*67*) and Goldstein and Meyerstein (*68*), an inner sphere mechanism is the best representation for Fenton reactions. Fe(II), when complexed to a ligand (L), forms the transient species $(L\text{-}Fe\cdot H_2O_2)^{2+}$ as in eq 1 (*67, 68*). The transient species formed in eq 1 could be broken down along several pathways to yield hydroxyl radical (eq 2), or an Fe(IV) specie (eq 3), or by directly oxidizing a substrate (R) as in eq 4. The relative importance of these reactions depends on the nature of the chelator and, by extension, the pH. These reactions could be similar to those used by some fungi in wood degradation.

$$L\text{-}Fe^{2+} + H_2O_2 \rightleftharpoons (L\text{-}Fe\cdot H_2O_2)^{2+} \quad (1)$$

$$(L\text{-}Fe\cdot H_2O_2)^{2+} \longrightarrow L\text{-}Fe^{3+} + {}^{\bullet}OH + OH^- \quad (2)$$

$$(L\text{-}Fe\cdot H_2O_2)^{2+} \longrightarrow L\text{-}Fe^{4+} + 2OH^- \quad (3)$$

$$(L\text{-}Fe\cdot H_2O_2)^{2+} + R \longrightarrow L\text{-}Fe^{3+} + OH^- + {}^{\bullet}ROH \quad (4)$$

Conversely, extracellular substances with chelating properties may function as transition metal scavengers to maintain fungal metabolism and produce extracellular metal-containing enzymes.

Iron Reducing Biomimetics

Our strategy has considered the use of the dihydroxybenzenes (DHBs), such as 1,2-benzenediol (CAT), 2,3-dihydroxybenzoic acid (2,3-DHBA) and 3,4-dihydroxybenzoic acid (3,4-DHBA), as simple models of fungal iron reducing compounds. These DHBs reduce Fe(III) to Fe(II) and promote the oxidation of the lignin model compound veratryl alcohol (VA) in the presence of H_2O_2 (*69*).

The kinetics and reaction mechanisms of Fe(III) reduction by 2,3-DHBA were reported by Xu and Jordan (*66*). The proposed mechanism considers first a reduction of free iron (III) in the Fe(III)/2,3-DHBA complex followed by two reversible electron–transfers to give the quinone product. Quinones play an important role in Fenton oxidation of aromatic compounds (*70*) and chelating agents have been found to drastically alter the iron catalytic efficiency in the Haber-Weiss reaction (*71*). It has been reported that catechols lead to an enhanced formation of hydroxyl radicals during ferric ion-hydrogen peroxide reactions (*72*). The participation of different Fe(III) reducing compounds in Fenton reactions and their role in increasing the oxidation potential of the Fe(II)/H_2O_2 system has been studied. Catechols (chlorogenic acid and caffeic acid) reduce Fe(III), leading to an enhancement of hydroxyl radical formation in the Fenton reaction (*72, 73*).

The increased oxidative activity in the DHBs-driven systems could be attributed to the formation of additional activated species that are produced in the Fenton reaction. Fenton chemistry alone, without DHB's, produces an insignificant amount of degradation of VA compared with reactions catalyzed by DHB's (*69*). Conversely, in the presence of MeOH (15%), used as an OH radical scavenger, the degradation of VA is completely inhibited in 2,3-DHBA and 3,4-DHBA systems, and partially inhibited in the CAT system.

The formation of the DHB-Fe(III)-H_2O_2 transient complex, can be detected by a fast decrease of the charge transference band (for catechol complex at 700 nm) according to Ito *et al.* (*74*). Figure 4 illustrates the proposed complex and the fast decreasing of the charge transference band by intramolecular oxidation of the catechol in the complex. It is suggested that this complex can react as in equations 1-4.

Chemiluminescence studies usually permit the quantification of all activated or radical species. Chemiluminescence produced in unchelated Fenton reactions is high, but decays in less than five minutes, whereas Fenton reactions in the presence of DHBs emit chemiluminescence for more than 120 minutes. The rate of VA degradation has been correlated with the period of chemiluminescence (CAT>2,3-DHB>3,4-DHB), and not with the integrated counts (*62*).

Parra *et al.* (*75*) reported the degradation of both lignin and pulp bleaching effluent by 2,3-DHBA and 3,4-dihydroxyphenylacetic acid (DHPAA) complexed with Fe(III). The 2,3-DHBA/Fe(III) and DHPAA/Fe(III) systems modified the fluorescence of dioxane lignin, decreasing its intensity by 100 and 85%, respectively. Also, a bleaching kraft effluent (first alkaline extraction) had its fluorescence intensity decreased by 80% after 2,3-DHBA/Fe(III) treatment.

Fenton chemistry has been proposed for use as an efficient option for effluent treatment. Through reverse-bioengineering researchers have discovered

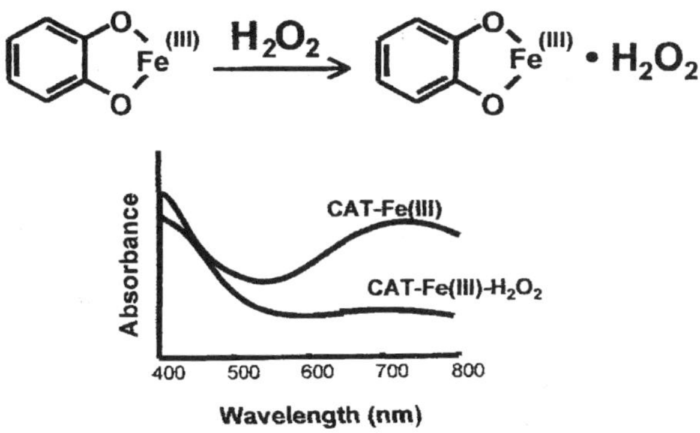

Figure 4. Proposed transient complex (Cat_n-$Fe^{3+}\cdot H_2O_2$) detected by fast decrease in the charge transfer band.

that the efficient Fenton system and mimetic analogs generated by wood degrading fungi can potentially be utilized for pulp and paper industry effluent treatment, and for the treatment of other effluents containing recalcitrant compounds. Understanding the biological pathways involved in these processes could allow new improvements in Fenton reactions making wastewater treatment with this type of system feasible. Using iron in small amounts could reduce the cost of the treatment, and iron oxide sludge production would be minimized. Advanced Oxidation Process (AOP) is the term commonly given to the generation and use of hydroxyl radicals in effluent treatment. Some examples are H_2O_2/UV, Fe/H_2O_2/UV, TiO_2/UV, etc. In this regard, the bioinspired process described here, which results in the production of OH radicals for effluent treatment, may be considered an AOP. The technical application of this mechanism for the production of OH radical in a cyclic system we have therefore termed the "Fungal Advanced Oxidation Process" (FAOP). The primary components of the system consist of Fe(III), an iron chelator, hydrogen peroxide and, eventually, an enzyme to recycles the chelator.

As a first approach, we have used catechol, 2,3-DHBA and 3,4-DHBA, all chelated with iron and hydrogen peroxide to treat a pulp bleaching effluent (*76*). The 2,3-DHBA/Fe(III)/H_2O_2 treatment at pH 4.0 decreases AOX by 95% compared to a 65% decrease with Fe(III), peroxide and no chelator. This clearly shows an enhancement of the Fenton reaction. A reduction in toxicity is also observed in samples treated with 2,3-DHBA at pH 4.0 (90%) when compared to the ferric iron and peroxide reaction (70%).

Degradation of 2-chlorophenol and 4-chlorophenol in aqueous solutions has also been examined. Complexes formed between DHBs and Fe(III) were more efficient in degrading chlorophenol than the DHB complexes formed with Fe(II) or in the absence of chelators. Comparing chelators in regard to their degradation abilities, catechol and 3,4-DHBA were more efficient than 2,3-DHBA.

The DHB–driven Fenton system has some advantages over non-DHB iron-peroxide reactions. These include the use of reduced amounts of iron, the higher pH of working solution, and the faster degradation rate. These promising results suggest that an improvement of the Fenton reaction using chelated-driven Fenton systems can be attained. Our experience has shown that, learning from fungal wood decay mechanisms, it is possible to mimic these processes and apply them in diverse fields. Our next step to further this work is to use fungal mycelium or extracellular enzymes to recycle the chelator.

Acknowledgements

The authors would like to express their appreciation for financial support received from FONDECYT (Grant 1010840) and FONDEF (Grant D97I2038) in Chile and CNPq, CAPES and FAPESP in Brazil for conducting multidisciplinary research projects related to wood biodegradation.

References

1. Kaim, W.; Schwederski, B. *Bioinorganic Chemistry: Inorganic Elements in the Chemistry of Life;* John Wiley & Sons: Chichester, 1994.
2. Carpita, N.; Sabularse, D.; Montezinos, D.; Delmer, D.P.; *Science* **1979,** *205,* 1144-1147.
3. Flournoy, D.S.; Kirk, T.K.; Highley, T. *Holzforschung* **1991,** *45,* 383-388.
4. Goodell, B.; Jellison, J.; Liu, J.; Daniel, G.; Paszczynski, A.; Fekete, F.; Krishnamurthy, S.; Jun, L.; Xu, G. *J. Biotechnol.* **1997,** *53,* 133-162.
5. Fekete, F.A.; Chandhoke, V.; Jellison, J. *Appl. Environ. Microbiol.* **1989,** *55,* 2720-2722.
6. Jellison, J.; Chandhoke, V.; Goodell, B.; Fekete, F. A. *Appl. Microbiol. Biotechnol.* **1991,** *35,* 805-809.
7. Enoki, A.; Itakura, S.; Tanaka, H. *J. Biotechnol.* **1997,** *53,* 265-272.
8. Lincoln, T.; Zeiger E. *Plant Physiology*; 2nd Edition. Sinauer Associates Inc. Publishers: Sunderland, MA. 1998.
9. Jellison, J.; Connolly, J.; Goodell, B.; Doyle, B.; Illman, B.; Fekete, F.; Ostrofsky A. *Int. Biodegrad. Biodet.* **1997,** *39,* 165-179.
10. Fengel, D.; Wegener, G. *Wood Chemistry, Ultrastructure, Reactions;* Walter de Gruyter, Berlin, 1989.
11. Baeza, J.; Freer, J. In *Wood and Cellulosic Chemistry;* Hon, D.; Shiraishi, N. Eds.; Marcel Dekker: New York, NY, 2001; pp 275-384.
12. Saka, S.; Mimori, R. *Mokuzai Gakkaishi* **1994,** *40,* 88-94.
13. Goodell, B.; Jellison, J. In: *Forest Products Biotechnology;* Bruce, A.; Palfreyman J. W. Eds.; Taylor & Francis: London, 1998, 235-249.
14. Agosin, E.; Blanchette, R.A.; Silva, H.; Lapierre, C.; Cease, K. R.; Ibach, E.; Rabad, A. R.; Muga, P. *Appl. Environ. Microbiol.* **1990,** *56,* 65-74.
15. Hartig, R. *Die Zersetzungserscheinungen des Holzes der Nadelbäume und der Eiche in forstlicher, botanischer und chemischer Richtung.* Springer, Berlin. 1878, 151 pp.
16. Boyce, J.S. *Forest Pathology.* 3rd Ed. McGraw-Hill Book Co., New York. 1961, 572 pp.
17. Blanchette, R.A. *Phytopathol.* **1984,** *74,* 725-730.

18. Martínez, A.T.; Barrasa, J.M.; Almendros, G.; González, A.E. In *Advances in Biological Treatment of Lignocellulosic Materials.*, Eds; M.P. Coughlan and M.T. Amaral-Collaço. Elsevier Appl. Sci., London. 1990, pp. 129-147.
19. Barrasa, J.M.; González, A.E.; Martínez, A.T. *Holzforschung* **1992,** *46,* 1-8.
20. Schmidt, O.; Schmitt, U.; Moreth, U.; Potsch, T. *Holzforschung* **1997,** *51,* 193-200.
21. Glenn, J.K.; Morgan, M.A.; Mayfield, M.B.; Kuwahara, M.; Gold, M.H. *Biochem. Biophys. Res. Comm.* **1983,** *114,* 1077-1083.
22. Tien, M.; Kirk, T.K. *Science* **1983,** *221,* 661-663.
23. Kuwahara, M.; Glenn, J.K.; Morgan, M.A.; Gold, M.H. *FEBS Letters* **1984,** *169,* 247-250.
24. Martínez, M.J.; Bockle, B.; Camarero, S.; Guillén, F.; Martínez, A.T. In *Enzymes for Pulp and Paper Processing;* Jeffries, T.W.; Viikari, L. Eds. ACS Symposium Series. ACS: Washington, DC, 1996; pp. 183-196.
25. Heinfling, A.; Bergbauer, M; Szewzyk, U. *Appl. Microbiol. Biotechnol.* **1997,** *48,* 261-266.
26. Palma, C.; Martínez, A.T.; Lema, J.M.; Martínez, M.J. *J. Biotechnol.* **2000,** *77,* 235-245.
27. Henriksson, G.; Johansson, G.; Pettersson G. *J. Biotechnol.* **2000,** *78,* 93-113.
28. Henriksson, G.; Zhang, L.; Li, J.; Ljungquist, P.; Reitberger, T.; Pettersson, G.; Johansson, G. *Biochim. Biophys. Acta* **2000,** *14,* 83-91.
29. Banci, L. *J. Biotechnol.* **1997,** *53,* 253-263.
30. Kersten, P.J.; Tien, M.; Kalyanaraman, B.; Kirk, T.K. *J. Biol. Chem.* **1985,** *260,* 2601-2612.
31. Kirk, T.K.; Cullen, D.. In *Environmentally Friendly Technologies for the Pulp and Paper Industry.* Young, R.A.; Akhtar M. Eds.; John Wiley & Sons: New York, NY, 1998; pp 273-308.
32. Wariishi, H.; Akileswaran, L.; Gold, MH. *Biochemistry* **1988,** *27,* 5365-5370.
33. Kapich, A.N.; Jensen, K.A.; Hammel, K.E. *FEBS Letters* **1999,** *461,*115-119.
34. Enoki, M.; Watanabe, T.; Nakagame, S.; Koller, K.; Messner, K.; Honda, Y.; Kuwahara, M. *FEMS Microbiol. Letters* **1999,** *180,* 205-211.
35. Thurston, C.F. *Microbiology* **1994,** *140,* 19-26.
36. Eggert, C.; Temp, U.; Dean, J.; Eriksson, K. *FEBS Letters* **1996,** *391,* 144-148.
37. Lia, K.; Horanyi, P.S.; Collins, R.; Phillips R.S.; Eriksson, K.-E. *Enzyme Microb. Technol.* **2001,** *28,* 301-307.

38. Gaskell, J.; Stewart, P.; Kersten, P.J.; Covert, S.F.; Reiser, J.; Cullen, D. *Biotechnol.* **1994,** *12,* 1372-1375.
39. Holzbaur, E.L.; Andrawis, A.; Tien, M. *Biochem. Biophys. Res. Commun.* **1988,** *155,* 626-633.
40. Cullen; D. *J. Bacteriol.* **1997,** *53,* 273-289.
41. Tello, M.; Corsini, G.; Larrondo, L.F.; Salas, L.; Lobos, S.; Vicuña, R. *Biochim. Biophys. Acta* **2000,** *1490,* 137-44.
42. Gettemy, J.M.; Ma, B.; Alic, M.; Gold, M.H. *Appl. Environ. Microbiol.* **1998,** *64,* 569-574.
43. Thiele, DJ. *Nucleic Acids Res* **1992,** *20,* 1183-1191.
44. Giedroc, DP; Chen, X; Apuy, JL. *Antioxid Redox Signal* **2001,** *3,* 577-596.
45. Li, D.; Alic, M.; Brown, J.A.; Gold, M.H. *Appl. Environ. Microbiol.* **1995,** *61,* 341-345.
46. Li, B.; Nagalla S.R.; Renganathan, V. *Appl. Environ. Microbiol.* **1997,** *63,* 796-799.
47. De Luca, N.G.; Wood, P.M. *Adv. Microbiol. Physiol.* **2000,** *43,* 39-74.
48. Yaver, D. S.; Xu, F.; Golightly, E. J.; Brown, K.M.; Brown S. H.; Rey, M. W; Schneider, P.; Halkier, T.; Mondorf, K.; Dalboge, H. *Appl. Environ. Microbiol.* **1996,** *62,* 834-841.
49. Mansur, M.; Suarez, T.; Fernandez-Larrea, J.; Brizuela, M.A.; Gonzalez, A.E. *Appl. Environ. Microbiol.* **1997,** *63,* 2637-2646.
50. Collins, P. J.; Dobson, A.D.W. *Appl. Environ. Microbiol.* **1997,** *63,* 3444–3450.
51. Palmieri, G; Giardina, P.; Bianco, C.; Fontanella, B.; Sannia, G. *Appl. Environ. Microbiol.* **2000,** *66,* 920-924.
52. Soden, DM; Dobson A.D. *Microbiology* **2001,** *147,* 1755-1763.
53. Hernandez-Macedo, M.L.; Ferraz, A.; Rodríguez, J.; Ottoboni, L.M.M.; De Mello, M.P. *Electrophoresis*, **2002,** *23,* In press.
54. Blanchette, R.A.; Krueger, E.W.; Haight, J.E.; Akhtar, M.; Akin, D.E. *J. Biotechnol.* **1997,** *53,* 203-213.
55. Machuca, A.; Ferraz A. *Enzyme Microb. Technol.* **2001,** *29,* 386-391.
56. Machuca, A.; Aoyama, H.; Duran, N. *Biochem. Biophys. Res. Commun.* **1999,** *256,* 20-26.
57. Paszczynski, A.; Crawford, R.; Funk, D.; Goodell, B. *Appl. Environ. Microbiol.* **1999,** *65,* 674-679.
58. Kerem, Z.; Jensen, K.A.; Hammel, K.E. *FEBS Letters* **1999,** *446,* 49-54.
59. Jensen, K.A.; Houtman, C.J.; Ryan, Z.C.; Hammel, K.E. *Appl. Environ. Microbiol.* **2001,** *67,* 2705-2711.
60. Milagres, A.M.F.; Arantes, V.; Medeiros, C.L.; Machuca, A. 8th International Conference on Biotechnology in the Pulp and Paper Industry. June 4-8 Helsinki, Finland. 2001; pp 305-306.

61. Ferraz, A.; Parra, C.; Freer, J.; Baeza, J.; Rodríguez, J.; *Int. Biodet. & Biodegradat.* **2001,** *47,* 203-208.
62. Schwyn, B.; Neilands, J.B. *Anal. Biochem.* **1987,** *160,* 47-56.
63. Csaky, T.Z. *Acta Chem. Scand.* **1948,** *2,* 450-454.
64. Arnow, L.E. *J. Biol. Chem.* **1937,** *118,* 531-537.
65. Chen, C.L.; Chang, H.M. In *Biosynthesis and Biodegradation of Wood Components;* Higuchi, T., Ed.; Academic Press: New York, N.Y., 1985; pp 535-556.
66. Xu, J.; Jordan, R.B. *Inorg. Chem.* **1988,** *27,* 4563-4566.
67. Winterbourn, C. *Toxicol. Letters* **1995,** *82/83,* 969-974.
68. Goldstein S.; Meyerstein D. *Acc. Chem. Res.* **1999,** *7,* 547-550.
69. Rodríguez, J.; Contreras, D.; Parra, C.; Freer, J.; Baeza, *J. Water Sci. Technol.* **2001,** *44,* 251–256.
70. Chen, R.; Pignatello J. *Environ. Sci. Technol.* **1997,** *31,* 2399-2406.
71. Buettner, G. *Bioelectrochem. Bioenergetics* **1987,** *18,* 29-36.
72. Iwahashi, H.; Morishita, H.; Ishii, T.; Sugata, R.; Kido, R. *J. Biochem.* **1989,** *105,* 429-434.
73. Moran, J.F.; Klucas, R.V.; Grayer, R.J.; Abian, J.; Becana, M. *Free Rad. Biol. & Med.* **1997,** *22,* 861-870.
74. Ito, S.; Ishikawa,Y.; Nishino S.; Kobayashi., T.; Ohba, S.; Nishida, Y. *Polyhedron* **1998,** *17,* 4379-4391.
75. Parra, C.; Rodriguez, J.; Baeza, J.; Freer, J.; Durán, N. *Biophys. Res. Commun.* **1998,** *251,* 399-402.
76. Rodríguez, J.; Contreras, D.; Parra, C.; Freer, J.; Baeza, J.; Durán N. *Water Sci. Technol.* **1999,** *40,* 351-355.

Chapter 10

Non-Enzymatic Iron, Manganese, and Copper Chemistry of Potential Importance in Wood Decay

William P. Henry

Department of Chemistry, Mississippi State University, Mississippi State, MS 39762

Iron, manganese, and copper may play an important role in the non-enzymatic fungal decay of wood. Specifically, these metals are believed to be involved in the generation of hydroxyl radicals or other oxidizing agents which, in turn, attack the wood structure. This review discusses the chemistry of iron, manganese, and copper and how this may relate to the fungal decay of wood. Particular attention is given to the coordination, oxidation-reduction, and Fenton chemistry of these metals.

Much of the destruction of the components in wood by fungi is enzymatic (*1*). Before these large enzymes can begin their digestion of the various components of wood, however, an initial disruption of the wood structure must be accomplished. This initial non-enzymatic attack creates openings through which the relatively large enzymes can diffuse. Various oxidizing agents, primarily the hydroxyl radical, have been implicated for this initial attack with hydrogen peroxide as the likely precursor to the hydroxyl radical (*2*).

Because radicals are both extremely reactive and non-specific in their reactivity they must be generated away from the fungus and in the vicinity of the wood component to be decomposed (*3*). If the radical is generated in close proximity to the fungus, the radical will have a deleterious effects on the fungus. With these factors in consideration, a generalized working hypothesis for the non-enzymatic degradation of wood is represented in Figure 1 (*3-11*).

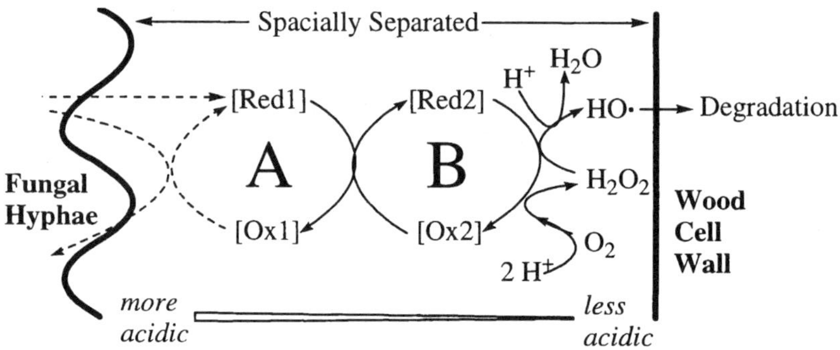

Figure 1. Possible process by which fungi generate hydroxyl radicals where A and B are redox cycles. Dashed arrows represent different possibilities for generation of Red1 (Reductant 1 in Redox Cycle A). Other oxidizing agents besides the hydroxyl radical may be involved in the fungal degradation mechanism.

The Fe^{II} cation (or other cationic metal species, as discussed later) reacts with hydrogen peroxide to generate a very reactive intermediate which subsequently oxidizes organic substrates in a process referred to as the Fenton reaction (*12-18*). The classic view is that the reactive intermediate is indeed the hydroxyl radical (eq 1). Thus, Fe^{II} (or Cu^{I}) can serve as Red2 in Figure 1, but iron in an aerobic environment exists in the trivalent oxidation state (Fe^{III}). As such, reduction of the Fe^{III} (or Cu^{II}) cation (Ox2) must occur for the Fenton process to operate. In addition, the Fenton process requires a hydrogen peroxide source. Alternatively, it is conceivable that iron, manganese, and/or copper can generate oxidizing agents through mechanisms besides the Fenton process, as will be discussed.

$$H_2O_2 + Fe^{2+} \longrightarrow Fe^{3+} + HO^- + HO\cdot \qquad (1)$$

Wood-rotting fungi also secrete oxalic acid (*7*). Thus, the region around the fungus is acidic and a good chelating agent, the oxalate dianion (ox^{2-}), is introduced into this environment. Recent work has determined that phenols and dihydroxybenzenes are also secreted by the fungus (*2,4,5,19,20*). In addition to being reducing agents, some of these species also coordinate to metal cations.

The above discussion suggests that both the oxidation-reduction chemistry that interconverts the common oxidation states of iron, manganese, and copper and the coordination chemistry of the various metal oxidation states with chelates present in the system could be important in the fungal degradation of wood. Therefore, these two topics, oxidation/reduction and coordination of iron, manganese and copper, are the focus of this review.

The Aqueous Chemistry of the Metal Cations

Iron

In addition to the metallic state there are two common oxidation states for iron, II and III. Other oxidation states found for iron complexes are –II, O, I, IV, V, and VI (*21*). In both the II and III oxidation states the coordination number tends to be 6 with an octahedral geometry (*22*). The structure of the hexaaquo complexes for both the Fe^{2+} and Fe^{3+} cations are represented below.

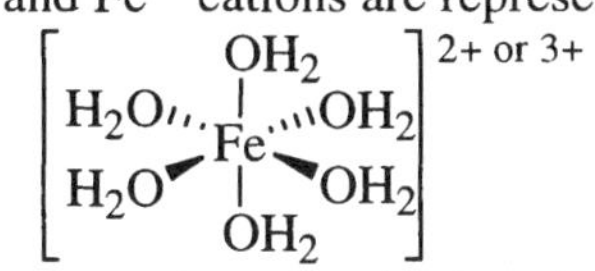

It is important to realize that the hydrated Fe^{3+} cation undergoes deprotonation (hydrolysis) according to eq 2 (*23*). The negative log of the equilibrium constant (pK) for this process at 25°C varies between 2.2 and 3.0, depending on the ionic strength of the solution.

$$[Fe(H_2O)_6]^{3+} \rightleftharpoons [Fe(H_2O)_5(OH)]^{2+} + H^+ \qquad (2)$$

This means that the pH must be approximately 1.5 or less for the hexaaquo complex to be the only species present in solution. At higher pH, hydroxy complexes are present as a significant amount of total iron. The hydroxy complexes react to form oxide-bridged dimers and other compounds. Eventually, precipitation of the red-brown solid with the empirical formula FeO(OH) occurs. In the absence of other ligands, for Fe^{3+} to remain solubilized the pH must be lower or equal to 1.5.

Like trivalent iron, Fe^{2+} undergoes hydrolysis (pK of 9.5). However, it is not necessary to decrease the acidity of the solution beyond what occurs upon dissolution of Fe^{2+} salts to prevent precipitation. Except under very acidic conditions some divalent iron exists as hydroxy complexes in solution (*21*).

The standard potential for the Fe^{3+}/Fe^{2+} couple (eq 3) is 0.77 V (*21*). This suggests that O_2 should readily oxidize Fe^{2+} to Fe^{3+} in aqueous solution.

However, because of the presence of hydroxy iron complexes the potential for the reduction of Fe^{III} to Fe^{II} is pH-dependent (*24*). The rate of autoxidation of Fe^{2+} is also pH-dependent and has been shown to be very slow under relatively acidic conditions. However, the presence of copper ions catalyzes the autoxidation of iron(II) such that the rate becomes reasonable even under acidic conditions (*25*). With an excess of Fe^{2+} the reaction proceeds such that 4 Fe^{2+} ions are required to reduce each oxygen molecule to H_2O. This reaction proceeds by an inner-sphere mechanism with hydrogen peroxide as an intermediate (*26*). However, when oxygen is present in excess over the Fe^{2+}, hydrogen peroxide is the main product (eq 4). As mentioned above, Fe^{2+} reacts with hydrogen peroxide to generate a very reactive oxidizing agent.

$$[Fe(H_2O)_6]^{3+} \underset{-\,e^-}{\overset{+\,e^-}{\rightleftharpoons}} [Fe(H_2O)_6]^{2+} \qquad (3)$$

$$2\,[Fe(H_2O)_6]^{2+} + O_2 \underset{-\,2\,H^+}{\overset{+\,2\,H^+}{\rightleftharpoons}} 2\,[Fe(H_2O)_6]^{3+} + H_2O_2 \qquad (4)$$

Manganese

Manganese compounds exist with oxidation states of –III to VII (*27*). Mn^{II} is the most common oxidation state, although Mn^{III} and Mn^{IV} complexes are relatively well-known in aqueous chemistry. The VI and VII oxidation states are extremely oxidizing and have been extensively employed in organic synthesis. Like the hydrated Fe^{2+} and Fe^{3+} cations, in the absence of other ligands Mn^{2+} in aqueous solution has an octahedral arrangement of water molecules in the coordination sphere.

In aqueous solution, Mn^{3+} disproportionates to Mn^{2+} and Mn^{4+} according to eq 5 (*28*). Mn^{4+} then precipitates as MnO_2 with the release of protons. As such, without other ligands present aqueous solutions of Mn^{3+} are unstable.

$$2\,Mn^{3+} \rightleftharpoons Mn^{2+} + Mn^{4+} \qquad (5)$$

Mn^{3+} is also slowly reduced to Mn^{2+} by H_2O. The reason for this instability is the extremely high Mn^{3+}/Mn^{2+} potential of approximately 1.56 V (*27*).

Copper

Copper has accessible oxidation states of 0, I, II, III, and IV (*29*). However, only the Cu^0, Cu^I, and Cu^{II} states are common. Cu^I compounds tend to be 4-coordinate, while many of the Cu^{II} compounds are 6-coordinate. Thus, in aqueous solution, Cu^{2+} exists as $[Cu(H_2O)_6]^{2+}$, although there is Jahn-Teller distortion in these d^9 metal systems. The Cu^0/Cu^+ redox couple has a potential of 0.52 V, while the potential of the Cu^+/Cu^{2+} couple is 0.153 V (*30*). These potentials are responsible for the fact that Cu^+ disproportionates to Cu^0 and Cu^{2+} in aqueous solution if no other ligands are present (eq 6).

$$2\,Cu^+ \rightleftharpoons Cu^{2+} + Cu^0 \qquad (6)$$

Oxalate Chemistry

As mentioned above, many decay fungi secrete significant oxalic acid (*7*). This acid is deprotonated in two steps (eqs 7 and 8). The pK_a's for the two processes are 1.1 and 3.6 (25°C, 1.0 ionic strength), respectively.

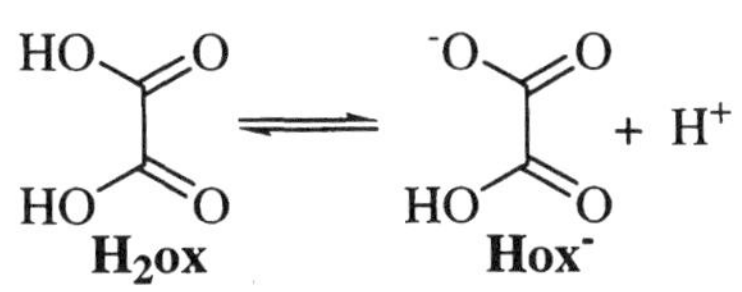

(7)

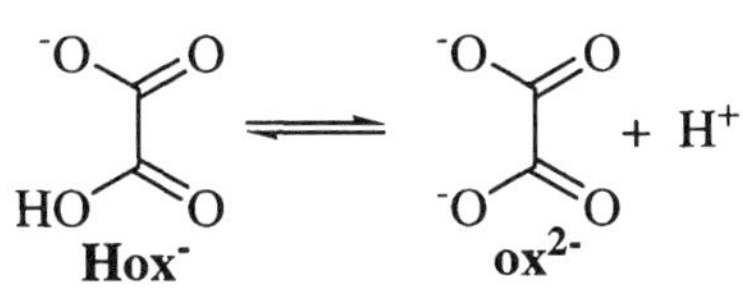

(8)

A review of the coordination of the oxalate dianion to transition metal cations has been published (*31*). In most instances the oxalate binds to the metal in a bidentate configuration resulting in a 5-membered ring (ox). However, the oxalate can also coordinate to a metal through only one oxygen. This binding configuration may involve protonation of the other carboxylate site (Hox). Two common bridging modes are also known for the oxalate group (μ-ox or μ-η^2,η^2-ox) (*32-40*).

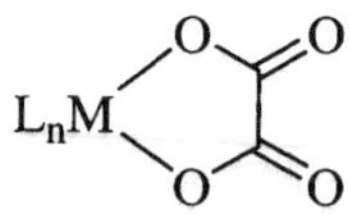

ox

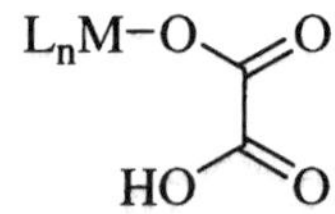

Hox

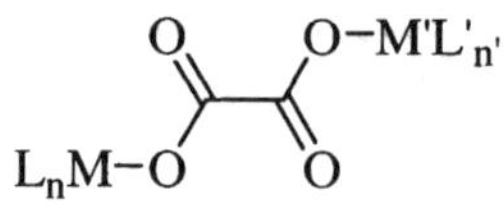

μ-ox

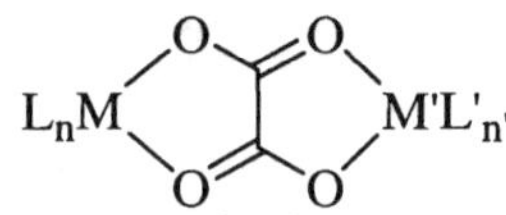

μ-η²,η²-ox

In aqueous solution the water molecules of aquo complexes of metal cations can be replaced by other ligands (*41*). These processes can be expressed by equilibrium constants for the individual processes (K_1, K_2, etc.) or as equilibrium constants for the overall process (β_1, β_2, etc.). For example, the oxalate dianion is an effective ligand toward metal cations of oxidation state "x", M^x (eqs 9-11).

$$M^x + ox^{2-} \rightleftharpoons [M(ox)]^{x-2} \tag{9}$$

$$[M(ox)]^{x-2} + ox^{2-} \rightleftharpoons [M(ox)_2]^{x-4} \tag{10}$$

$$[M(ox)_2]^{x-4} + ox^{2-} \rightleftharpoons [M(ox)_3]^{x-6} \tag{11}$$

$$K_1 = \frac{[M(ox)^{x-2}]}{[M^{x+}][ox^{2-}]}; K_2 = \frac{[M(ox)_2{}^{x-4}]}{[M(ox)^{x-2}][ox^{2-}]}; K_3 = \frac{[M(ox)_3{}^{x-6}]}{[M(ox)_2{}^{x-4}][ox^{2-}]}$$

$$\beta_1 = \frac{[M(ox)^{x-2}]}{[M^{x+}][ox^{2-}]} = K_1; \beta_2 = \frac{[M(ox)_2{}^{x-4}]}{[M^{x+}][ox^{2-}]^2} = K_1K_2;$$

$$\beta_3 = \frac{[M(ox)_3{}^{x-6}]}{[M^{x+}][ox^{2-}]^3} = K_1K_2K_3$$

Knowing the values of the various equilibrium constants (K_1, β_2, and β_3) and those for the deprotonation of the ligand (K_{a1} and K_{a2} for oxalic acid) allows for a determination of which species are present in an aqueous solution of known initial concentrations and pH.

Besides being an excellent chelating ligand, oxalate is also a reducing agent. At temperatures well beyond those experienced during wood decay processes solid samples of metal oxalate complexes rapidly undergo thermal decomposition to produce carbon dioxide (and, under some conditions, carbon monoxide) (*42*). Concomitant with gas evolution is a reduction of the metal center. A general chemical equation for this process is given in eq 12 where n is the number of oxalate ligands coordinated to the metal center.

$$2\,[M^{x}(ox)_n]^{x-2n}(s) \xrightarrow{\Delta} 2\,[M^{x-I}(ox)_{n-1}]^{x-2n+1} + 2\,CO_2 + C_2O_4^{2-} \qquad (12)$$

A similar reaction occurs in aqueous solution under photolytic conditions (eq 13 and 14) (*31*). While the photolytic conditions required to drive these reactions would not generally be present during wood decay processes, it is important to be aware that they occur (*3*). The general mechanism for this process involves the formation of the radical anion of carbon dioxide ($CO_2^{\cdot-}$) from one metal oxalate complex. This radical anion is an extremely effective reducing agent that reduces a second higher oxidation state metal center by one electron.

$$[M^{x}(ox)_n]^{x-2n} \xrightarrow{h\nu} [M^{x-I}(ox)_{n-1}]^{x-2n+1} + CO_2 + CO_2^{\cdot-} \qquad (13)$$

$$[M^{x}(ox)_n]^{x-2n} + CO_2^{\cdot-} \longrightarrow [M^{x-I}(ox)_{n-1}]^{x-2n+1} + CO_2 + C_2O_4^{2-} \qquad (14)$$

Iron

Table I compares the values for K_1, β_2, and β_3 for oxalate coordination to Fe^{II} and Fe^{III}. Apparent from these values is that the oxalate dianion coordinates to the Fe^{III} cation much more strongly than it does to the Fe^{II} species. This difference in speciation for the two iron oxidation states may be extremely important in the decomposition of wood (*3*).

Table I. Stability constants for coordination of oxalate anion to metal cations at 25°C[a]

Cation	*Log K_1*	*Log β_2*	*Log β_3*
Fe^{II}	3.05 [1.0]	5.08 [1.0]	5.22 [0.5][b]
Fe^{III}	7.58 [1.0]	13.81 [1.0]	18.6 [1.0]
Mn^{II}	3.0 [0.1]	4.4 [0.1]	
Mn^{III}	9.98 [2.0]	16.57 [2.0]	19.42 [2.0]
Cu^{II}	4.85 [0.1]	8.85 [0.1]	

a) Except where noted, all values are from the NIST database of stability constants (*43*). The ionic strength is given in brackets. b) reference (*3*).

With the possibility for different numbers of oxalate ligands bound to iron, the different Fe^{II}/Fe^{III} potentials should be considered. The potential for the tris-oxalate complex is reported to be –0.120 V, while for the bis- and mono-oxalate complexes the values are 0.181 V and 0.468 V, respectively (*3*). As expected, the oxidation of the Fe^{II} complex becomes more favorable with an increase in the number of oxalate ligands coordinated.

The Fe^{III} complex of tris-oxalate will undergo reduction thermally in the solid state with the evolution of carbon dioxide (eq 12; $M^x = Fe^{III}$, n = 3) (*42*). Likewise, in aqueous solution the complex will photolytically form the Fe^{II} oxalate species (eqs 13 and 14; $M^x = Fe^{III}$, n = 3) (*44-46*).

It has been suggested that iron(III) can be reduced by oxalate under normal fungal decay conditions (*7*). However, to date there is no substantial support for this hypothesis; iron(III) reduction by oxalate, with no other organic present, only occurs with elevated temperatures or in the presence of light. It should be emphasized, however, that these systems only included iron(III) and oxalate and were thus very simple models; a situation not found under "real life" fungal decay conditions where a wide variety of other organics and metal co-exist. It is possible that ligands or other agents present in the wood decay environment could mediate this reaction even at ambient temperatures. Therefore, caution must be exercised when extrapolating laboratory results to the biological system.

Manganese

As can be seen from Table I, the oxalate dianion binds to the Mn^{III} much more readily than it does to Mn^{II}. Indeed, the coordination by oxalate stabilizes the Mn^{III} to preclude the disproportionation reaction (eq 5) (*28,47*). In aqueous solution manganese(III) oxalate complexes undergo an oxidation/reduction reaction as depicted below (eq 15) (*48-50*). Mn^{II} complexes, carbon dioxide and the carbon dioxide radical are formed. The carbon dioxide radical can react with either another Mn^{III} to yield a second equivalent of Mn^{II} and carbon dioxide (eq 16) or O_2 to give the superoxide radical anion, $O_2^{\cdot-}$ (eq 17). The carbon dioxide radical would also reduce any Fe^{III} present to Fe^{II}. Subsequently, $O_2^{\cdot-}$ can react with Mn^{II} to produce hydrogen peroxide (eq 18). Manganese oxide with oxalate has been shown to degrade lignin under acidic conditions (*51*). Following reduction of the MnO_2 to Mn^{III} by oxalate, presumably some or all of the reactions listed occur.

$$[Mn^{III}(ox)_3]^{3-} \longrightarrow [Mn^{II}(ox)_2]^{2-} + CO_2 + CO_2^{\cdot-} \qquad (15)$$

$$[Mn^{III}(ox)_3]^{3-} + CO_2^{\cdot-} \longrightarrow [Mn^{II}(ox)_2]^{2-} + CO_2 + C_2O_4^{2-} \qquad (16)$$

$$O_2 + CO_2^{\cdot-} \longrightarrow O_2^{\cdot-} + CO_2 \qquad (17)$$

$$[Mn^{II}(ox)_2]^{2-} + O_2^{\bullet -} \xrightarrow[2\,H^+]{} [Mn^{III}(ox)_2]^- + H_2O_2 \quad (18)$$

Thus, unlike the iron oxalate chemistry discussed above manganese oxalates can react in the absence of light at room temperature to give a reduced metal and a reactive radical. Like the iron(III) complexes, Mn^{III} oxalate complexes undergo thermal decomposition in the solid state to yield Mn^{II} complexes (eq 12; $M^x = Mn^{III}$, n = 3) (*42*). The oxalate complexes of Mn^{III} also react under photolytic conditions as described in eq 13 ($M^x = Mn^{III}$, n = 3) (*31,52,53*). MnO_2/oxalate systems have been shown to degrade the cell wall of various woods (*54*).

Copper

Solid copper(II) oxalate undergoes internal redox chemistry at elevated temperatures similar to the other metal oxalate complexes (*42*). The stability constants for the binding of oxalate to Cu^{II} are given in Table I. The magnitude of these values are a little higher than those observed for the other metals in the same oxidation state.

Catechol/*p*-Hydroquinone Chemistry

In addition to oxalic acid, both 1,2- and 1,4-dihydroxybenzenes and their oxidized (quinone) forms have been found in the extracellular regions around fungi (*2,4,5,19,55,56*). 1,2-Dihydroxybenzenes (catechols) are well known to chelate to metals (*57-62*). In fact, the siderophore enderobactin binds iron(III) cations through three catecholate groups (*63-69*). Examples of the modes of coordination for catechols with metals are given below.

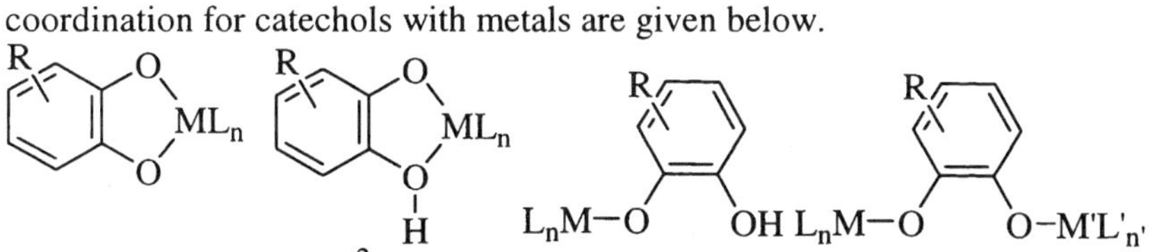

cat η^2-**Hcat** **Hcat** μ-**cat**

An extremely interesting aspect of the coordination chemistry of the catechols is represented in eq 19. Catechols can undergo a valence tautomerism to yield a reduced metal center with an ortho-semiquinone ligand (*60,70-92*). This is a form of inner-sphere redox chemistry. The determination

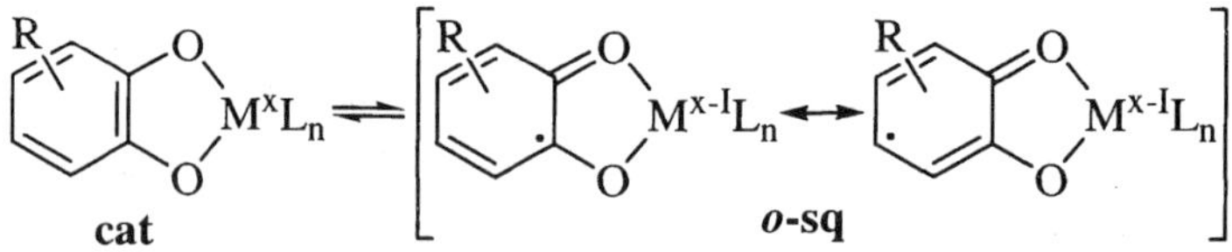

(19)

of whether a particular complex is a catechol or semiquinone can be deduced from structural information, ESR spectroscopy, or Mössbauer spectroscopy.

Catechols and *p*-hydroquinones are reducing agents (eqs 20 and 21). They can be oxidized by two one-electron steps to yield quinones (*83,93,94*). The intermediate species in both processes is referred to as a semiquinone radical.

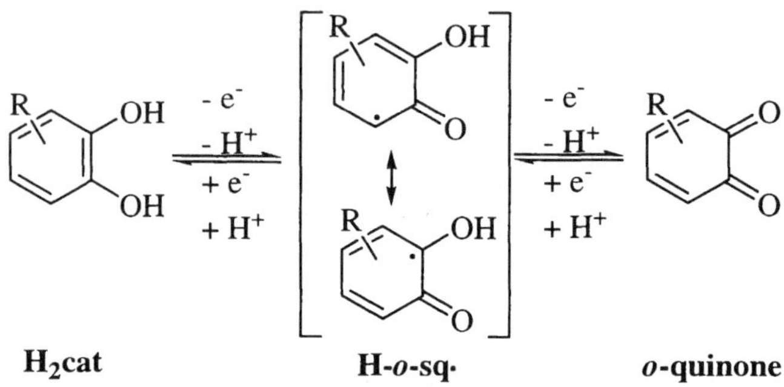

(20)

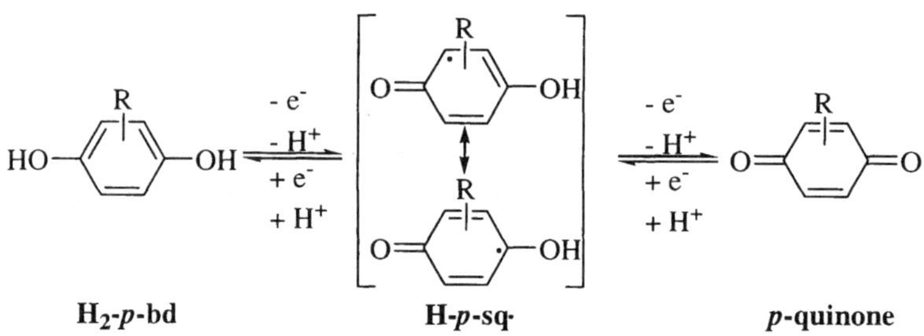

(21)

As mentioned above, for oxalate complexation the equilibrium constants for the deprotonation of the acid, or protonation of the base, must be known to use the stability constants for speciation calculations (eq 22 and 23).

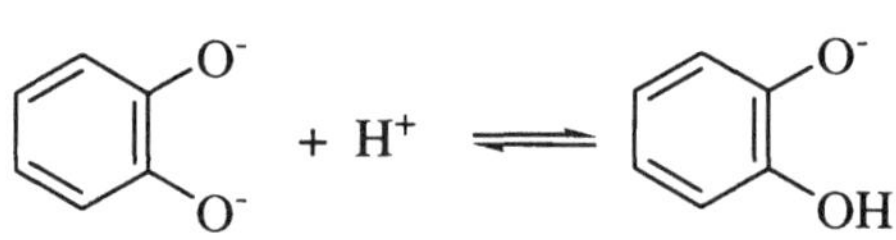

(22)

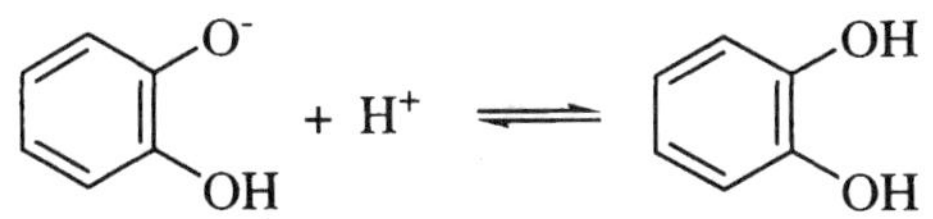

(23)

The log K_1 is reported to be 13.1 (25°C, 1.0 ionic strength) while log K_2 is 9.27 under the same conditions (*43*).

For the coordination of catechol to metal ions, the data in the NIST database is reported for eqs 24-26 (*43*).

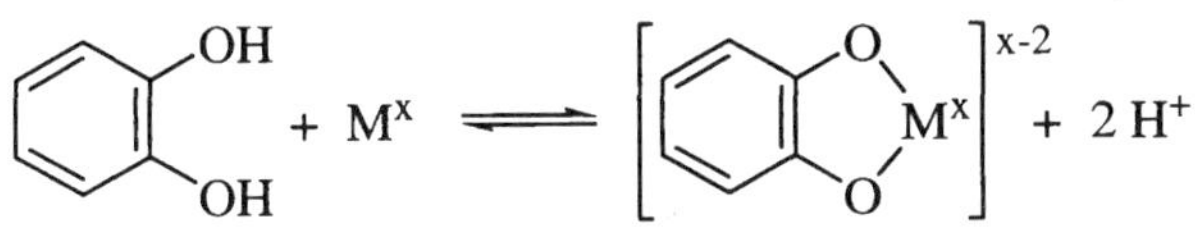

(24)

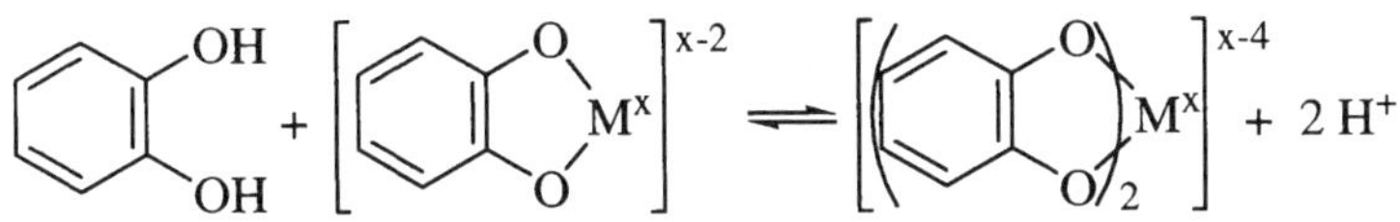

(25)

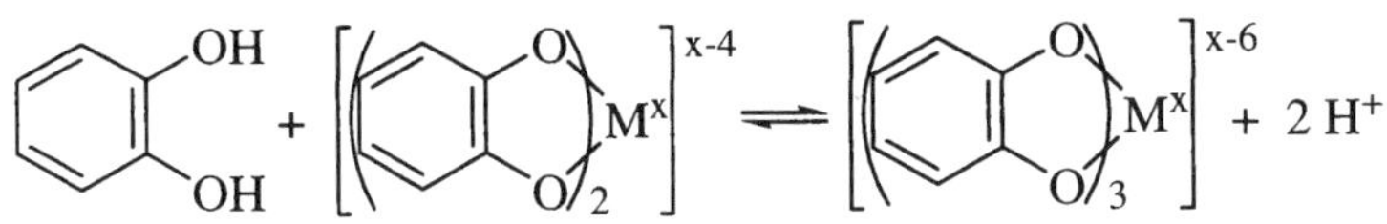

(26)

Thus, the equilibrium constants for complexation of catecholate to metals include deprotonation of the ligand, in contrast to that reported for the oxalate ligand.

Catechols undergo another set of reactions that might occur under the conditions surrounding decay fungi. Dioxygenation reactions are known to occur enzymatically, but have also been demonstrated for relatively simple model systems (*95*). The two possible reactions are intradiol and extradiol cleavage, shown respectively in eqs 27 and 28. For the intradiol cleavage a muconic acid derivative is formed. In the extradiol reaction a carboxylic acid with a formyl group is the product.

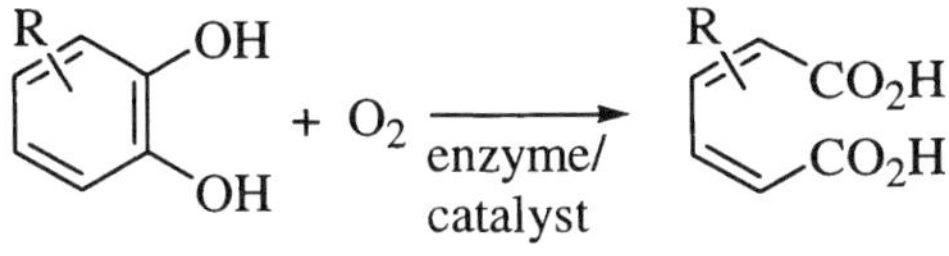

(27)

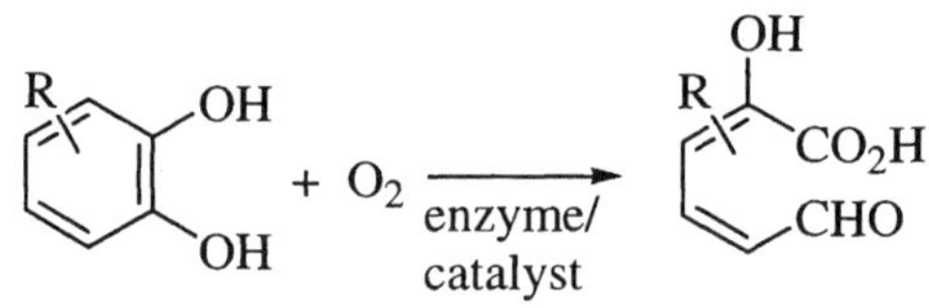

(28)

Iron

Mentasti studied the kinetics and determined the equilibrium constant for complexation of one catechol with iron(III) (eq 24) (*57*). Using stopped-flow techniques he determined the log K_1 to be –1.36 while the second order rate constant for the reaction of catechol with $[(H_2O)_5Fe(OH)]^{2+}$ was reported to be 3.1×10^3 l mol^{-1} s^{-1}. The values of log K_1, log K_2, and log K_3 for the chelation of catechol with Fe^{3+} and Fe^{2+} are given in Table II.

Table II. Equilibrium constants for the coordination of catechol to metal cations at 25°C[a]

Cation	*Log K_1*	*Log K_2*	*Log K_3*
Fe^{II}	-14.33 [1.0]	-16.7 [1.0]	
Fe^{III}	-1.36 [1.0]	-7.53 [0.1]	-13.16 [0.1]
Mn^{II}	-14.81 [1.0]	-17.0 [1.0]	
Cu^{II}	-8.64 [1.0]	-10.64 [1.0]	

a) All values are from the NIST database of stability constants (*43*). The ionic strength is given in brackets.

Mentasti and coworkers went on to study the redox reaction between catechol and iron(III) (*96*). They postulate that the catechol first complexes to Fe^{III} in a monodentate binding mode (eq 29). This complex tautomerizes to give an Fe^{II} metal center coordinated to a semiquinone (eq 30). The semiquinone then quickly dissociates (eq 31) and reacts with another Fe^{III} by an outer-sphere mechanism to yield a second equivalent of Fe^{II} and *o*-quinone (eq 32). A similar mechanism has been postulated for the oxidation of *p*-hydroquinone by metal ions.

$$[(H_2O)_5Fe(OH)]^{2+} + H_2cat \rightleftharpoons [(H_2O)_5Fe^{III}(Hcat)]^{2+} + H_2O \quad (29)$$

$$[(H_2O)_5Fe^{III}(Hcat)]^{2+} \rightleftharpoons [(H_2O)_5Fe^{II}(H\text{-}o\text{-}sq\cdot)]^{2+} \quad (30)$$

$$[(H_2O)_5Fe^{II}(H\text{-}o\text{-}sq\cdot)]^{2+} + H_2O \rightleftharpoons [Fe(H_2O)_6]^{2+} + o\text{-}sq\cdot \qquad (31)$$

$$[Fe(H_2O)_6]^{3+} + o\text{-}sq\cdot \rightleftharpoons [Fe(H_2O)_6]^{2+} + o\text{-quinone} + H^+ \qquad (32)$$

Some iron catecholate and ortho-semiquinone complexes react with O_2 to give intradiol (eq 27) (*74,75,96-102*) and extradiol cleavage (eq 28) (*74,75,103*) of the ring. Other iron systems have been demonstrated to catalyze ring cleavage reactions (*104-106*). In addition, iron complexes are known to catalyze the ortho-hydroxylation of phenols to give catechols (*73,107,108*).

Manganese

Like Fe^{II}, the equilibrium constants for the coordination of catechol to Mn^{II} are low (see Table II). The NIST database does not report stability constants for the reaction of Mn^{III} with catechol. However, some very interesting chemistry is observed for manganese catechol complexes. Some complexes undergo valence tautomerization as depicted in eq 19 (*81,91*) for the Mn^{III}(cat)/Mn^{II}(*o*-sq) pair. Even more interesting is that in one system, the form of the complex for these two possibilities is dependent upon the associated counterion (*78*). For the tris(catecholate)Mn^{IV} complexes there is no valence tautomerism to the $Mn^{III}(cat)_2$(*o*-sq) complex (*109*). The electrochemical interconversion of catecholate complexes of Mn^{II}, Mn^{III} and Mn^{IV} has been reported (*93*).

Copper

There has been extensive research on the copper chemistry of catechols. Part of this research was motivated by the desire to elucidate the mechanism by which copper-containing enzymes oxidize phenol to catechol, and catechol to *o*-quinone, under aerobic conditions (*110-113*). In addition, copper complexes that model the catechol dioxygenase's ability to cleave the aromatic ring of catechols (eq 24) have been reported (*76,114,115*).

Both semiquinone and catechol derivatives of copper are known (*60,72,76,77,79-84*). The tautomeric form is very dependent upon the ancillary ligands coordinated to the copper center (*77,79,84*). The semiquinone tautomer has been implicated as the form that is attacked by dioxygen during the intradiol cleavage reaction (*76*).

Fenton Chemistry

As mentioned earlier, the Fenton process is the reaction between a low valent metal complex, such as copper(I) or iron(II), with hydrogen peroxide to generate a species which effectively oxidizes organic molecules (*12*). Traditionally, the reaction for the Fe^{2+} system is written as in eq 1, with the oxidizing intermediate as the hydroxyl radical. However, the identity of the oxidizing intermediate has been the subject of much discussion (*13-18,116,117*). A variation in product distribution between the Fe^{II}/H_2O_2 system and the radiolytically-generated hydroxyl radical reaction is the main reason for the postulate that HO· is not the reactive intermediate formed in the Fenton process (*13*). To explain the various postulates, a description of the two possible mechanisms for electron transfer of an electron from the metal to hydrogen peroxide, inner-sphere and outer-sphere, will be outlined.

Outer-Sphere Electron Transfer

A simplified approach to this mechanism suggests that for electron transfer to occur, the oxidant and reductant need only come into close proximity. Of course, this assumes that the potentials are suitable for the oxidation/reduction process to occur. For the reaction of an Fe^{II} complex with hydrogen peroxide, the hydrogen peroxide radical anion would form but would decompose rapidly to hydroxide and the hydroxyl radical.

$$[L_nFe^{II}]^y + H_2O_2 \rightleftharpoons [L_nFe^{III}]^{y+1} + H_2O_2^{\cdot-} \qquad (33)$$

$$H_2O_2^{\cdot-} \xrightarrow{\text{v. fast}} OH^- + HO\cdot \qquad (34)$$

Inner-Sphere Electron Transfer

In this mechanism, the hydrogen peroxide would initially enter the coordination sphere of the Fe^{II}. This step in the process is referred to as ligand substitution (or ligand association, if the starting complex is coordinatively unsaturated). An iron(II) hydrogen peroxide or hydroperoxide complex would be formed as shown in Figure 2. It has been postulated that one of these peroxide iron complexes is the species which oxidizes the organic substrate in Fenton chemistry (path a) (*13*). For free hydroxyl radicals to be formed, one of the peroxide species would decompose to an iron(III) hydroxide and OH radical

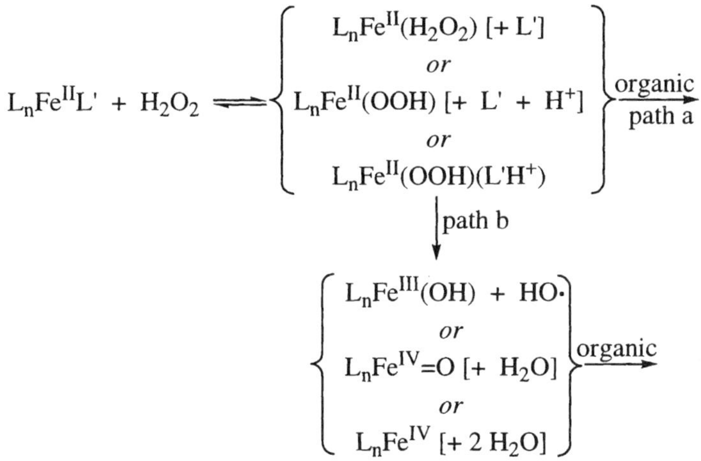

Figure 2. Postulated inner-sphere mechanistic pathways and intermediates for the Fenton process.

which could then oxidize the organic substrate (path b) (*17*). If this is the path for generation of hydroxyl radicals, it is equivalent to the outer-sphere mechanism (eqs 30 and 31). Alternately, instead of generating a hydroxyl radical, the peroxide intermediate could react further to give an iron(IV) (ferryl) complex which would serve as an oxidizing agent for the organic substrate.

Recent work has suggested that different oxidizing species can operate in the same system, with the oxidizing species dependent upon the ancillary ligands coordinated to the iron(II) center (*118*). A theoretical study suggests that an iron(IV) species is the oxidizing agent in the Fenton process where only water is coordinated to the metal center (*119*). However, an iron(IV) oxalate complex is unlikely to be stable. As is the case with manganese(III) oxalate (eq 15), the oxalate ligand would be oxidized and the Fe^{IV} metal center reduced to Fe^{III}.

Non-enzymatic Fungal Decay Mechanisms

As mentioned in the Introduction, the production of hydroxyl radicals from hydrogen peroxide (or other oxidizing agents) can be represented as in Figure 1. In wood decay processes it has been hypothesized that the reducing agent which converts hydrogen peroxide to hydroxide ion and hydroxyl radical is an iron(II) or other reduced metal species (*3-7,9,10,55,120*). A number of different methods for the reduction of iron(III) to iron(II) have been reported. Other metals, including copper(I), also react with hydrogen peroxide to give a hydroxyl radical. Extracellular catechols/hydroquinones have been isolated and shown to act as a reducing agent for iron(III) (*2,4,5,19,20,55*). Enoki has postulated that the iron is coordinated to a glycopeptide and reduction of the iron(III) involves the conversion of the glycopeptide from the reduced to oxidized forms (*6*). Schmidt et al. suggested that the oxalate dianion acts as the reducing agent (*7*), although this likely was photolytically-mediated reaction (*3*). Another possibility is that the carbon dioxide radical anion formed by the manganese oxalate chemistry (eq 17) reduces iron(III) to iron(II). Wood claims that the major reducing agent for the iron(III) complex is the cellobiose dehydrogenase enzyme (*3*).

Reduction of O_2 to give hydrogen peroxide is another important step shown in Figure 1. One hypothesized mechanism suggests the reducing agent for this process is NADH or ascorbate catalyzed by an iron glycopeptide (*6,10*). As mentioned, Fe(II) complexes will reduce O_2 to H_2O_2; this is one postulate for the generation of hydrogen peroxide (*3*). As mentioned above, copper ion catalyzes this process, even in the acidic conditions in wood. An alternative route is that the intermediate semiquinone formed when catechol is oxidized (H-*o*-sq·) reacts with O_2 to generate a peroxy radical which decomposes to give HOO· which is further reduced by iron(II) or some other reducing agent to H_2O_2 (*55*). A third possibility is that the manganese oxalate chemistry could result in hydrogen peroxide (eq 18). Contradicting Figure 1, the secretion of hydrogen peroxide by the fungus is also postulated (*2,7*).

Conclusions

Tremendous progress has been made in the study of fungal processes for the decay of wood. However, research must continue so that we can better understand the processes employed by the fungus for the metal-catalyzed non-enzymatic degradation of wood. At the same time, further research concerning the metal coordination and oxidation/reduction chemistry of the various extracellular fungal substances present in the system should be undertaken.

References

1. Eaton, R. A.; Hale, M. D. C. *Wood: Decay, Pests and Protection*; Chapman & Hall: London, 1993; pp 160-186.
2. Xu, G.; Goodell, B. *J. Biotechnol.* **2001**, *87*, 43-57.
3. Hyde, S. M.; Wood, P. M. *Microbiology* **1997**, *143*, 259-266.
4. Goodell, B.; Jellison, J.; Liu, J.; Daniel, G.; Paszczynski, A.; Fekete, F.; Krishnamurthy, S.; Jun, L.; Xu, G. *J. Biotechnol.* **1997**, *53*, 133-162.
5. Jensen, K. A. Jr.; Houtman, C. J.; Ryan, Z. C.; Hammel, K. E. *Appl. Environ. Microbiol.* **2001**, 2705-2711.

6. Hirano, T.; Tanaka, H.; Enoki, A. *Mokuzai Gakkaishi* **1995**, *41*, 334-341.
7. Schmidt, C. J.; Whitten, B. K.; Nicholas, D. D. *Proc. Am. Wood-Preserv. Assoc.* **1981**, *77*, 157-164.
8. Kapich, A. N.; Jensen, K. A.; Hammel, K. E. *FEBS Lett.* **1999**, *461*, 115-119.
9. Tanaka, H.; Itakura, S.; Enoki, A. *Mater. Org.* **2000**, *33*, 91-105.
10. Enoki, A.; Hirano, T.; Tanaka, H. *Mater. Org.* **1992**, *27*, 247-261.
11. Enoki, A.; Tanaka, H.; Fuse, G. *Wood Sci. Technol.* **1989**, *23*, 1-12.
12. Walling, C. *Acc. Chem. Res.* **1975**, *8*, 125-131.
13. Sawyer, D. T.; Sobkowiak, A.; Matsushita, T. *Acc. Chem. Res.* **1996**, *29*, 409-416.
14. Walling, C. *Acc. Chem. Res.* **1998**, *31*, 155-157.
15. MacFaul, P. A.; Wayner, D. D. M.; Ingold, K. U. *Acc. Chem. Res.* **1998**, *31*, 159-162.
16. Sawyer, D. T. *Coord. Chem. Rev.* **1997**, *165*, 297-313.
17. Goldstein, S.; Meyerstein, D. *Acc. Chem. Res.* **1999**, *32*, 547-560.
18. Winterbourn, C. C. *Toxicol. Lett.* **1995**, *82/83*, 969-974.
19. Paszczynski, A.; Crawford, R.; Funk, D.; Goodell, B. *Appl. Environ. Microbiol.* **1999**, *65*, 674-679.
20. Goodell, B.; Jellison, J. in *Forest Products Biotechnology*; Bruce, A.; Palfreyman, J. W., Eds.; Taylor & Francis: London, 1998; pp 235-249.
21. Hawker, P. N.; Twigg, M. V. in *Comprehensive Coordination Chemistry*; Wilkinson, G.; Gillard, R. D.; McCleverty, J. A., Eds.; Pergamon: Oxford, 1987; Vol. 4, pp 1179-1288.
22. Cotton, F. A.; Wilkinson, G. *Advanced Inorganic Chemistry*, 5th ed.; Wiley & Sons: New York, NY, 1988; pp 709-724.
23. Flynn, C. M. Jr. *Chem. Rev.* **1984**, *84*, 31-41.
24. Miller, D. M.; Buettner, G. R.; Aust, S. D. *Free Radical Biol. Med.* **1990**, *8*, 95-108.
25. Huffman, R. E.; Davidson, N. *J. Am. Chem. Soc.* **1956**, *78*, 4836-4842.
26. Feig, A. L.; Lippard, S. J. *Chem. Rev.* **1994**, *94*, 759-805.
27. Cotton, F. A.; Wilkinson, G. *Advanced Inorganic Chemistry*, 5th ed.; Wiley & Sons: New York, NY, 1988; pp 697-709.
28. Chiswell, B.; McKenzie, E. D.; Lindoy, L. F. in *Comprehensive Coordination Chemistry*; Wilkinson, G.; Gillard, R. D.; McCleverty, J. A., Eds.; Pergamon: Oxford, 1987; Vol. 4, pp 1-122.
29. Hathaway, B. J. in *Comprehensive Coordination Chemistry*; Wilkinson, G.; Gillard, R. D.; McCleverty, J. A., Eds.; Pergamon: Oxford, 1987; Vol. 5, pp 533-774.
30. Cotton, F. A.; Wilkinson, G. *Advanced Inorganic Chemistry*, 5th ed.; Wiley & Sons: New York, NY, 1988; pp 755-775.
31. Krishnamurty, K. V.; Harris, G. M. *Chem. Rev.* **1961**, *61*, 213-246.

32. Cotton, F. A.; Donahue, J. P.; Lin, C.; Murillo, C. A. *Inorg. Chem.* **2001**, *40*, 1234-1244.
33. Castillo, O.; Luque, A.; Roman, P.; Lloret, F.; Julve, M. *Inorg. Chem.* **2001**, *40*, 5526-5535.
34. Decurtins, S.; Schmalle, H. W.; Schneuwly, P.; Oswald, H. R. *Inorg. Chem.* **1993**, *32*, 1888-1892.
35. Glerup, J.; Goodson, P. A.; Hodgson, D. J.; Michelsen, K. *Inorg. Chem.* **1995**, *34*, 6255-6264.
36. Kou, H.-Z.; Liao, D.-Z.; Yang, G.-M.; Cheng, P.; Jiang, Z.-H.; Yan, S.-P.; Wang, G.-L.; Yao, X.-K.; Wang, H.-G. *Polyhedron* **1998**, *17*, 3193.
37. Larionova, J.; Mombelli, B.; Sanchiz, J. ; Kahn, O. *Inorg. Chem.* **1998**, *37*, 679-684.
38. Pellaux, R.; Schmalle, H. W.; Huber, R.; Fischer, P.; Hauss, T.; Ouladdiaf, B.; Decurtins, S. *Inorg. Chem.* **1997**, *36*, 2301-2308.
39. Kitagawa, S.; Okubo, T.; Kawata, S.; Kondo, M.; Katada, M.; Kobayashi, H. *Inorg. Chem.* **1995**, *34*, 4790-4796.
40. Wrobleski, J. T.; Brown, D. B. *Inorg. Chem.* **1979**, *18*, 2738-2749.
41. Martell, A. E.; Motekaitis, R. J. *Determination and Use of Stability Constants, 2nd Edition*; Wiley-VCH: New York, NY, 1992.
42. Nagase, K. *Bull. Chem. Soc. Jpn.* **1973**, *46*, 144-146.
43. Smith, R. M.; Martell, A. E.; Motekaitis, R. J. *NIST Critically Selected Stability Constants of Metal Complexes Database;* Version 6.0 ed. National Institute of Standards and Technology: Gaithersburg, MD, 2001.
44. Zepp, R. G.; Faust, B. C.; Hoigne, J. *Environ. Sci. Technol.* **1992**, *26*, 313-319.
45. Vincze, L.; Papp, S. *J. Photochem.* **1987**, *36*, 289-296.
46. Loginov, A. V.; Katenin, S. B.; Voyakin, I. V.; Shagisultanova, G. A. *Koord. Khim.* **1986** , *12*, 1621-1626.
47. Davies, G. *Coord. Chem. Rev.* **1969**, *4*, 199-224.
48. Kimura, M.; Ohota, M.; Tsukahara, K. *Bull. Chem. Soc. Jpn.* **1990**, *63*, 151-155.
49. Mukhopadhyay, S.; Chaudhuri, S.; Das, R.; Banerjee, R. *Can. J. Chem.* **1993**, *71*, 2155-2159.
50. Luther, G. W. I.; Ruppel, D. T.; Burkhard, C. *ACS Symp. Ser.* **1998**, *715*, 265-280.
51. Hames, B. R.; Kurek, B.; Pollet, B.; Lapierre, C.; Monties, B. *J. Agric. Food Chem.* **1998**, *46*, 5362-5367.
52. Ionescu, S. G.; Oncescu, T. *J. Photochem.* **1982**, *20*, 299-304.
53. Harriman, A. *Coord. Chem. Rev.* **1979**, *28*, 147-175.
54. Kurek, B.; Hames, B. R.; Lequart, C.; Ruel, K.; Pollet, B.; Lapierre, C.; Gaudard, F.; Monties, B. *ACS Symp. Ser.* **2001**, *785*, 272-285.
55. Kerem, Z.; Jensen, K. A.; Hammel, K. E. *FEBS Lett.* **1999**, *446*, 49-54.

56. Fekete, F. A.; Chandhoke, V.; Jellison, J. *Appl. Environ. Microbiol.* **1989**, *55*, 2720-2722.
57. Mentasti, E.; Pelizzetti, E. *J. Chem. Soc., Dalton Trans.* **1973**, 2605-2608.
58. Tyson, C. A.; Martell, A. E. *J. Am. Chem. Soc.* **1968**, *90*, 3379-3386.
59. Mentasti, E.; Pelizzetti, E.; Baiocchi, C. *J. Inorg. Nucl. Chem.* **1976**, *38*, 2017-2021.
60. Pierpont, C. G.; Buchanan, R. M. *Coord. Chem. Rev.* **1981**, *38*, 45-87.
61. Jameson, R. F.; Wilson, M. F. *J. Chem. Soc., Dalton Trans.* **1972**, 2614-2616.
62. Jameson, R. F.; Wilson, M. F. *J. Chem. Soc., Dalton Trans.* **1972**, 2617-2621.
63. Harris, W. R.; Weitl, F. L.; Raymond, K. L. *J. Chem. Soc., Chem. Commun.* **1979**, 177-178.
64. Salama, S.; Stong, J. D.; Neilands, J. B.; Spiro, T. G. *Biochemistry* **1978**, *17*, 3781-3785.
65. Raymond, K. N.; Carrano, C. J. *Acc. Chem. Res.* **1979**, *12*, 183-190.
66. Raymond, K. N.; Isied, S. S.; Brown, L. D.; Fronczek, F. R.; Nibert, J. H. *J. Am. Chem. Soc.* **1976**, *98*, 1767-1774.
67. Isied, S. S.; Kuo, G.; Raymond, K. N. *J. Am. Chem. Soc.* **1976** , *98*, 1763-1767.
68. Harris, W. R.; Carrano, C. J.; Cooper, S. R.; Sofen, S. R.; Avdeef, A. E.; McArdle, J. V.; Raymond, K. N. *J. Am. Chem. Soc.* **1979**, *101*, 6097-6104.
69. Raymond, K. N. *Adv. Chem. Ser.* **1977**, *162*, 33-54.
70. Manos, M. J.; Tasiopoulos, A. J.; Raptopoulou, C.; Terzis, A.; Woollins, J. D.; Slawin, A. M. Z.; Keramidas, A. D.; Kabanos, T. A. *J. Chem. Soc., Dalton Trans.* **2001**, 1556-1558.
71. Pierpont, C. G. *Inorg. Chem.* **2001**, *40*, 5727-5728.
72. Balogh-Hergovich, E.; Speier, G. *Inorg. Chim. Acta* **1988**, *147*, 51-54.
73. Funabkiki, T.; Yokomizo, T.; Suzuki, S.; Yoshida, S. *Chem. Commun.* **1997**, 151-152.
74. Funabiki, T.; Ishikawa, M.; Nagai, Y.; Yorita, J.; Yoshida, S. *J. Chem. Soc., Chem. Commun.* **1994**, 1951-1952.
75. Funabiki, T.; Tada, S.; Yoshioka, T.; Takano, M.; Yoshida, S. *J. Chem. Soc., Chem. Commun.* **1986**, 1699-1700.
76. Speier, G. *New J. Chem.* **1994** , *18*, 143-147.
77. Rall, J.; Wanner, M.; Albrecht, M.; Hornung, F. M.; Kaim, W. *Chem.-Eur. J.* **1999**, *5*, 2802-2809.
78. Caneschi, A.; Dei, A. *Angew. Chem., Int. Ed.* **1998**, *37*, 3005-3007.
79. Rall, J.; Kalm, W. *J. Chem. Soc., Faraday Trans.* **1994**, *90*, 2905-2908.
80. Thompson, J. S.; Calabrese, J. C. *J. Am. Chem. Soc.* **1986**, *108*, 1903-1907.
81. Pierpont, C. G.; Lange, C. W. *Prog. Inorg. Chem.* **1994**, *41*, 331-442.

82. Thompson, J. S.; Calabrese, J. C. *Inorg. Chem.* **1985**, *24*, 3167-3171.
83. Harmalker, S.; Jones, S. E.; Sawyer, D. T. *Inorg. Chem.* **1983**, *22*, 2790-2794.
84. Speier, G.; Tisza, S.; Tyeklar, Z.; Lange, C. W.; Pierpont, C. G. *Inorg. Chem.* **1994**, *33*, 2041-2045.
85. Attia, A. S.; Bhattacharya, S.; Pierpont, C. G. *Inorg. Chem.* **1995**, *34*, 4427-4433.
86. Shoner, S. C.; Power, P. P. *Inorg. Chem.* **1992**, *31*, 1001-1010.
87. Ding, Z.; Bhattacharya, S.; McCusker, J. K.; Hagen, P. M.; Hendrickson, D. N.; Pierpont, C. G. *Inorg. Chem.* **1992**, *31*, 870-877.
88. Lauffer, R. B.; Heistand, R. H. I.; Que, L. Jr. *Inorg. Chem.* **1983**, *22*, 50-55.
89. Attia, A. S.; Conklin, B. J.; Lange, C. W.; Pierpont, C. G. *Inorg. Chem.* **1996**, *35*, 1033-1038.
90. Dei, A.; Gatteschi, D.; Pardi, L. *Inorg. Chem.* **1993**, *32*, 1389-1395.
91. Attia, A. S.; Pierpont, C. G. *Inorg. Chem.* **1998**, *37*, 3051-3056.
92. Scott, S. L.; Bakac, A.; Espenson, J. H. *J. Am. Chem. Soc.* **1992**, *114*, 4605-4610.
93. Jones, S. E.; Chin, D.-H.; Sawyer, D. T. *Inorg. Chem.* **1981**, *20*, 4257-4262.
94. Sawyer, D. T.; Sobkowiak, A.; Roberts, J. L. Jr. *Electrochemistry for Chemists*, 2nd ed.; Wiley & Sons: New York, NY, 1995; pp 446-451.
95. Que, L. Jr.; Ho, R. Y. N. *Chem. Rev.* **1996**, *96*, 2607-2624.
96. Mentasti, E.; Pelizzetti, E.; Saini, G. *J. Chem. Soc., Dalton Trans.* **1973**, 2609-2614.
97. Yamahara, R.; Ogo, S.; Watanabe, Y.; Funabiki, T.; Jitsukawa, K.; Masuda, H.; Einaga, H. *Inorg. Chim. Acta* **2000**, *300-302*, 587-596.
98. Duda, M.; Pascaly, M.; Kreba, B. *Chem. Commun.* **1997**, 835-836.
99. Koch, W. O.; Krueger, H.-J. *Angew. Chem., Int. Ed. Engl.* **1995**, *34*, 2671-2674.
100. Cox, D. D.; Que, L. Jr. *J. Am. Chem. Soc.* **1988**, *110*, 8085-8092.
101. Jang, H. G.; Cox, D. D.; Que, L. Jr. *J. Am. Chem. Soc.* **1991**, *113*, 9200-9204.
102. Mialane, P.; Tchertanov, L.; Banse, F. ; Sainton, J.; Girerd, J.-J. *Inorg. Chem.* **2000**, *39*, 2440-2444.
103. Ogihara, T.; Hikichi, S.; Akita, M.; Moro-oka, Y. *Inorg. Chem.* **1998**, *37*, 2614-2615.
104. Weiner, H.; Hayashi, Y.; Finke, R. G. *Inorg. Chim. Acta* **1999**, *291*, 426-437.
105. Viswanathan, R.; Palaniandavar, M.; Balasubramanian, T.; Muthiah, T. P. *Inorg. Chem.* **1998**, *37*, 2943-2951.

106. Goodell, B.; Qian, Y.; Jellison, J.; Richard, M.; Qi, W. in *Biotechnology in the Pulp and Paper Industry*; Viikari, L.; Lantto, R., Eds.; Elsevier: Amsterdam, 2002; pp 37-47.
107. Liu, C.; Ye, X.; Zhan, R.; Wu, Y. *J. Mol. Catal. A: Chem.* **1996**, *112*, 15-22.
108. Hage, J. P.; Sawyer, D. T. *J. Am. Chem. Soc.* **1995**, *117*, 5617-5621.
109. Hartman, J. R.; Foxman, B. M.; Cooper, S. R. *Inorg. Chem.* **1984**, *23*, 1381-1387.
110. Mandal, S.; Macikenas, D.; Protasiewicz, J. D.; Sayre, L. M. *J. Org. Chem.* **2000**, *65*, 4804-4809.
111. Malachowski, M. R.; Huynh, H. B.; Tomlinson, L. J.; Kelly, R. S.; Furbeejun, J. W. *J. Chem. Soc., Dalton Trans.* **1995**, 31-36.
112. Speier, G. *J. Mol. Catal.* **1986**, *37*, 259-267.
113. Demmin, T. R.; Swerdloff, M. D.; Rogic, M. M. *J. Am. Chem. Soc.* **1981**, *103*, 5795-5804.
114. Rogic, M. M.; Demmin, T. R. *J. Am. Chem. Soc.* **1978**, *100*, 5472-5487.
115. Brown, D. G.; Beckmann, L.; Ashby, C. H.; Vogel, G. C.; Reinprecht, J. T. *Tetrahedron Lett.* **1977**, 1363-1364.
116. Sugimoto, H.; Sawyer, D. T. *J. Am. Chem. Soc.* **1984**, *106*, 4283-4285.
117. Sawyer, D. T.; Kang, C.; Llobet, A.; Redman, C. *J. Am. Chem. Soc.* **1993**, *115*, 5817-5818.
118. Mekmouche, Y.; Menage, S.; Toia-Duboc, C.; Fontecave, M.; Galey, J.-B.; Lebrun, C.; Pecaut, J. *Angew. Chem., Int. Ed.* **2001**, *40*, 949-952.
119. Ensing, B.; Buda, F.; Blochl, P.; Baerends, E. J. *Angew. Chem., Int. Ed.* **2001**, *40*, 2893-2895.
120. Tanaka, H.; Itakura, S.; Hirano, T.; Enoki, A. *Holzforschung* **1996**, *50*, 541-548.

Chapter 11

The Role of Enzymes and Mediators in White-Rot Fungal Degradation of Lignocellulose

Kaichang Li

Department of Wood Science and Engineering, Oregon State University, Corvallis, OR 97331

White-rot fungi can selectively and efficiently degrade lignin through the action of secreted enzymes. Major lignin-degrading enzymes include lignin peroxidase (LiP), manganese peroxidase (MnP) and laccase. LiP and MnP require hydrogen peroxide (H_2O_2) for their activities, but they are inactivated by a high concentration of H_2O_2. There are various H_2O_2–producing enzymes in white-rot fungi. LiP is able to oxidize both phenolic and non-phenolic lignin substructures, whereas MnP and laccase are only able to degrade phenolic lignin substructures. An unsaturated fatty acid or a thio-containing compound enables MnP to degrade non-phenolic lignin substructures. Several *in vitro* studies reveal that some enzymes can synergistically degrade lignin. As lignin-degrading enzymes are too bulky to penetrate plant cell walls for lignin degradation, the role of various small organic/inorganic compounds, serving as redox mediators for lignin-degrading enzymes in lignin degradation, is discussed in detail.

Introduction

Wood is composed mainly of cellulose, hemicelluloses, and lignin. The intimate associations among these three components make the selective removal of them, especially lignin, a difficult task, whereas selective degradation of lignin has great implications on the pulp and paper industry. White-rot fungi are the only microorganisms that can substantially and selectively degrade the phenolic and non-phenolic lignin ring structures. Extensive studies have shown that enzymes play a very important role in the fungal degradation of lignocellulose (*1*). Because enzymes are too bulky to penetrate plant cell walls for selective degradation of lignin, small organic or inorganic compounds are believed to serve as redox mediators to transfer electrons from lignin to enzymes. The enzymes and redox mediators that could be involved in lignin degradation are reviewed in this chapter.

Enzymes Associated with Lignin Degradation

Lignin is a structurally complicated phenolic polymer. The most salient characteristic of lignin structures is that they lack repeating linkage types such as are found in protein and carbohydrates. Even though it has long been known that certain fungi degrade wood including lignin, the concept that a fungus degrades lignin through enzymes was not proposed until the 1970s. In 1976, Ander and Eriksson demonstrated for the first time that phenol oxidase activity was closely correlated with lignin degradation by the white-rot fungus *Sporotrichum pulverulentum* (*2*). The first lignin-degrading enzyme, now known as lignin peroxidase (LiP), was purified and characterized from a white-rot fungus *Phanerochaete chrysosporium* in 1983 (*3,4*). In the same year, another type of lignin-degrading enzyme, a peroxidase whose activity was greatly dependent upon a manganese ion, now known as manganese peroxidase (MnP), was also purified and characterized from *Phanerochaete chrysosporium* (*5*). It is worthwhile noting that *Sporotrichum pulverulentum* and *Phanerochaete chrysosporium* are actually the same fungus. Both LiP and MnP are heme-containing glycoproteins. Since their isolation and characterization, these two enzymes have been extensively studied. Their abilities to oxidize lignin model compounds and various lignin preparations have been investigated in great detail (*6-12*). Catalyses of these two enzymes have been studied and genes encoding them have been cloned and sequenced (*13-17*). The X-ray crystal structures of both LiP and MnP have also recently been elucidated (*18-20*).

In the laboratory, LiP activity is normally assayed by exploiting the enzyme's ability to oxidize veratryl alcohol to veratryl aldehyde (*21*). In other words, a fungal culture that is able to oxidize veratryl alcohol to veratryl aldehyde is said to contain LiP. MnP activity is characteristic of its dependence on Mn(II). However, a peroxidase that can oxidize veratryl alcohol but is also dependent upon Mn(II) for the oxidation has recently been purified and characterized (*22*). Thus, the boundary between LiP and MnP may not be as clear as previously thought.

Both LiP and MnP, as peroxidases, require hydrogen peroxide for their activities. However, a high concentration of hydrogen peroxide quickly inactivates both enzymes. The inactivation mechanisms have been investigated and ways have been devised to prevent their inactivation from a high concentration of hydrogen peroxide *in vitro* (*13,23*). In the *in vivo* degradation of lignin by a white-rot fungus, it is believed that a low concentration of hydrogen peroxide is supplied continuously through a H_2O_2-producing enzyme.

There are many ways that a white-rot fungus could generate hydrogen peroxide for LiP and MnP. Extracellular H_2O_2-producing enzymes that have been studied include aryl alcohol oxidase (AAO) (*24-31*), glyoxal oxidase (*32,33*), pyranose 2-oxidase (*34*), and cellobiose dehydrogenase (CDH) (*35,36*). AAO oxidizes aryl alcohol to aryl aldehyde and concomitantly produces hydrogen peroxide from oxygen. Glyoxal oxidase oxidizes glyoxal to oxalic acid while reducing oxygen to hydrogen peroxide. Pyranose 2-oxidase oxidizes a pyranose, such as glucose, to a ketone at the 2-position of a pyranose with the concomitant reduction of oxygen to hydrogen peroxide. CDH oxidizes cellobiose to cellobionic acid. A quinone, or ferric ion, rather than oxygen, is the preferred oxidant for CDH (*36*). However, because a CDH from *P. chrysosporium* can indeed reduce oxygen to hydrogen peroxide, it is considered a H_2O_2-producing enzyme (*36*).

Several intracellular H_2O_2-producing enzymes from white-rot fungi have also been studied, but to a lesser extent than the extracellular enzymes described above. The extracellular enzymes include glucose 1-oxidase (*37*), pyranose 2-oxidase (*38,39*), and methanol oxidase (*40*). Many H_2O_2-producing enzymes could possibly supply LiP or MnP with H_2O_2 for lignin degradation. It is still poorly understood which enzyme plays the primary role in supplying LiP or MnP with H_2O_2.

Another phenoloxidase is laccase, a copper-containing glycoprotein. Laccase oxidizes phenol to quinone with the concomitant reduction of oxygen to water. By means of recombinant DNA techniques, laccase catalysis has been studied in great detail in recent years (*41-43*). The crystal structure of a T2-copper deleted laccase has been reported (*44,45*). However, a crystal structure of a laccase with four copper atoms has yet to be resolved.

Contributions of Individual Enzymes to the Fungal Degradation of Lignin

Contribution of LiP

Lignin is a free-radical-coupling product of one or more of three monolignols: *p*-hydroxycinnamic alcohol, coniferyl alcohol, and syringyl alcohol. In softwood, lignin is derived mainly from polymerization of coniferyl alcohol, whereas hardwood lignin is derived from both coniferyl and syringyl alcohols. In softwood lignin, the major linkages between two coniferyl alcohols include β-O-4, β-β, 4-O-5, 5-5, and β-1 (*46*). A lignin substructure containing a phenolic hydroxyl group is called a phenolic lignin substructure; all other lignin substructures are called non-phenolic lignin substructures. It is known that about 10% of substructures in softwood lignin are phenolic (*46*).

It has been demonstrated that LiP activity is closely correlated to lignin degradation by *P. chrysosporium*, thus indirectly showing that LiP is involved in lignin degradation (*3,4,47,48*). Purified LiP from *P. chrysosporium* has been shown to degrade both phenolic and non-phenolic lignin model compounds (*49-53*). Depolymerization of lignin preparations by purified LiP has also been demonstrated (*54,55*). All these results suggest that LiP plays an important role in the fungal degradation of lignin. However, repolymerization of lignin preparations by purified LiP has also been demonstrated (*56,57*). The use of purified LiP to degrade lignin in unbleached kraft pulp, however, has not resulted in success. Although discrepant results on LiP's ability to degrade lignin *in vitro* do not discredit the important role of LiP in the fungal degradation of lignin *in vivo*, the process by which a fungus uses LiP for depolymerization of lignin *in vivo* is still poorly understood.

Contribution of MnP

In the catalysis of MnP, MnP oxidizes Mn(II) to Mn(III) and this oxidation state of manganese is then chelated by an organic ligand such as tartaric acid (*13*). The chelated Mn(III) has been proposed to be the actual oxidant in lignin degradation (*13*). It has been demonstrated that MnP-less mutants from *Trametes versicolor* are unable to degrade lignin, which implies that MnP plays an important role in lignin degradation (*58*). The white-rot fungi *Dichomitus squalens* and *Ceriporiopsis subvermispora* degraded lignin efficiently without expressing detectable LiP, and MnP was demonstrated as a key enzyme in the degradation of lignin by these two fungi (*59-62*). MnP is widely distributed in white-rot fungi. A large body of evidence now suggests that MnP is one of the major phenoloxidases responsible for the fungal degradation of lignin in white rot fungi.

Contribution of laccase

P. chrysosporium is one of the most extensively studied white-rot fungi. When degrading lignin, the fungus secretes both LiP and MnP. As discussed above, we can be reasonably sure that a combination of LiP and MnP plays a very important role in the complete degradation of lignin. However, a recent survey has shown that only 40% of white-rot fungi studied so far produce a combination of LiP and MnP, whereas a combination of MnP and laccase is much more common (*63*). In an extreme case, a white-rot fungus, *Pycnoporus cinnabarinus*, produced only laccase, lacking both LiP and MnP (*63*). However, the fungus degrades lignin very efficiently. A laccase-less mutant of *P. cinnabarinus* was not able to degrade synthetic lignin and lignin in unbleached kraft pulp, thus demonstrating that laccase is essential for the fungus to degrade lignin (*64,65*). Because most white-rot fungi secrete laccase under ligninolytic conditions, it is now commonly accepted that laccase plays an important role in the fungal degradation of lignin.

Contribution of other enzymes

Cellobiose dehydrogenase (CDH) is widely distributed in the wood-degrading fungi. Because it oxidizes cellobiose with the concomitant reduction of quinones, CDH has been proposed to link cellulose hydrolysis and lignin degradation (*35,66,67*). The catalytic features of CDH suggest that it could have various biological functions in the fungal degradation of plant components (*36*). Possible roles of CDH in lignin degradation may include detoxification of quinones for fungal growth, generation of hydroxyl free radicals via Fenton reaction, prevention of aromatic free radicals from repolymerization, and synergistic actions with MnP in lignin degradation (*68-76*). However, the exact role of CDH in the fungal degradation of lignin *in vivo* is still poorly understood.

H_2O_2-producing enzymes could be involved in the fungal degradation of lignin by supplying LiP or MnP with hydrogen peroxide. Hydrogen peroxide produced by these enzymes could react with Fe (II) to form hydroxyl free radicals ($HO^{\cdot}$) that are so reactive that they can degrade both carbohydrates and lignin.

Synergistic Actions of Enzymes in Lignin Degradation

Laccase and MnP

Both laccase and MnP are able to oxidize phenolic substructures only in lignin. It is interesting to note that more than 60% of white-rot fungi studied so far produce laccase and MnP at the same time (*63*). It is perplexing why a fungus produces two phenoloxidases to tackle the same problem. An *in vitro* study has shown that laccase and MnP display synergistic effects in the depolymerization of lignin preparations (*77*). Differential expression of MnP and laccase in white-rot fungi has also been detected in the presence of manganese or aromatic compounds (*78*). However, the specific roles of laccase and MnP in the synergistic action are still poorly understood.

MnP and CDH

It has been demonstrated that CDH can reduce insoluble MnO_2 in wood to Mn(II), thus supplying MnP with Mn(II) (*73*). It has also been demonstrated that cellobionic acid generated by CDH is a good chelator for Mn(III) (*73*). Close associations of MnP and CDH in lignin degradation have been discussed in detail (*73,75*).

Hydrolytic enzymes and oxidative enzymes

Hydrolytic enzymes such as cellulases and xylanases in white-rot fungi have been studied now for several decades. A recent study has shown that patterns of hydrolytic enzyme production in white-rot fungi could differ significantly from one to another (*79,80*). For example, *C. subvermispora* was found to lack cellobiose hydrolase, thus explaining why the fungus has a weak ability to degrade cellulose (*80*). Because a white-rot fungus cannot live on lignin alone, hydrolytic enzymes obviously play a very important role in fungal growth and lignin degradation (*1*). In a recent study, MnP and xylanase were shown to have superior synergistic effects on delignification of unbleached kraft pulp (*81*). Overall, much more research is needed to establish the relationship between hydrolytic enzymes and oxidative enzymes in lignocellulose degradation.

Mediators Involved in Lignocellulose Degradation

It has been demonstrated that enzymes are generally too bulky to penetrate plant cell walls for lignin degradation. Thus, some small inorganic or organic compounds are commonly believed to serve as diffusible oxidants for lignin degradation.

In LiP-based lignin degradation, veratryl alcohol has been shown to enhance the ability of LiP to oxidize various lignin model compounds and to mineralize ^{14}C-labled lignin (*82-85*). In this context, LiP was proposed to first oxidize veratryl alcohol to a cation free radical that subsequently diffuses to, and oxidizes lignin. In other words, veratryl alcohol was proposed to serve as a redox mediator. However, it is debatable whether veratryl alcohol is able to serve as a redox mediator in lignin degradation because the cation free radical of veratryl alcohol has a fairly short lifetime. The possible roles of veratryl alcohol in lignin degradation have been discussed in detail (*86*).

Chelated Mn(III) is the mediator for MnP. Unfortunately, under physiological conditions, MnP appears to be able to oxidize only phenolic lignin substructures. It is hard to imagine that MnP would play a major role in breaking down polymeric lignin if it is only able to oxidize about 10% of the lignin structures. As part of the effort to expand the role of MnP in lignin degradation, two different systems have been demonstrated *in vitro* to enable MnP to degrade non-phenolic lignin model compounds (Schemes 1 and 2) (*8,10*). In Scheme 1, a thio free radical (RS-) was proposed to be the actual oxidant for oxidation of

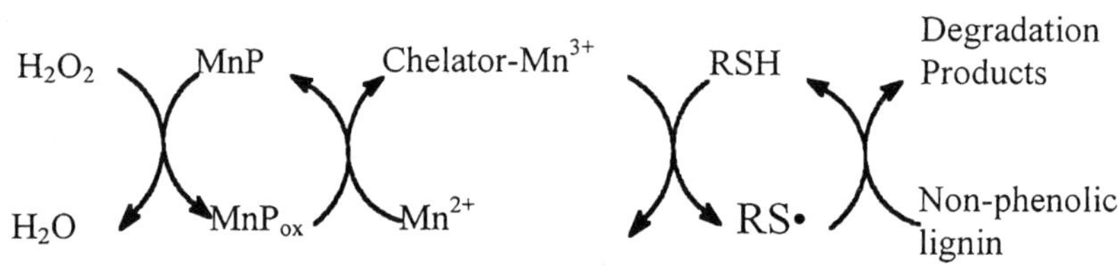

Scheme 1. Thio-mediated degradation of non-phenolic lignin

non-phenolic lignin model compounds (*8*). In Scheme 2, a peroxyl free radical (ROO-) was proposed to be responsible for the oxidation (*10,61*). A white-rot fungus could use either or both of these systems for lignin degradation because various thio-containing compounds such as glutathione and unsaturated fatty

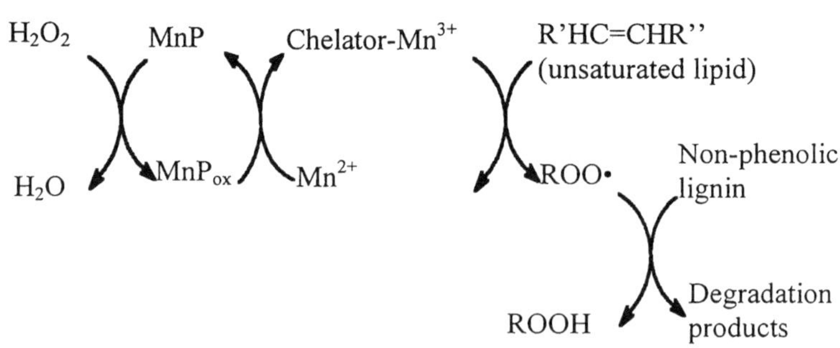

Scheme 2. Lipid-mediated degradation of non-phenolic lignin

acids are present during lignin degradation. The question is how these thio-compounds or unsaturated fatty acids are transported to the lignin degradation sites. It is worth noting that MnP was successfully used to degrade lignin in unbleached kraft pulp in the presence of Tween 80, a surfactant containing unsaturated fatty acid chains (*81,87-89*).

Laccase is able to oxidize only phenolic lignin substructures. However, in the presence of a redox mediator, laccase is able to efficiently oxidize non-phenolic lignin substructures (*90*). In recent years, extensive efforts have been devoted to developing a laccase/mediator system for bleaching unbleached kraft pulp (*91,92*). Several synthetic organic and inorganic compounds that enable laccase to effectively degrade lignin have been developed (*91-95*). Most organic laccase mediators contain a N-OH functional group with the N attached to an aromatic ring (*91*). Phenothiazine derivatives are also effective laccase mediators and have been used commercially to bleach denim (*96,97*). Mechanisms by which a laccase mediator system degrades lignin have been extensively studied (*98-102*). The relationship between the chemical structures of laccase mediators and the effectiveness of a laccase mediator system has also been investigated (*93,103-106*). The importance of laccase mediator systems in lignin degradation has recently been reviewed (*91,92,107*). In a sharp contrast to our well developed understanding of laccase mediator systems for use in pulp bleaching, the process by which laccase is used for lignin degradation *in vivo* is still poorly understood.

As mentioned previously, a white-rot fungus, *P. cinnabarinus*, degrades lignin while producing laccase as the sole phenoloxidase. Genes encoding the only two laccase isozymes have been cloned and sequenced (*108,109*). The major laccase isoform from the fungus has been purified and characterized (*110*). However, the laccase from *P. cinnabarrinus* has characteristics typical of blue laccases from other white-rot fungi (*99,110*). In other words, the superior ability of the fungus to degrade lignin is not because its laccase is substantially different from other laccases. A fungal metabolite, 3-hydroxyanthranilic acid (3-HAA), enabled laccase to oxidize selected lignin model compounds and a synthetic lignin preparation, and was proposed to be a natural laccase mediator (*111*). However, an in-depth study has demonstrated that 3-HAA does not play an important role in the fungal degradation of lignin (*112*). Several natural phenolic and thio-containing compounds were demonstrated to serve as laccase mediators to oxidize polycyclic aromatic hydrocarbons (*113*). However, degradation of lignin by laccase and one of these natural products has yet to be demonstrated. Their actual involvement as laccase mediators in fungal degradation of lignin is thus debatable. It has been shown that laccase could oxidize Mn(II) to Mn(III) in the presence of appropriate phenolic substances (*114*). Chelated Mn(III) can indeed serve as an oxidant for degradation of phenolic lignin substructures and further degrade non-phenolic lignin substructures via a thio free radical or a peroxy free radical as described for

MnP. However, Mn(II) is not the preferred substrate for laccase. The oxidation of Mn (II) to Mn(III) might not be the primary function of laccase because MnP alone can efficiently utilize Mn(II) available for lignin degradation.

There is another possible process whereby a white-rot fungus could effectively use laccase for lignin degradation. Non-phenolic lignin substructures may first be converted to phenolic substructures via a hydroxyl or peroxyl free radical. Laccase then fragments the phenolic lignin substructures. Hydroxyl free radical is produced primarily via Fenton reaction, i.e., the reaction of hydrogen peroxide with Fe(II). Production of hydroxyl radical by the synergistic action of fungal laccase and aryl alcohol oxidase has also been demonstrated (*115*). It has been shown that oxidation of methoxyhydroquinones by laccase could activate oxygen to form superoxide anion free radicals that have been proposed to be involved in lignin degradation (*116*). However, all possible explanations for how a white-rot fungus uses laccase for the degradation of lignin *in vivo* are still more or less speculative. Further study on this issue could shed light on how to improve the effectiveness of enzymatic bleaching of unbleached kraft pulp.

Acknowledgement

Parts of the works cited here and this paper were supported by a grant from the National Research Initiative Competitive Grants Program of the USDA (Award No. 2001-35103-10149).

References

1. Eriksson, K.-E.; Blanchette, R. A.; Ander, P. *Microbial and enzymatic degradation of wood and wood components*; Springer-Verlag: New York, 1990.
2. Ander, P.; Eriksson, K.-E. *Arch. Microbiol.* **1976**, *109*, 1-8.
3. Tien, M.; Kirk, T. K. *Science* **1983**, *221*, 661-663.
4. Tien, M.; Kirk, T. K. *Proc. Natl. Acad. Sci. USA* **1984**, *81*, 2280-2284.
5. Glenn, J. K.; Morgan, M. A.; Mayfield, M. B.; Kuwahara, M.; Gold, M. H. *Biochem. Biophys. Res. Commun.* **1983**, *114*, 1077-1083.
6. Wariishi, H.; Huang, J.; Dunford, H. B.; Gold, M. H. *J. Biol. Chem.* **1991**, *266*, 20694-20699.
7. Wariishi, H.; Valli, K.; Gold, M. H. *Biochemistry* **1989**, *28*, 6017-6023.
8. Wariishi, H.; Valli, K.; Renganathan, V.; Gold, M. H. *J. Biol. Chem.* **1989**, *264*, 14185-14191.
9. Wariishi, H.; Valli, K.; Gold, M. H. *Biochem. Biophys. Res. Commun.* **1991**, *176*, 269-275.

10. Bao, W.; Fukushima, Y.; Jensen, K. A., Jr.; Moen, M. A.; Hammel, K. E. *FEBS Lett.* **1994**, *354*, 297-300.
11. Schmidt, H. W. H.; Haemmerli, S. D.; Shoemaker, H. E.; Leisola, M. S. A. *Biochemistry* **1989**, *28*, 1776-1783.
12. Banci, L.; Ciofi-Baffoni, S.; Tien, M. *Biochemistry* **1999**, *38*, 3205-3210.
13. Wariishi, H.; Dunford, H. B.; MacDonald, I. D.; Gold, M. H. *J. Biol. Chem.* **1989**, *264*, 3335-3340.
14. Wariishi, H.; Akileswaran, L.; Gold, M. H. *Biochemistry* **1988**, *27*, 5365-5370.
15. Tien, M.; Tu, C. P. *Nature* **1987**, *326*, 520-523.
16. Brown, A.; Sims, P. F.; Raeder, U.; Broda, P. *Gene* **1988**, *73*, 77-85.
17. Pribnow, D.; Mayfield, M. B.; Nipper, V. J.; Brown, J. A.; Gold, M. H. *J. Biol. Chem.* **1989**, *264*, 5036-5040.
18. Edwards, S. L.; Raag, R.; Wariishi, H.; Gold, M. H.; Poulos, T. L. *Proc. Natl. Acad. Sci. USA* **1993**, *90*, 750-754.
19. Sundaramoorthy, M.; Kishi, K.; Gold, M. H.; Poulos, T. L. *J. Biol. Chem.* **1994**, *269*, 32759-32767.
20. Sundaramoorthy, M.; Kishi, K.; Gold, M. H.; Poulos, T. L. *J. Biol. Chem.* **1997**, *272*, 17574-17580.
21. Orth, A. B.; Royse, D. J.; Tien, M. *Appl. Environ. Microbiol.* **1993**, *59*, 4017-4023.
22. Camarero, S.; Sarkar, S.; Ruiz-Duenas, F. J.; Martinez, M. J.; Martinez, A. T. *J. Biol. Chem.* **1999**, *274*, 10324-10330.
23. Timofeevski, S. L.; Reading, N. S.; Aust, S. D. *Arch. Biochem. Biophys.* **1998**, *356*, 287-295.
24. de Jong, E.; Cazemier, A. E.; Field, J. A.; de Bont, J. A. *Appl. Environ. Microbiol.* **1994**, *60*, 271-277.
25. Farmer, V. C.; Hendrerson, M. E. K.; Russell, J. D. *Biochem. J.* **1960**, *74*, 257-262.
26. Bourbonnais, R.; Paice, M. G. *Biochem. J.* **1988**, *255*, 445-450.
27. Sannia, P.; Limongi, P.; Cocca, E.; Buonocore, F.; Nitti, G.; Giardina, P. *Biochim. Biophys. Acta* **1991**, *1073*, 114-119.
28. Guillén, F.; Martínez, A. T.; Martínez, M. J. *Eur. J. Biochem.* **1992**, *209*, 603-611.
29. Guillén, F.; Martínez, A. T.; Martínez, M. J. *Appl. Microbiol. Biotechnol.* **1990**, *32*, 465-469.
30. Muheim, A.; Waldner, R.; Sanglard, D.; Reiser, J.; Schoemaker, H. E.; Leisola, M. S. A. *Eur. J. Biochem.* **1991**, *195*, 369-375.
31. Muheim, A.; Waldner, R.; Leisola, M. S. A.; Fiechter, A. *Enzyme Microb Technol* **1990**, *12*, 204-209.
32. Kersten, P. J. *Proc. Natl. Acad. Sci. USA* **1990**, *87*, 2936-2940.
33. Kersten, P. J.; Kirk, T. K. *J. Bacteriol.* **1987**, *169*, 2195-2201.

34. Daniel, G.; Volc, J.; Kubatova, E. *Appl. Environ. Microbiol.* **1994**, *60*, 2524-2532.
35. Ayers, A. R.; Ayers, S. B.; Eriksson, K. E. *Eur. J. Biochem.* **1978**, *90*, 171-181.
36. Henriksson, G.; Johansson, G.; Pettersson, G. *J. Biotechnol.* **2000**, *78*, 93-113.
37. Kelley, R. L.; Reddy, C. A. *Methods Enzymol.* **1988**, *161B*, 307-316.
38. Volc, J.; Eriksson, K.-E. L. *Methods Enzymol.* **1988**, *161B*, 316-322.
39. Volc, J.; Kubárocá, E.; Sedmera, P.; Daniel, G.; Gabriel, J. *Arch. Microbiol.* **1991**, *156*, 297-301.
40. Eriksson, K.-E. L.; Nishida, A. *Methods Enzymol.* **1988**, *161B*, 322-326.
41. Xu, F. In *Encyclopedia of bioprocess technology: fermentation, biocatalysis, and bioseparation*; Flickinger, M. C., Drew, S. W., Eds.; John Wiley & Sons, Inc.: 1999, p 1545-1554.
42. Xu, F.; Berka, R. M.; Wahleithner, J. A.; Nelson, B. A.; Shuster, J. R.; Brown, S. H.; Palmer, A. E.; Solomon, E. I. *Biochem. J.* **1998**, *334*, 63-70.
43. Xu, F.; Palmer, A. E.; Yaver, D. S.; Berka, R. M.; Gambetta, G. A.; Brown, S. H.; Solomon, E. I. *J. Biol. Chem.* **1999**, *274*, 12372-12375.
44. Ducros, V.; Brzozowski, A. M.; Wilson, K. S.; Brown, S. H.; Østergaard, P.; Schneider, P.; Yaver, D. S.; Pedersen, A. H.; Davies, G. J. *Nat. Struct. Biol.* **1998**, *5*, 310-316.
45. Ducros, V.; Brzozowski, A. M.; Wilson, K. S.; Østergaard, P.; Schneider, P.; Svendson, A.; Davies, G. J. *Acta Crystallogr. D Biol. Crystallogr.* **2001**, *57*, 333-336.
46. Adler, E. *Wood Sci. Technol.* **1977**, *11*, 169-218.
47. Perez, J.; Jeffries, T. W. *Appl. Environ. Microbiol.* **1990**, *56*, 1806-1812.
48. Schoemaker, H. E.; Leisola, M. S. A. *J. Biotechnol.* **1990**, *13*, 101-109.
49. Kersten, P. J.; Kalyanaraman, B.; Hammel, K. E.; Reinhammar, B.; Kirk, T. K. *Biochem. J.* **1990**, *268*, 475-480.
50. Kersten, P. J.; Tien, M.; Kalyanaraman, B.; Kirk, T. K. *J. Biol. Chem.* **1985**, *260*, 2609-2612.
51. Schoemaker, H. E.; Harvey, P. J.; Bowen, R. M.; Palmer, J. M. *FEBS Lett.* **1985**, *183*, 7-12.
52. Kirk, T. K.; Farrell, R. L. *Annu. Rev. Microbiol.* **1987**, *41*, 261-266.
53. Hammel, K. E.; Kalyanaraman, B.; Kirk, T. K. *Proc. Natl. Acad. Sci. USA* **1986**, *83*, 3708-3712.
54. Hammel, K. E.; Jensen, K. A.; Mozuch, M. D.; Landucci, L. L.; Tien, M.; Pease, E. A. *J. Biol. Chem.* **1993**, *268*, 12274-12281.
55. Hammel, K. E.; Moen, M. A. *Enzyme Microb. Technol.* **1991**, *13*, 15-18.
56. Haemmerli, S. D.; Leisola, M. S.; Fiechter, A. *FEMS Microbiol. Lett.* **1986**, *35*, 33-36.
57. Odier, E.; Mozuch, M. D.; Kalyanaraman, B.; Kirk, T. K. *Biochimie* **1988**, *70*, 847-852.

58. Addleman, K.; Dumonceaux, T.; Paice, M. G.; Bourbonnais, R.; Archibald, F. S. *Appl. Environ. Microbiol.* **1995**, *61*, 3687-3694.
59. Périé, F. H.; Gold, M. H. *Appl. Environ. Microbiol.* **1991**, *57*, 2240-2245.
60. Périé, F. H.; Sheng, D.; Gold, M. H. *Biochim. Biophys. Acta* **1996**, *1297*, 139-148.
61. Jensen, K. A., Jr.; Bao, W.; Kawai, S.; Srebotnik, E.; Hammel, K. E. *Appl. Environ. Microbiol.* **1996**, *62*, 3679-3686.
62. Srebotnik, E.; Kenneth A. Jensen, J.; Hemmel, K. E. *Proc. Natl. Acad. Sci. USA* **1994**, *91*, 12794-12797.
63. Eggert, C.; Temp, U.; Eriksson, K.-E. L. In *ACS Symposium Series 655, Enzymes for Pulp and Paper Processing;* Jeffries, T. W., Viikari, L., Eds.; American Chemical Society: Washington, D. C., 1996, p 130-150.
64. Eggert, C.; Temp, U.; Eriksson, K.-E. L. *FEBS Lett.* **1997**, *407*, 89-92.
65. Bermek, H.; Li, K.; Eriksson, K.-E. L. *J. Biotechnol.* **1998**, *66*, 117-124.
66. Ayers, A. R.; Eriksson, K. E. *Methods Enzymol.* **1982**, *89*, 129-135.
67. Duarte, J. C.; Costa-Ferreira, M.; Sena-Martins, G. *Bioresource Technol.* **1999**, *68*, 43-48.
68. Kremer, S. M.; Wood, P. M. *Eur. J. Biochem.* **1992**, *208*, 807-814.
69. Ander, P.; Mishra, C.; Farrell, R. L.; Eriksson, K. E. L. *J. Biotechnol.* **1990**, *13*, 189-198.
70. Temp, U.; Eggert, C. *Appl. Environ. Microbiol.* **1999**, *65*, 389-395.
71. Westermark, U.; Eriksson, K.-E. *Acta Chem. Scand. B* **1974**, *28*, 204-208.
72. Westermark, U.; Eriksson, K.-E. *Acta Chem. Scand.* **1974**, *B28*, 209-214.
73. Roy, B. P.; Paice, M. G.; Archibald, F. S.; Misra, S. K.; Misiak, L. E. *J. Biol. Chem.* **1994**, *169*, 19745-19750.
74. Wood, P. M. *FEMS Microbiol. Rev.* **1994**, *13*, 313-320.
75. Ander, P. *FEMS Microbiol. Rev.* **1994**, *13*, 297-312.
76. Eriksson, K.-E. L.; Habu, N.; Samejima, M. *Enzyme Microb. Technol.* **1993**, *15*, 1001-1008.
77. Galliano, H.; Gas, G.; Seris, J. L.; Boudet, A. M. *Enzyme Microb. Technol.* **1991**, *13*, 478-482.
78. Scheel, T.; Hofer, M.; Ludwig, S.; Holker, U. *Appl. Microbiol. Biotechnol.* **2000**, *54*, 686-691.
79. Sethuraman, A.; Akin, D. E.; Eriksson, K. E. *Appl. Microbiol. Biotechnol.* **1999**, *52*, 689-697.
80. Sethuraman, A.; Akin, D. E.; Eriksson, K. E. L. *Biotechnol. Appl. Biochem.* **1998**, *27*, 37-47.
81. Bermek, H.; Li, K.; Eriksson, K.-E. L. *Tappi J.* **2000**, *83*, 69.
82. Shoemaker, H. E.; Leisola, M. S. A. *J. Biotechnol.* **1990**, *13*, 101-109.
83. Harvey, P. J.; Schoemaker, H. E.; Palmer, J. M. *FEBS Lett.* **1986**, *195*, 242-246.
84. Goodwin, D. C.; Aust, S. D.; Grover, T. A. *Biochemistry* **1995**, *34*, 5060-5065.

85. Koduri, R. S.; Tien, M. *J. Biol. Chem.* **1995**, *270*, 22254-22258.
86. Zapanta, L. S.; Tien, M. *J. Biotechnol.* **1997**, *53*, 93-102.
87. Paice, M. G.; Reid, I. D.; Bourbonnais, R.; Archibald, F. S.; Jurasek, L. *Appl. Environ. Microbiol.* **1993**, *59*, 260-265.
88. Harazono, K.; Kondo, R.; Sakai, K. *Appl. Environ. Microbiol.* **1996**, *62*, 913-917.
89. Kondo, R.; Karazono, K.; Sakai, K. *Appl. Environ. Microbiol.* **1994**, *60*, 4359-4363.
90. Bourbonnais, R.; Paice, M. G. *FEBS Lett.* **1990**, *267*, 99-102.
91. Li, K.; Eriksson, K.-E. L. *Ind. J. Microbiol.* **2000**, *40*, 149-261.
92. Call, H. P.; Mücke, I. *J. Biotechnol.* **1997**, *53*, 163-202.
93. Li, K.; Prabhu, N. G.; Cooper, D. A.; Xu, F.; Elder, T.; Eriksson, K.-E. L. In *ACS symposium series 785, Oxidative Delignification Chemistry: Fundamentals and Catalysis*; Argyropoulos, D. S., Ed.; American Chemical Society: Washington DC, 2001, p 400-412.
94. Amann, M. In *9th International Symposium on Wood and Pulping Chemistry* Montreal, Canada, 1997, p F(4-1) - F(4-5).
95. Bourbonnais, R.; Rochefort, D.; Paice, M. G.; Renaud, S.; Leech, D. *Tappi J.* **2000**, *83*, 68.
96. Kulys, J.; Krikstopaitis, K.; Ziemys, A. *J. Bio. Inorg. Chem.* **2000**, *5*, 333-340.
97. Schneider, P.; Pedersen, A. H.; PCT world patent, 1995, p WO 95/01426.
98. Li, K.; Helm, R. F.; Eriksson, K.-E. L. *Botechnol. Appl. Biochem.* **1998**, *27*, 239-243.
99. Li, K.; Xu, F.; Eriksson, K. E. *Appl. Environ. Microbiol.* **1999**, *65*, 2654-2660.
100. Majcherczyk, A.; Johannes, C.; Hüttermann, A. *Appl. Microbiol. Biotechnol.* **1999**, *51*, 267-276.
101. Muheim, A.; Fiechter, A.; Harvey, P. J.; Schoemaker, H. E. *Holzforschung* **1992**, *46*, 121-126.
102. Bourbonnais, R.; Leech, D.; Paice, M. G. *Biochim. Biophys. Acta* **1998**, *1379*, 381-390.
103. Xu, F.; Kulys, J. J.; Duke, K.; Li, K.; Krikstopaitis, K.; Deussen, H. J.; Abbate, E.; Galinyte, V.; Schneider, P. *Appl. Environ. Microbiol.* **2000**, *66*, 2052-2056.
104. Bourbonnais, R.; Paice, M. G.; Freiermuth, B.; Bodie, E.; Borneman, S. *Appl. Environ. Microbiol.* **1997**, *63*, 4627-4632.
105. Xu, F.; Deussen, H.-J. W.; Lopez, B.; Lam, L.; Li, K. *Eur. J. Biochem.* **2001**, *268*, 4169-4176.
106. Xu, F.; Li, K.; Elder, T. *Enzyme Microb. Technol.* In press.
107. Archibald, F. S.; Bourbonnais, R.; Jurasek, L.; Paice, M. G.; Reid, I. D. *J. Biotechnol.* **1997**, *53*, 215-236.

108. Eggert, C.; LaFayette, P. R.; Temp, U.; Eriksson, K. E.; Dean, J. F. *Appl. Environ. Microbiol.* **1998**, *64*, 1766-1772.
109. Temp, U.; Zierold, U.; Eggert, C. *Gene* **1999**, *236*, 169-177.
110. Eggert, C.; Temp, U.; Eriksson, K.-E. L. *Appl. Environ. Microbiol.* **1996**, *62*, 1151-1158.
111. Eggert, C.; Temp, U.; Dean, J. F.; Eriksson, K.-E. L. *FEBS Lett.* **1996**, *391*, 144-148.
112. Li, K.; Horanyi, P.; Collins, R.; Phillips, R. S.; Eriksson, K.-E. L. *Enzyme Microb. Technol.* **2001**, *28*, 301-307.
113. Johannes, C.; Majcherczyk, A. *Appl. Environ. Microbiol.* **2000**, *66*, 524-528.
114. Archibald, F.; Roy, B. *Appl. Environ. Microbiol.* **1992**, *58*, 1496-1499.
115. Guillén, F.; Gomez-Toribio, V.; Martinez, M. J.; Martinez, A. T. *Arch. Biochem. Biophys.* **2000**, *383*, 142-147.
116. Guillén, F.; Muñoz, C.; Gómez-Toribio, V.; Martínez, A. T.; Jesus Martínez, M. *Appl. Environ. Microbiol.* **2000**, *66*, 170-175.

Chapter 12

Enzyme Processes for Pulp and Paper: A Review of Recent Developments

William R. Kenealy and Thomas W. Jeffries

Institute for Microbial and Biochemical Technology, U.S. Department of Agriculture, Forest Service, Forest Products Laboratory, Madison, WI 53705

The pulp and paper industry is applying new, ecologically sound technology in its manufacturing processes. Many interesting enzymatic applications have been proposed in the literature. Implemented technologies tend to change the existing industrial process as little as possible. Commercial applications include xylanases in prebleaching kraft pulps and various enzymes in recycling paper. In the future, value-added products could be built around enzyme processes. When new applications are proposed that do not fit into existing practices, either the process or the enzyme must be altered. The decision depends on process economics and the feasibility of changing the technology. We review here new applications of enzymes in the pulp and paper industry and how they might be changed to implement the technology on an industrial scale. Also, this review suggests how existing enzymes may be used by process engineers to improve the efficiency of unit operations, the pulp products, or both.

This is a difficult time for the pulp and paper industry. Consumer standards are high, and manufacturing is competitive. Pulp production is increasingly derived from tropical regions (*1*). Cost reduction pressures are causing consolidation of companies through mergers and acquisitions; many research and development laboratories are being downsized, closed, or directed toward short-term objectives and opportunities; and profitability is being constrained by external factors including globalization, environmental concerns, and competition (*2*). Pulp and paper is a large industry that is highly capital-intensive and has been periodically affected by overcapacity (*1*). In the United States, more than 300 kg of paper are consumed annually per capita (*3*). To maintain this level of supply, we need to find new ways to use our forest resources more efficiently and with fewer environmental consequences. Emerging technologies based on sustainable use of renewable resources hold promise for the rejuvenation and growth of the pulp and paper industry.

Enzyme Applications in Pulp and Paper

Biotechnology has the potential to increase the quality and supply of feedstocks for pulp and paper, reduce manufacturing costs, and create novel high-value products. Novel enzyme technologies can reduce environmental problems and alter fiber properties. Because the pulp and paper industry is capital-intensive with facilities specific to the tasks (*1*), new technology must either reduce expenses or fit easily into the existing process design. Pulp and paper companies are reluctant to build or expand plants when the overall industry has enough capacity to satisfy market demands (*1*). Nevertheless, the industry has embraced enzymes for use in the paper-making process.

Environmental and Manufacturing Benefits

While the size and commodity nature of many of its products make changes difficult, the pulp and paper industry has responded to environmental and economic concerns. It implements new technology as required and when economics dictate change (*4*). For example, in response to environmental concerns and regulations, the industry has greatly reduced chlorinated aromatic byproducts that can be formed during pulp bleaching (*5-7*), first by reducing the amount of residual lignin in pulps and second by turning to other bleaching agents (*8*). An enzyme technology based on microbial xylanases has helped to achieve this goal by reducing or even eliminating the need for chlorine in the manufacture of elemental chlorine free (ECF) and totally chlorine free (TCF) printing and writing paper grades (*8-11*). Enzymes have helped meet environmental goals in other ways as well. By reducing costs involved in deinking (*12*), enzymes have increased the ability of manufacturers to recycle fiber, thereby placing fewer demands on timber resources. Enzymes have been used commercially to reduce paper manufacturing costs or improve the product. Lipases can control the accumulation of pitch during the production of paper from pulps with high resin content, such as sulfite and mechanical pulps from pine (*13*). Enzymes also help remove contaminants in the recycle stream. They can reduce the accumulation of adhesives and pitch residues, called stickies, on machines (*14*). They can facilitate the deinking of recycled paper and improve pulp drainage, which is particularly important as the amount of recycled fiber in the feedstock stream increases. With higher drainage rates, paper machines are able to operate faster, which again saves capital costs (*5,13,15,16*).

Xylanases have saved chemical costs for the industry (*7*) without interfering with the existing process. This technology has increased the bleaching speed in both TCF and ECF processes (*17*) and, in the case of chlorine dioxide bleaching, has actually increased the throughput of the plant due to debottlenecking at the chlorine dioxide generator (*7*). Developments of this last type are viewed very favorably since they enable the industry to make better use of its existing capital equipment.

Innovation and Implementation

Many other enzyme applications are possible based on properties demonstrated in the laboratory. These include eliminating caustic chemicals for cleaning paper machines, enhancing kraft pulping, reducing refining time, decreasing vessel picking, facilitating retting, selectively removing fiber components, modifying fiber properties, increasing fiber flexibility, and covalently linking side chains or functional groups. Commercial development of these applications will require better knowledge of enzyme mechanisms and actions on fibers, development of improved processes for their use, and changing enzymes to function better under existing operating conditions.

Efficacy is critical. The effect must be sufficient and reliable enough to warrant a change in process technology, or it will not be used. Economy of scale can greatly reduce the expense of enzyme production but unless the benefit is substantial, a paper maker will not implement the technology. Technical barriers can also block implementation. The use of an enzyme in a process step might call for adjusting operating parameters such as pH, adding chemicals such as surfactants, or making capital expenditures such as mixing tanks. In addition to the cost to implement the enzyme technology, the changes required for altered downstream processing might create expenses. For these reasons, innovative enzyme processes are most readily accepted when they can be incorporated into existing practices.

Often the biological or biochemical basis for an enzyme application is not well understood and must be developed through an empirical approach. For example, the ambient pH and temperature of kraft pulps dictate that xylanases should exhibit both alkaline activity and thermostability, but other factors make similar enzymes better suited to enhance bleachability. Xylanase use must be tested and levels of enzyme, the source of the enzyme, and location of enzyme treatment in the process are all critical factors to the successful deployment of the technology at the mill (*18*). The particular pulp or fiber might require the use of specific enzymes or doses. The pH optimum, the temperature of operation, and the sensitivity of the enzymes to other components of the process might all have to be modified.

Twenty years ago, the only means to make these changes would be to find new organisms and new enzymes. Today, these traditional means are available, and in addition, recombinant DNA has allowed the cloning of enzymes from known producers, difficult-to-culture microorganisms, and even unculturable organisms. These cloned enzymes can be modified for temperature, pH, and stability using similar techniques. Random mutagenesis, gene shuffling, directed evolution, and site-specific changes in the active site or supporting structure have all allowed the biotechnologist to change the natural enzyme to fit the need. The advent of *in silico* design adds another dimension to the possibilities by allowing the testing and design of enzymes that are not presently found in nature.

Pulp and Paper Processes

This review will present selected applications of enzymes in pulp and paper processing. Pulping starts with the conversion of wood or agricultural materials into a flexible fiber that can be made into paper. Depending on the ultimate application, many pulping processes can be used. The most common, kraft pulping, is a hot alkaline sulfide digestion of wood that removes most of the lignin and leaves a cellulosic fiber that is flexible yet strong. Acid sulfite pulping is also used to produce printing and writing paper grades. These chemical pulping processes form the basis for the production of printing, writing, and packaging paper grades. Mechanical, thermomechanical, and chemo-thermomechanical pulping produces fibers in much higher yield but with lower strength and optical properties. Such fibers can be incorporated into newsprint or magazine stock. After pulping, the fiber is washed and then bleached if the end use calls for white paper. Many other manipulations such as sizing, addition of fillers, and color addition can take place after this to produce the final paper product.

In addition to primary fiber, recycled fiber derived from recovered postconsumer paper is a major fiber source. Basic steps in paper recycling are pulping, washing, screening and flotation (to remove ink and other contaminants), bleaching, and wastewater cleanup. Enzymes can be useful in each of these process steps. The remainder of this review will describe enzyme uses in these processes and will also touch on the alteration of enzymes to better fit the process needs.

Enzyme Use in Pulping

Enzymes used in pulping can increase the yield of fiber, lessen further refining energy requirements, or provide specific modifications to the fiber. Cellulases have been used in many processes in the paper industry. Enzyme pretreatments using cellulase, hemicellulase, and pectinase have been shown to enhance the kraft pulping of sycamore chips and other pulp sources (*19-21*). This enzyme mixture allowed for better delignification of the pulp and savings in bleaching chemicals without altering the strength of the paper. The cost of the enzymes and questions about the effectiveness of a large enzyme aiding the low molecular weight pulping chemicals (*16*) has led to skepticism about the implementation of cellulases to enhance the kraft process.

Cellulases have been tested on mechanical pulps as well. In this case, there are conflicting results shown by a benefit in brightness and an increase in energy required for refining (*22*) using a crude cellulase on radiata pine compared with reduced energy required after a cellobiohydrolase treatment on spruce mechanical pulps (*23,24*). Other enzymes such as laccase and protease have been reported to reduce energy requirements in mechanical pulping (*16*). Laccase treatment has the additional benefit of increasing fiber bonding, which enhances the strength of the paper (*25*).

The use of enzymes in the refining of virgin fibers has been ongoing for decades. Kraft pulp has been treated with cellulases and xylanases, and both enzymes have reduced the energy required for further refining (*26*). The cellulases must be used carefully so they do not reduce the strength of the fibers (*27*). Xylanase treatments are more effective on unbleached pulps than they are on bleached kraft pulps (*16*).

The yield of thermomechanical pulp can also be increased by the use of enzymes. De-esterifying the soluble O-acetyl-galactoglucomannans of Norway spruce using an acetyl esterase was shown to precipitate the galactoglucomannans onto the fiber and increase the yield of fiber from the process (*28*).

Dissolving pulps are derived from pulps that contain a high level of cellulose. These pulps are treated to form soluble reactive carbohydrate chains that are then extruded into fibers or films. Endoglucanase treatment decreased the viscosity and chain length and increased the reactivity of a pulp made from eucalyptus and acacia (*29*). The endoglucanases were more efficient at hydrolyzing the pulp than were cellobiohydrolases at the same protein dosage (*29*). The use of cellulases in a two step process, where the alkali insoluble material was recovered, treated with enzymes, and recombined with the first extraction, resulted in a dissolving pulp that was more soluble in alkali than pulp treated directly with enzymes followed by alkali. Although when compared at the same level of hydrolysis, the fibers from the two stop process were actually slightly lower in solubility (*30*). Alkali extraction of dissolving pulps made from bleached hardwood kraft fiber and recycled paper rich in hardwood fiber, followed by xylanase treatment and a second alkali extraction, provided a reduction in the hemicellulose content of the pulp, acceptable viscosity, and alkali solubility (*31*).

Enzyme Use in Bleaching

Pulp bleaching has been repeatedly targeted as an application for enzymes in pulp and paper processing. The goal of bleaching is to whiten the pulp by changing or removing colored components. Initially, lignin-degrading enzymes were considered most important. In 1986, however, Viikari et al. (*10*) demonstrated that xylanases were effective as a prebleaching agent for pulps. In more recent years, laccases or manganese peroxidases, either alone or in combination with low molecular weight mediators, have been examined for their usefulness in pulp bleaching.

Xylanases in Prebleaching

The use of xylanases for prebleaching kraft pulp has been one of the greatest success stories of enzymes in the pulp and paper industry. Enzyme use helps to solve some of the environmental concerns associated with the use of chlorine in bleaching. The subject has been reviewed many times (*5,6,9,15,32-35*) and will not be covered here except for more recent developments. The mechanism by which xylanase assists in bleaching has come under study (*9,11*). Xylanase treatment can improve lignin extraction, alter carbohydrate and lignin associations, or cleave redeposited xylan (*9,36,37*).

One UV absorbing material, hexenuronic acid, is formed during kraft pulps from 4-methyl-glucuronic acid residues present on the xylan (*38-40*). The cleavage of the xylan allows the removal of such groups, thus saving on chemicals that would be needed to bleach those residues (*41*). An additional mild acid treatment will also selectively remove these colored components (*42,43*). The removal of the hexenuronic acid by xylanase treatments also helped to prevent brightness reversion of the treated kraft pulps (*44*).

Other studies on fiber modification have indicated that xylanase treatment allows alkali better access to the lignin (*45*). This enables more of the lignin to be removed and improves the bleaching efficiency for the remaining lignin in the fiber (*46*). The xylanase could be removing xylan that blocks access to the lignin, or xylan could precipitate on the surface of the fiber during kraft pulping, thereby preventing lignin extraction. This improved lignin removal and the removal of the bleach-consuming hexenuronic acid might explain the mechanism of xylanase prebleaching (*47*). However, there is evidence that xylan does not necessarily reprecipitate on the surface in all pulps (*48*), and our understanding of the xylanase bleaching mechanism at this point does not provide a means of predicting if bleaching will be enhanced by a given xylanase. Even the demonstration of carbohydrate (xylan) removal does not always correlate with bleaching efficacy (*49*).

Novel Xylanases

The potential applications of xylanases and the need for enzymes that might be more appropriate for use in pulp bleaching has caused many researchers to look for xylanases in different organisms and environments (*50*). Because the kraft process results in pulps that are alkaline and at higher temperatures, enzymes that do not require adjustment of temperature or pH are better suited to the process. Thermophilic (*50-55*) and alkaline sources (*56-58*) for these enzymes have been recently reviewed. Substantial amounts of work have been devoted to isolating and cloning new xylanases from all sources. Using xylanase as a search term in the scientific bibliographic databases, we found an average of 24 papers per year from 1982 to 1990. The same search from 1991 to 2000 revealed an average of 188 papers per year. Not all of these were papers about new xylanases, but a better indicator may be the number of xylanase-encoding DNA sequences that have been entered into data banks. A recent computer analysis (BLAST) of the nucleotide sequences with representatives of the Family 10 and 11 xylanases revealed more than 380 closely related protein-encoding sequences in the public databases.

Knowledge of the structure of a protein and its amino acid sequence might be used to explain why some xylanases have activity in alkaline conditions (*59*) while others are active only in acidic conditions (*60*). Structural analysis of the *Bacillus circulans* xylanase provided insight into the unusually high pKa of an active site glutamate (*59*). The *Aspergillus niger* xylanase was active in acidic conditions due to the presence of an aspartic acid residue whereas other alkaline active xylanases have an asparagine residue (*60*). The knowledge gained from these studies can be used to better adapt xylanases to a given function. Alterations of xylanases to increase their activity under process conditions will be covered in a later section.

Variation in Applications

Xylanases have been used on a wide variety of pulp fibers. Pulps from softwoods (radiata pine, Douglas-fir, western hemlock, redcedar, loblolly pine, slash pine, black spruce) (*61-68*), hardwoods (aspen, sweetgum, oak, tupelo) (*65,66,69-71*), bagasse (*72,73*), eucalyptus (*67,74-80*), ramie fibers (*81*), bamboo (*82*), khar grass (*79*), and wheat straw (*83*) have all been treated with xylanases and the treatment appears useful in assisting the bleaching of the pulp.

Not all xylanases work with all pulp sources. Indeed, there have been several reports of some enzymes working better than others, and synergy among xylanases or other carbohydrases has also been shown (*84,85*). In general, most of the xylanases used for bleaching belong to the Family G (or 11) enzymes, which are smaller (*49,86-89*) than those in Family F (or 10). Some xylanases act better with other xylanases (*84,85*) or other carbohydrases (*90-94*). The method of pulp production may be important for the activity of other carbohydrases to improve pulp bleaching (*95*).

The source of the fiber is not the only variable in determining the use of xylanases in bleaching. One variable is the type of pulp being produced. Kraft, sulfite, dissolving, mechanical, and thermomechanical pulps are all possible. Xylanase treatment of kraft pulps is quite common and is well represented in the mentioned uses. Kraft pulping followed by oxygen delignification complicates the use of xylanase since the levels of enzymes must be empirically adjusted (*96*). Oxygen delignification makes more xylan accessible, and the enzyme doses for bleaching are actually lower than for kraft pulp.

Xylanase treatments have had mixed success with sulphite pulps. Repeated xylanase and alkaline oxygen treatments provided some increase in the ability to bleach sulphite pulp (*97*). The need for repeated enzymatic treatments without some method of enzyme recycle would make this economically unworkable. Xylanase treatment was tested on sulphite pulps bleached with hydrogen peroxide (*98*) and no benefit was shown. Reducing the residual pentose content of dissolving pulps was possible using xylanase, but success depended upon having a low level of pentosans present in the pulp and repeating treatment and alkali extraction (*99*). Xylanase was able to enhance the brightness of a fungal-pretreated dissolving pulp (*100*).

There is also variability in the stage of the process where the enzyme treatment is performed. It could be pre- (*19*) or post-pulping, before or after alkaline treatment (*101*), or before or after the bleaching treatment. Alkaline leaching and enzyme treatment work by different, noncomplementary mechanisms, so it is not useful to combine the processes. The method of bleaching is also a variable, and improving bleaching with ozone, chlorates, peracetic acid, and peroxide is important for the industry (*102,103*). Use of enzyme treatments may have to be reoptimized when using a different bleaching method.

Other Enzymes Used for Pulp Bleaching

In general, the effect of xylanase on final pulp brightness is considered indirect. It removes reactants with the bleaching agent or obstacles to the bleaching action. A similar effect was shown for cellulase treatment and color removal in recycled yellow pages (*104*). There are some reports of direct brightening with xylanase treatments (*105,106*), but most of the direct bleaching with enzymes is done using oxidative enzymes that directly attack the color-producing compounds in the lignin.

The enzymes that attack the lignin components of the fiber are oxidative. Peroxidases use hydrogen peroxide, and laccases use oxygen and eventually react with the lignin-derived moieties. The fungi *Trametes versicolor* and *Phanerochaete chrysosporium* degrade lignin in kraft pulp and enhance its brightness (*107*). *P. chrysosporium* produces manganese peroxidase (MnP) and lignin peroxidase (LiP). *T. versicolor* makes LiP, MnP, and laccase. Cellobiose dehydrogenase is also made by both fungi, and its role in wood degradation and potential applications has recently been reviewed (*108-112*).

Treating pulp with fungal enzymes partially enhanced brightness but did not brighten the pulp to the same extent as fungal growth on the pulp (*107,113,114*). However, the use of oxidative enzymes rather than whole organisms does avoid the problem of cellulose hydrolysis. *T. versicolor* produces laccase and MnP at the highest levels when the greatest increase in brightening occurs. Both laccase and MnP can reduce the content of lignin in pulp provided the conditions are correct (*107,113,114*).

MnP requires hydrogen peroxide and Mn^{++} for full activity. Laccase requires oxygen and a low molecular weight compound, termed a mediator, for effective bleaching. Some pulps do not require addition of Mn^{++} (*115*). Either they already have sufficient Mn^{++} or chelators, such as oxalate, are able to liberate sufficient Mn^{++} for the enzyme to have an effect (*115*). As yet, no natural mediators have been identified from the fungal cultures involved in bleaching, and some have thought that MnP is more important than laccase (*116*). Some evidence also suggests that extractives from Norway spruce might contain a natural mediator for laccase (*117*).

MnP and LiP both require hydrogen peroxide for activity, yet they are inactivated when it is present at moderately high levels. Both enzymes also contain heme, which must be inserted properly prior to secretion. Fungal cultures make these enzymes in relatively low yield. The cost of the enzymes, their requirements for hydrogen peroxide, and their sensitivity to hydrogen peroxide limited the early use of these enzymes (*114,118*). Enzymes are now more available, but their sensitivity to hydrogen peroxide still exists. Continuous low level addition of hydrogen peroxide (*116,118*), or a system to generate it (glucose oxidase and glucose), can provide the needed hydrogen peroxide (*115,119*).

The use of MnP and xylanase along with caustic extraction and hydrogen peroxide bleaching has been proposed as a TCF method of pulp bleaching (*118*). Hydrogen peroxide bleaching after MnP treatment has been found effective. In one instance, MnP increased the final brightness by 10 points (*116*). The simultaneous use of xylanase and MnP had a synergistic effect during a prolonged bleaching with continuous addition of hydrogen peroxide (*120*). The use of xylanase prior to laccase–mediator treatment also allowed better removal of the lignin. However, the simultaneous treatment with laccase was ineffective (*121*).

As noted, the initial proposed use of laccase for bleaching (*122*) was limited by the availability of the enzyme and the cost of the mediators needed for lignin destruction (*114*). Laccase and MnP can both provide equivalent levels of bleaching (*113*), but laccase is much easier to produce in quantity and is commercially available. Both laccase and MnP cause an initial darkening of the pulp that is subsequently removed during bleaching and washing (*113*). The search for new laccases and new mediators has met with some success. New laccases (*123*) have been reported from various sources (*123,124*), which now include bacteria such as *Azospirillum lipoferum* (*125*) and *Bacillus subtilis* (*126*). New mediators are being tested (*123,127,128*) including transition metal complexes (*129*).

The more common laccase mediators, 1-hydroxybenzotriazole (1-HBT) and 2,2'-azinobis-(3-ethylbenzthiazoline-6-sulfonate) (ABTS), have been used by many researchers comparing effects with various laccases (*130-136*). 1-HBT and ABTS were better mediators for delignification than other mediators tested (*137*). However, both are expensive (*137*). Other mediators such as violuric acid can be used, but they suffer from inactivating the laccase as does 1-HBT (*138*). While laccase and mediators allow brightening of low lignin pulps, on high lignin pulps, they still reduce the lignin content (*139*). However, the brightening effect is lost (*131*).

Different laccases have also been shown to react differently with mediators (*137,138*). Combinations have to be tested prior to use. A single treatment of pulp with laccase and mediators will only remove a percentage of the lignin (*140,141*). The extraction of the pulp with alkali, and a second treatment with laccase, will remove more lignin (*141*).

While there is great promise in using oxidative enzymes in the bleaching of pulp, the process is not yet economical. The cost of the enzyme will always be a factor. However, the sensitivity of MnP to hydrogen peroxide and the need for an inexpensive, effective, nontoxic mediator for laccase treatment are critical remaining problems.

Enzyme Use in Fiber Recycling

The primary objectives in recycling paper are to remove ink and other contaminants while retaining optical and strength properties of the fibers. Enzymes can be used to enhance dewatering (drainage) rates, facilitate contaminant removal, and increase bond strength in recycled fibers. Drainage resistance of secondary fibers adversely affects sheet formation, slows operation of the paper machine, and increases drying energy (*142*). Deinking and contaminant removal likewise can benefit from enzyme treatments that facilitate separation of fibers in the washing and flotation processes.

Refining and Drainage

When they are properly applied, endoglucanases can enhance the drainage rates for recycled fibers beyond what can be attained by polymer addition (*143-146*). Enzyme treatments require additional retention time and sufficient mixing to affect the fiber surface structure. At optimum pH, temperature, and consistency and with appropriate mixing, treatments can be achieved in as little as 15 minutes. The enzyme dose and treatment time depends on the preparation and its activity. Enzyme addition prior to refining can lower the energy required to meet strength specifications or can improve strength properties at a fixed refining level (*145*). Enzyme addition after refining increases freeness, thereby enabling operation of the paper machine at a faster rate (*146*). Enzyme-assisted refining can enable lower head box consistency for better paper formation, improved strength, decreased basis weight, and increased use of recycled fibers.

Mooney et al. (*147*) reported evidence indicating that increased drainage results from selective digestion of the smaller fiber fragments. However, this effect also could be attributable to removal of cellulose microfibrils from paper fiber surfaces (*148,149*). In either case, care must be exercised to keep treatments at a low dose because endoglucanases that attack amorphous cellulose also cause rapid loss of fiber strength (*150,151*). In an attempt to avoid strength loss, Pala et al. (*152*) examined the use of cellulose binding domains (CBDs) isolated from cellulase preparations following proteolytic digestion. At low doses, CBDs increased both drainage rates and paper strength properties, but at higher dosage rates, the beneficial effects on strength indices were less pronounced. Pala et al. (*152*) hypothesized that the beneficial effect on strength was attributable to an increase in the microfibrilation of the fiber surface. Increased drainage, however, was also attributable to residual hydrolytic activity because CBD preparations in which some reducing sugar release was detectable increased drainage to a greater extent than preparations without detectable activity.

The efficacy of enzyme treatment depends on the fiber type. Mechanical fibers are much more resistant than chemical fibers to cellulase activity. Low doses of cellulases can increase handsheet density and reduce coarseness while having minimal effects on strength. Such treatments are most effective with kraft fiber (*153*). Because bond strength restoration is most important with secondary fibers and because recycled fibers are often filled with calcium carbonate, alkaline cellulase activity might be a desirable characteristic (*154*). However, highly active alkaline cellulases have not yet been described.

Starches used in sizings can accumulate in the treatment water of fiber recycling mills and interfere with drainage. In such instances, alpha-amylase can improve drainage properties, presumably by reducing the viscosity of the backwater (*155,156*). Amylase treatments have been shown to increase the drainage of recycled paper pulp, allowing the paper machines to run faster (*155*).

Recovered paper is recycled from various sources. Mixed office waste (MOW), old newspapers, and old corrugated containers (OCC) are the main recycled paper streams. The adhesives, glues, coatings, and binders present in recycled paper can cause the accumulation of stickies in the pulp slurry (*157*). When stickies accumulate into larger particles, they cause problems with the paper production and need to be removed from the system. An undefined esterase was reported to be effective in reducing the size and number of stickies in the recycled paper sources (*157*).

Cellulase and hemicellulase mixes have been reported to improve drainage of OCC pulps and liner boards and to allow a lower amount of polymeric additives to be used to strengthen the paper (*143,144,158*). The use of cellulases and xylanases can be a problem with anionic surfactants, which are often used during recycling (*159*). The use of cationic or nonionic surfactants actually enhances the activity of cellulases and xylanases (*159*).

Deinking

One of the main applications of enzymes to recycled fibers has been to remove print (*160,161*). More than 70% of MOW paper consists of uncoated papers that are printed with copy and laser printer toners, which are often difficult to remove by conventional, alkaline deinking processes (*162*). Cellulases are particularly effective in facilitating the removal of toners from office waste papers (*163*). Both cellulase and a nonionic surfactant are required for effective deinking (*164*). Recently, Park and Park (*165*) showed that chemically modified cellulases were more effective for recycled fibers than unmodified cellulases. These researchers added polyethylene oxide copolymer derivatives and maleic anhydride to the amino groups of the enzyme and then used the modified enzyme to treat recycled fibers. Paper made from fibers treated with modified cellulase had better physical properties, such as tensile strength and internal bond formation, than fibers made from unmodified cellulase. The modified cellulase was more effective in separating ink particles from fiber during flotation deinking (*166*). Paper freeness, whiteness, and tensile strength increased by 31%, 13%, and 24%, respectively.

Treimanis et al. (*167*) showed that cellulase decreases the interaction of toner and ink particles with fibers. This treatment is more effective with alkaline sized papers after adjusting the pH to 3.0 and the addition of a hydrocarbon surfactant (*168*). Based on enzyme trials and microscopic observations, Dinus and Welt (*161*) concluded that the primary role of cellulases in deinking involves separating ink–fiber agglomerates and dislodging or separating ink particles and fibrous materials in response to mechanical action during disintegration. Alpha amylase also appears to have a positive effect on deinking of recycled fibers (*169*), but it is not clear whether this mechanism is the same as that observed with cellulase.

Other Enzyme Uses in the Pulp and Paper Industry

Deposit Removal

Water used for the paper process often contains nutrients and is kept in conditions conducive to microbial growth. Microbial growth or deposits can be a problem when mills have closed their water loops, providing the possibility of concentrating organisms and nutrients (*170*). The microorganisms that cause these problems can be many different species, both aerobic and anaerobic (*170,171*). Slime formation on equipment is often the result of such conditions. Slime will cause many problems such as pump failure and breaks and blockages in paper machines. The first approach to this problem is to prevent slime formation by adding biocides. Once slime has formed, its removal is an important operation in the paper making industry and often requires shut down and caustic boil outs to properly clean the line.

Much of slime, or biofilm, is composed of carbohydrate and protein polymers. Enzymes such as amylases and proteases have been found useful in enzymatic boil out of slime encrusted equipment (*172-174*). One approach to the removal of slime from equipment is to produce sufficient slime from organisms isolated from the source, set up enrichments, and select organisms that will grow on it (*175*). Organisms were found that could degrade these polymers depending on the polysaccharide and organism used to make it. This approach may provide a source of new enzymes to be used for cleaning such deposits (*175*).

A similar approach was taken with colloidal material that could be reduced by culture filtrates of a fungus *T. versicolor* grown on the colloidal material (*176-178*). Dissolved and colloidal substances are a problem especially when mills close their water loops to conserve water use (*179-181*). Lipase has been used to remove lipid-based materials in these suspensions for a number of years (*182-186*). Removal of extractives by laccase-based catalysis of polymer formation and lipases provided additional removal of organic materials (*177*). Carbohydrate components of the colloidal material (*187*) could be removed by esterase treatment, which results in the carbohydrates being deposited onto the fiber (*28*).

Enzyme Use in Fiber Modifications

Interest has been growing in the enzyme-assisted modification of fibers. Enzymes such as peroxidases and laccases have been used to polymerize or co-polymerize materials with wood-based fibers (*188*). Attachment of guaiacol sulfonate using laccase made lignin more water soluble (*189*). The attachment of 4-hydroxyphenylacetic acid was also demonstrated, but no differences in lignin solubility could be distinguished (*189*). High kappa pulps were modified with 4-hydroxyphenylacetic acid in the presence of laccase (*190*). This modification increased the carboxylic acid groups, water retention, tensile strength, and burst strength of the resulting paper. The same treatment with tyrosine and guaiacol sulfonate had only minimal effects (*190*). Laccase catalyzed the polymerization of acrylic compounds onto lignosulfonates (*191*). Laccase in the presence of specific peroxides also grafted acrylamide onto lignin (*192*). Incubation of lignin with wood-based fibers in the presence of laccase formed covalent attachments with carbohydrate polymers (*193*). Laccase also assisted in enhancing the bonding in fiberboard made from Norway spruce (*117*) and beech fibers (*194*). Water soluble components of the spruce thermomechanical pulp aided in this bonding, and if removed, a mediator could replace their effects (*117*). Mediators also improved the strength of unbleached kraft pulp modified with laccase (*195*).

Peroxidases and other oxidases have been used to modify polymers as well (*188*). Peroxidases have been used to attach gallate esters to chitosan (*196*). However, it was difficult to measure what reactions actually took place. Low molecular weight lignin fragments were precipitated by treatment with horseradish peroxidase or potato-polyphenoloxidase, thus removing them from solution (*197*). The use of manganese peroxidase in aqueous organic solvents allowed the polymerization of guaiacol and other phenolic compounds and aromatic amines (*198*). The ability to function in organic solvents and modify polymeric compounds allows many reactions to proceed via oxidative enzymes, since many of the substrates that the oxidative enzymes react with are sparingly soluble in water (*199*). While these enzymes are able to catalyze reactions efficiently, they operate by free radical addition, and the bond formation can be predicted but not necessarily controlled and directed. This is especially true when the polymer being used is complex like lignin and wood-based fibers.

Cellulases can increase fiber flexibility and thereby improve pulp properties. This can be particularly important for coarse fibers with thick walls such as those derived from Douglas-fir. An enzyme charge of 1 mg cellulase protein/g of ovendry pulp can improve handsheet density and smoothness while increasing freeness and reducing coarseness (*153*). Excessive enzyme treatment, however, erodes the fiber surface and reduces pulp strength (*200*). Degradation is most apparent with short fibers, and refining is least effective with the long, coarse fibers (*201*), so if the pulps are fractionated into various lengths prior to enzyme treatment, the results are more satisfactory (*202*). By treating the fiber fractions separately, each under optimal conditions, and then recombining the fractions, Mansfield and Saddler (*203*) were able to increase tensile strengths of the resulting pulps by 25% to 35%. Cellulase and xylanase treatments can increase the density of the resulting handsheets made from kraft pulps (*204*). The beneficial effect of cellulases and xylanases on pulp properties appears to result from a collapsing of the coarse fiber and a resulting increase in the surface area available for interfiber bonding (*201*). Mechanical pulps are much more resistant to enzyme treatment. Laccases can increase bonding of mechanical fibers, but this probably results from a different mechanism than that observed with cellulases and xylanases (*25*).

Cellulose Binding Domains

Enzymes that attack cellulose and xylan often have portions (domains) of the protein that bind to cellulose (*205,206*). Binding domains specific for xylan have been identified (*207*), but a single amino acid change can alter this specificity from xylan to cellulose. Xylanase catalytic subunits with CBDs as part of their overall tertiary structure are common. The biochemical role of a CBD is to keep the enzyme catalytic unit close to the substrate surface. The need for the CBD in fiber processing might depend on the uses of the enzyme. The roles of CBDs in the hydrolysis and modification of chemical pulps have been examined (*208*). The presence or absence of CBDs did not affect the action of enzymes against soluble substrates. However, enzymes with a CBD enhanced fiber beating to a greater extent than did enzymes without the CBD. Since beating is thought to increase microfibrillation, the CBD could play a role in opening up the cellulose crystallite structure.

Opening the structure of cellulose or swelling of fiber has recently been attributed to swollenin (*209*), a protein made by *Trichoderma reesei* that has homology to CBD and expansin (*210*). The latter is a plant protein involved in separating cellulose fibrils during wall expansion and cell growth. Thus, CBD might be viewed as a domain that makes substrates more accessible (*210*). The recombinant-derived CBD of *Cellulomonas fimi* actually blocked the access of cellulase to the substrate rather than making it more available (*210*). Deletion of the CBD of xylanases from *Pseudomonas fluorescens* and *C. fimi* did not affect the ability of the enzymes to attack pulp xylan but had variable effects on reducing the lignin content and bleaching (*211*). The *P. fluorescens* enzyme had the same effect on bleaching with or without the CBD, whereas the *C. fimi* enzyme was marginally better with the CBD. Similar findings on the need for the carbohydrate binding module for the hydrolysis of insoluble xylan by the xylanase from *Clostridium stercorarium* have also been reported (*212*). The xylanase might have dual functionality being both a cellulose binding protein and a xylanase (*213*).

Some proteins with an ability to bind to cellulose do not have a specific definable CBD. The cellobiose dehydrogenase from *P. chrysosporium* binds less strongly than the cellobiohydrolase of *T. reesei* (*214*). Attempts to localize a CBD from the cellobiose dehydrogenase via proteolysis indicated that the structure is probably internal to the structure of the protein. The proteolysis of a cellulase complex from *Penicillium verruculosum* provided a fraction that was able to hydrolyze soluble xylan but did not hydrolyze insoluble xylan (*215*). The xylanase from this organism was a major component of the cellulase complex and presumably contains a strong CBD (*215*). New CBD sequences are being discovered, and there may be multiplicity (*216*) in the binding of carbohydrases to the substrates they attack.

Some of the more interesting uses of CBDs might be to alter fiber surfaces without actually hydrolyzing the fiber. For example, Kitaoka and Tanaka (*217*) have recently described a CBD-based additive that enhances paper strength.

Modifying Enzymes to Attain Activity Under Specific Conditions

The cloning of specific enzymes from described organisms and DNA sequences is becoming routine. Many new xylanases (*34,50-55,67,68,218-232*), cellulases (*50,51,53-55,233,234*), and other fiber modifying enzymes (*50,53-55,92,235,236*) have been recently cloned, produced, and characterized as to their activities. Often these enzymes can be produced without ever having to deal with the natural host. With the plethora of genomic sequences being discerned, many new enzymes might be available by searching the new sequences for regions that code for specific enzymes. This directed approach has provided new potential enzymes for use.

Degenerate primers coding for conserved regions of enzymes have been used in PCR studies to produce gene fragments encoding these enzymes. This technique has been used to produce novel enzymes from unknown organisms present in the environment. A recent example of this is the cloning of a 2,5-diketo-D-gluconic acid reductase from an unknown organism (*237*). The resultant enzyme had much higher catalytic activity than existing enzymes from known organisms. Although there may be some limitations on what might be made by recombinant organisms, there are many enzymatic targets that might benefit from this approach. Thermophilic organisms must have active enzymes for all of their cellular functions. The nature of the environment in which these organisms are found dictates that these enzymes must be thermostable. Many enzymes have been cloned from the DNA of these organisms. Although a thermophilic enzyme can tolerate higher temperatures, it will act slower at lower temperatures. Thus, a thermophilic enzyme should be sought when a process has to remain hot or requires the use of solvents, but many other alternatives exist when the process does not have to be heated and is at a nondenaturing pH. In fact, using a thermophilic enzyme might cost more for a given treatment if the fiber has to be specifically heated for the enzyme to be effective.

When an enzyme has been studied to the extent that the crystal structure is known, some specific planned changes can be made and the response in the activity of the protein predicted. The introduction of disulfide bonds through site-specific mutations have increased the thermostability of a *B. circulans* xylanase (*238*). All of the designed changes provided increased thermostability. However, only one out of eight changes provided an increase in the optimal temperature of activity (*238*). Site-specific changes were also introduced into the MnP of *P. chrysosporium*, which allowed the enzyme to tolerate higher concentrations of hydrogen peroxide than the wild type (*239*). The increased tolerance was somewhat offset by a decrease in the affinity of the enzyme for hydrogen peroxide.

Site-specific changes require knowledge of the sequence of the protein and some concept of how changes can be imparted. Comparison of the sequence of *Streptomyces lividans* xylanase to thermostable xylanases from the same family provided a road map to increasing the thermostability of the *S. lividans* enzyme (*229*). The expression of xylanases in *E. coli* and a screening assay for thermostability also helped to identify more thermostable variants of the *S. lividans* xylA xylanase (*221*). Domains that have thermostabilizing activity have been identified (*33,240*). The first 29 amino acids of the *Thermomonospora fusca* xylanase A enzyme imparted increased thermostability to *S. lividans* xylanase B (*241*) and also *T. reesei* xylanase II (*242,243*). An increase in optimal operating temperature was found in variants of *S. lividans* xylanase B by further random gene shuffling using random fragmentation of the *T. fusca* and *S. lividans* genes (*241*). The thermostability and thermal activity of the modified *T. reesei* enzyme has been incorporated into the BioBrite HB60C, which shows improved activity in ECF bleaching (*242,244*).

Other domains from the family 10 xylanases have also been shown to have thermostabilizing effects. Domains from *C. fimi* (*227,245*), *Thermomonospora alba* (*218*), *Neocallimastix frontalis* (*228*), *Streptomyces olivaceoviridis* (*218,245*), and *Fibrobacter succinogenes* (*246*) have all been studied. The results from these studies indicate that the domains often convey their function to new chimeric proteins. This could be the ability to bind to cellulose, bind to other proteins (*246*), change active site domains, and change substrate binding domains (*245*). The shuffling of modules or domains can again produce proteins with altered characteristics. An example of modified enzyme would be the combination of cellulase activity with that of (1-3,1-4)-β-glucanase activity (*247*). This multienzyme was able to degrade the specific linkages present in barley glucan.

The altering of an enzyme using random mutations is powerful and does not require detailed structural information. The *Coprinus cinereus* peroxidase was modified using both site directed and random methods (*248*). The manual combination of individual site directed mutants resulted in an enzyme with 110 times the thermal stability and 2.8 times the oxidative stability. However, when *in vivo* shuffling of mutations through yeast homologous recombination was used, an enzyme with 174 times the thermostability and 100 times the oxidative stability was obtained (*248*).

The directed evolution of an enzyme can provide a powerful tool to attain enzyme optimization. The technique requires a useful host organism and a screen to select desired clones (*249*). Directed evolution increased the alkaline activity and thermostability of a *Neocallimastix patriciarum* xylanase (*250*). Error prone PCR, mutator strains, DNA shuffling, and domain swapping can all be used when the selection is powerful and specific (*249*). Clearly, knowledge and random changes can be applied, and this allows nature to provide a range of possibilities for enzyme modification.

In silico design of an enzyme may lead to further development in this area (*251,252*). Intimate structural knowledge of an enzyme's active site is required for these types of modifications. However, the alterations provided by nature are rather finite and work within a context of what was already present and working. With an enzyme designed by *in silico* methods, the principles of protein folding, bond angles, hydrophobic cores, hydrogen bonding solvation, and other forces (*253,254*) can provide us with so many possibilities (many of which have not been produced by nature) that they cannot be all tested. Designing thermophilic variants of proteins with this technology is already possible (*255*). The temperature limit for enzyme activity might be as high as 200°C (*256*). Narrowing these possible structural variants to what can be readily tested is the challenge (*257*).

Acknowledgment

The Forest Products Laboratory is maintained in cooperation with the University of Wisconsin. This article was written and prepared by U.S. Government employees on official time, and it is therefore in the public domain and not subject to copyright.

References

1. Yin, R. *Tappi J.* **1998**, *81*, 69-74.
2. Raymond, D. R. *Solutions!* **2001**, *1*, 64-67.
3. Meadows, D. G.; Synovec, M. M. *Tappi J.* **1997**, *80*, 104-108.
4. Phillips, R. *Tappi J.* **2000**, *83*, 42-46.
5. Bajpai, P. *Biotechnol. Prog.* **1999**, *15*, 147-157.
6. Beg, Q. K.; Kapoor, M.; Mahajan, L.; Hoondal, G. S. *Appl. Microbiol. Biotechnol.* **2001**, *56*, 326-338.
7. Yee, E.; Tolan, J. S. *Pulp and Paper Canada* **1997**, *98*, T370-T337.
8. Mcdonough, T. J. *Tappi J.* **1995**, *78*, 55-62.
9. Viikari, L.; Kantelinen, A.; Sundquist, J.; Linko, M. *FEMS Microbiol. Rev.* **1994**, *13*, 335-350.
10. Viikari, L.; Ranva, M.; Kantelinen, A.; Sanquist, J.; Linko, M. *Third international conference in biotechnology in the pulp and paper industry,* **1986**, Stockholm, Sweden, 67-69.
11. Dence, C. W.; Reeve, D. W. In *Pulp bleaching-principles and practice*; **1996**, Tappi Press, Atlanta, pp 365-377.

12. Bajpai, P; Bajpai, P. K. *Tappi J.* **1998**, *81*, 111-117.
13. Irie, Y.; Matsukura, T.; Hata, K. **1989,** Jap. Patent 160,997/90.
14. Gutierrez, A.; Del Rio, J. C.; Martinez, M. J.; Martinez, A. T. *Trends Biotechnol.* **2001**, *19*, 340-348.
15. Kirk, T. K.; Jeffries, T. W. In *Enzymes for pulp and paper processing*; **1996**, American Chemical Society: Atlanta, pp 1-14.
16. Wong, K. K. Y.; Mansfield, S. D. *Appita J.* **1999**, *52*, 409-418.
17. Atkinson, D.; Moody, D.; Gronberg, V. *Inv. Tec. Papel* **1998**, *136*, 199-209.
18. Tolan, J. S.; Guenette, M. *Adv. in Biochem. Engin.* **1997**, *57*, 289-231.
19. Jacobs, C. J.; Venditti, R. A.; Joyce, T. W. *Proceedings of the Pulping Conference* **1997**, Tappi Press, San Francisco, 593-602.
20. Jacobs-Young, C. J.; Gustafson, R. R.; Heitmann, J. A. *Paper and Timber* **2000**, *82*, 114-119.
21. Jacobs, C. J.; Venditti, R. A.; Joyce, T. W. *Tappi J.* **1998**, *81*, 260-266.
22. Richardson, J. D.; Wong, K. K. Y.; Clark, T. A. *J. Pulp Pap. Sci.* **1998**, *24*, 125-129.
23. Pere, J.; Liukkonen, S.; Siika–Aho, M.; Gullichsen, J.; Viikari, L. *Proceedings of the TAPPI Pulping Conference* **1996**, Nashville, 693-696.
24. Pere, J.; Siika-Aho, M.; Viikari, L. *Tappi J.* **2000**, *82*, 1-8.
25. Wong, K. K. Y.; Richardson, J. D.; Mansfield, S. D. *Biotechnol. Prog.* **2000**, *16*, 1025-1029.
26. Dickson, A. R.; Wong, K. K. Y.; Mansfield, S. D. *Tappi J.* **2000**, 83, (7) 64.
27. Seo, Y. B.; Shin, Y. C.; Jeon, Y. *In Tappi J.* **2000**, 83, (11) p 64.
28. Thornton, J.; Tenkanen, M.; Ekman, R.; Holmbom, B.; Viikari, L. *Holzforschung* **1994**, *48*, 436-440.
29. Rahkamo, L.; Siika-aho, M.; Vehvilainen, M.; Dolk, M.; Viikari, L.; Nousiainen, P.; Buchert, J. *Cellulose* **1996**, *3*, 153-163.
30. Rahkamo, L.; Viikari, L.; Buchert, J.; Paakkari, T.; Suortti, T. *Cellulose* **1998**, *5*, 79-88.
31. Jackson, L. S.; Heitmann, J. A.; Joyce, T. W. *Tappi J.* **1998**, *81*, 171-178.
32. Bajpai, P.; Bajpai, P. K. *Process Biochem.* **1992**, *27*, 319-325.
33. Kulkarni, N.; Shendye, A.; Rao, M. *FEMS Microbiol. Rev.* **1999**, *23*, 411-456.
34. Sunna, A.; Antranikian, G. *Crit. Rev. Biotechnol.* **1997**, *17*, 39-67.
35. Suurnakki, A. T. M.; Buchert, J.; Viikari, L. *Adv. in Biochem. Engin.* **1997**, *57*, 261-287.
36. Gliese, T.; Kleemann, S.; Fischer, K. *Pulp Pap. Can.* **1998**, *99*, 171-174.
37. Sivaswamy, S. N.; Udayachandran, R.; Venkataraman, T. S. *IPPTA* **1998**, *10*, 1-9.
38. Davis, M.; Rosin, B.; Landucci, L. L.; Jeffries, T. W. *Tappi Biological Symposium* **1997**, Tappi Press, Atlanta, San Francisco, 435-442.
39. Jiang, Z.–H.; Van Lierop, B.; Berry, R. *Tappi J.* **2000**, *83*, 167-175.

40. Laine, J.; Buchert, J.; Viikari, L.; Stenius, P. *Holzforschung* **1996**, *50*, 208-214.
41. Suurnakki, A.; Heijnesson, A.; Buchert, J.; Westermark, U.; Viikari, L. *J. Pulp Pap. Sci.* **1996**, *22*, J91-J96.
42. Vuorinen, T.; Fagerstrom, P.; Buchert, J.; Tenkanen, M.; Teleman, A. *J. Pulp Pap. Sci.* **1999**, *25*, 155-162.
43. Chakar, F. S.; Allison, L.; Ragauskas, A. J.; McDonough, T. J.; Sezgi, U. O. *Tappi J.* **2000**, 83, 62.
44. Buchert, J.; Bergnor, E.; Lindblad, G.; Viikari, L.; Ek, M. *Tappi J.* **1997**, *80*, 165-171.
45. Maiti, B.; Whitmire, D. *Chem. Eng. Commun.* **1997**, *162*, 169-178.
46. Suurnakki, A.; Li, T–Q.; Buchert, J.; Tenkanen, M.; Viikari, L.; Vuorinen, T.; Odberg, L. *Holzforschung* **1997**, *51*, 27-33.
47. De Jong, E.; Wong, K. K. Y.; Saddler, J. N. *Holzforschung* **1997**, *51,* 19-26.
48. Suurnakki, A.; Heijnesson, A.; Buchert, J.; Viikari, L.; Westermark, U. *J. Pulp Pap. Sci.* **1996**, *22*, J43-J47.
49. Christov, L.; Biely, P.; Kalogeris, E.; Christakopoulos, P.; Prior, B. A.; Bhat, M. K. *J. Biotechnol.* **2000**, *83*, 231-244.
50. Sunna, A.; Moracci, M.; Rossi, M.; Antranikian, G. *Extremophiles* **1997**, *1*, 2-13.
51. Bergquist, P. L.; Gibbs, M. D.; Morris, D. D.; Te'o, V. S.; Saul, D. J.; Moran, H. W. *FEMS Microbial Ecology* **1999**, *28*, 99-110.
52. Bergquist, P. L.; Gibbs, M. D.; Morris, D. D.; Thompson, D. R.; Uhl, A. M.; Daniel, R. M. *Methods Enzymol.* **2001**, *330*, 301-319.
53. Bauer, M. W.; Driskill, L. E.; Kelly, R. M. *Cur. Opin. in Biotech.* **1998**, *9*, 141-145.
54. Maheshwari, R.; Bharadwaj, G.; Bhat, M. K. *Microbiol. Mol. Biol. Rev.* **2000**, *64*, 461-488.
55. Bruins, M. E.; Janssen, A. E. M.; Boom, R. M. *Appl. Biochem. Biotechnol.* **2001**, *90*, 155-186.
56. Horikoshi, K. *FEMS Microbiol. Rev.* **1996**, *18*, 259-270.
57. Dimitrov, P. L.; Kambourova, M. S.; Mandeva, R. D.; Emanuilova, E. I. *FEMS Microbiol. Lett.* **1997**, *157*, 27-30.
58. Nakamura, S.; Nakai, R.; Wakabayashi, K.; Ishiguro, Y.; Aono, R.; Horikoshi, K. *Biosci. Biotechnol. Biochem.* **1994**, *58*, 78-81.
59. Davoodi, J.; Wakarchuk, W. W.; Campbell, R. L.; Carey, P. R.; Surewicz, W. K. *Eur. J. Biochem.* **1995**, *232*, 839-843.
60. Krengel, U.; Dijkstra, B. W. *J. Mol. Biol.* **1996**, *263*, 70-78.
61. Saltin, G.; Edwards, L. *Adv. Pulp Papermaking* **1995**, *91*, 53-61.
62. Vicuna, R.; Yeber, M. C.; Osses, M. *J. Biotechnol.* **1995**, *42*, 69-73.
63. Vicuna, R.; Osses, M.; Jara, A. *Tappi Biological Symposium* **1997**, Tappi Press, Atlanta, San Francisco, 155-161.

64. Wong, K. K. Y.; Nelson, S. L.; Saddler, J. N. *J. Biotechnol.* **1996**, *48*, 137-145.
65. Hamilton, J.; Senior, D. J.; Rodriguez, A.; Santiago, D.; Szwec, J.; Ragauskas, A. J. *Tappi J.* **1996**, *79*, 231-234.
66. Chen, C.-C.; Adolphson, R.; Dean, J. F. D.; Eriksson, K-E. L.; Adams, M. W. W.; Westpheling, J. *Enzyme Microb. Technol.* **1997**, *20*, 39-45.
67. Morris, D. D.; Gibbs, M. D.; Chin, C. W. J.; Koh, M. H.; Wong, K. K. Y.; Allison, R. W.; Nelson, P. J.; Bergquist, P. L. *Appl. Environ. Microbiol.* **1998**, *64*, 1759-1765.
68. Saul, D. J.; Williams, L. C.; Reeves, R. A.; Gibbs, M. D.; Bergquist, P. L. *Appl. Environ. Microbiol.* **1995**, *61*, 4110-4113.
69. Valchev, V.; Valchev, I.; Christova, E. *Cell Chem. Technol.* **1998**, *32*, 457-462.
70. Duran, N.; Milagres, A. M. F.; Esposito, E.; Haun, M. In *Enzymatic Degradation of Insoluble Carbohydrates* **1995** Saddler, J. N.; Penner, M. H., Eds.; American Chemical Society: Washington, DC, pp 332-338.
71. Shah, A. K.; Cooper, A.; Adolphson, R. B.; Li, X-L.; Chen, H.; Ljungdahl, L. G.; Eriksson, K–E. L. *Tappi J.* **2000**, *83*, 95.
72. Madlala, A. M.; Bissoon, S.; Singh, S.; Christov, L. *Biotechnol. Lett.* **2001**, *23*, 345-351.
73. Prasad, D. Y.; Rajesh, K. S.; Praburaj, T.; Rao, N. R. M.; Joyce, T. W. *Cell. Chem. Technol.* **1996**, *30*, 463-472.
74. Roncero, M. B.; Torres, A. L.; Colom, J. F.; Vidal, T. *Process Biochem.* **2000**, *36*, 45-50.
75. Ryynanen, H.; Nelson, P. J.; Chin, C. W. J. *Appita J.* **1995**, *48*, 440-444.
76. Vicuna, R.; Escobar, F.; Osses, M.; Jara, A. *Biotechnol. Lett.* **1997**, *19*, 575-578.
77. Vidal, T.; Torres, A. L.; Colom, J. F.; Siles, J. *Appita J.* **1997**, *50*, 144-147.
78. Christov, L. P.; Szakacs, G.; Balakrishnan, H. *Process Biochem.* **1999**, *34*, 511-517.
79. Gupta, S.; Bhushan, B.; Hoondal, G. S. *J. Appl. Microbiol.* **2000**, *88*, 325-334.
80. Beg, Q. K.; Bhushan, B.; Kapoor, M.; Hoondal, G. S. *Enzyme Microb. Technol.* **2000**, *27*, 459-466.
81. Zheng, L. S.; Du, Y. M.; Zhang, J. Y. *Biotechnol. Lett.* **2000**, *22*, 1363-1367.
82. Bajpai, P.; Bajpai, P. K. *Tappi J.* **1996**, *79*, 225-230.
83. Jimenez, L.; Navarro, E.; Garcia, J. C.; Perez, I. *Tappi J.* **2000**, *83*, 65.
84. Elegir, G.; Sykes, M.; Jeffries, T. W. *Enzyme Microb. Technol.* **1995**, *17*, 954-959.
85. Fernandes, A. C.; Fontes, C.; Gilbert, H. J.; Hazlewood, G. P.; Fernandes, T. H.; Ferreira, L. M. A. *Biochem. J.* **1999**, *342*, 105-110.

86. Georis, J.; Giannotta, F.; De Buyl, E.; Granier, B.; Frere, J. M. *Enzyme Microb. Technol.* **2000**, *26*, 178-186.
87. Clarke, J. H.; Rixon, J. E.; Ciruela, A.; Gilbert, H. J.; Hazlewood, G. P. *Appl. Microbiol. Biotechnol.* **1997**, *48*, 177-183.
88. Baraznenok, V. A.; Becker, E. G.; Ankudimova, N. V.; Okunev, N. N. *Enzyme Microb. Technol.* **1999**, *25*, 651-659.
89. Morris, D. D.; Bergquist, P. L. *Tappi Biological Symposium* **1997**, Tappi Press, Atlanta, San Francisco, 181-187.
90. Clarke, J. H.; Davidson, K.; Rixon, J. E.; Halstead, J. R.; Fransen, M. P.; Gilbert, H. J.; Hazlewood, G. P. *Appl. Microbiol. Biotechnol.* **2000**, *53*, 661-667.
91. Saake, B.; Clark, T.; Puls, J. *Holzforschung* **1995**, *49*, 60-68.
92. Sunna, A.; Gibbs, M. D.; Chin, C. W. J.; Nelson, P. J.; Bergquist, P. L. *Appl. Environ. Microbiol.* **2000**, *66*, 664-670.
93. Gomes, J.; Gomes, I.; Terler, K.; Gubala, N.; Ditzelmuller, G.; Steiner, W. *Enzyme Microb. Technol.* **2000**, *27*, 414-422.
94. Montiel, M. D.; Rodriguez, J.; Perez–Leblic, M. I.; Hernandez, M.; Arias, M. E.; Copa–Patino, J. L. *Appl. Microbiol. Biotechnol.* **1999**, *52*, 240-245.
95. Suurnakki, A.; Clark, T. A.; Allison, R. W.; Viikari, L.; Buchert, J. *Tappi J.* **1996**, *79*, 111-117.
96. Allison, R. W.; Clark, T. A.; Ellis, M. J. *Appita J.* **1996**, *48*, 201-206.
97. Christov, L. P.; Prior, B. A. *Enzyme Microb. Technol.* **1996**, *18*, 244-250.
98. Buchert, J.; Kantelinen, A.; Suurnakki, A.; Viikari, L.; Janson, J. *Holzforschung* **1995**, *49*, 439-444.
99. Bajpai, P.; Bajpai, P. K. *7th International Conference on Biotechnology in the Pulp and Paper Industry* **1998**, Vancouver, C3-C6.
100. Christov, L. P.; Akhtar, M.; Prior, B. A. *Holzforschung* **1996**, *50*, 579-582.
101. Li, J.; Paice, M. G.; MacLeod, J. M.; Jurasek, L. *J. Pulp Pap. Sci.* **1996**, *22*, J207-J212.
102. Lockie, M. *Pulp Paper Internat.* **1998**, Oct. 44-46.
103. Whiteside, J. A. B. *Pulp Paper Internat.* **1998**, Oct. 46-51.
104. Serra, J. A.; Pastor, F. I. J.; Blanco, A.; Colom, J. F.; Vidal, T. *ISWPC* **1997**, 100-101, 100-104.
105. Wong, K. K. Y.; Martin, L. A.; Gama, F. M.; Saddler, J. N.; de Jong, E. *Biotechnol. Bioeng.* **1997**, *54*, 312-318.
106. Wong, K. K. Y.; James, C. S.; Campion, S. H. *J. Pulp Pap. Sci.* **2000**, *26*, 377-383.
107. Paice, M. G.; Archibald, F. S.; Bourbonnais, R.; Jurasek, L.; Reid, I. D.; Charles, T.; Dumonceaux, T. In *Enzymes for Pulp and Paper Processing* **1996**, Jeffries, T. W.; Viikari, L., Eds.; American Chemical Society: Washington, DC.
108. Henriksson, G.; Johansson, G.; Pettersson, G. *J. Biotechnol.* **2000**, *78*, 93-113.
109. Ander, P. *FEMS Microbiol. Rev.* **1994**, *13*, 297-312.

110. Duarte, J. C.; Costa-Ferreira, M.; Sena-Martins, G. *Bioresour. Technol.* **1999**, *68*, 43-48.
111. Burke, R. M.; Cairney, J. W. G. *New Phytology* **1998**, *139*, 637-645.
112. Cameron, M. D.; Aust, S. C. *Enzyme Microb. Technol.* **2001**, *28*, 129-138.
113. Paice, M. G.; Bourbonnais, R.; Reid, I. D. *Tappi J.* **1995**, *78*, 161-169.
114. Paice, M. G.; Bourbonnais, R.; Reid, I. D.; Archibald, F. S.; Jurasek, L. *J. Pulp Pap. Sci.* **1995**, *21*, J280-J284.
115. Harazono, K.; Kondo, R.; Sakai, K. *Appl. Environ. Microbiol.* **1996**, *62*, 913-917.
116. Paice, M. G.; Archibald, F. S.; Bourbonnais, R.; Reid, I. D.; Renaud, S. *Tappi Biological Symposium* **1997**, Tappi Press, Atlanta, San Francisco, 346-348.
117. Hassingboe, J.; Lawther, J. M.; Felby, C. *7th International Conference on Biotechnology in the Pulp and Paper Industry* **1998**,Vancouver, A125-A128.
118. Harazono, K.; Kondo, R.; Sakai, K.; Tsuchikawa, K. *Kami Pa Gikyoshi* **1996**, *50*, 1292-1298.
119. Reid, I. D.; Paice, M. G. *Appl. Environ. Microbiol.* **1998**, *64*, 2273-2274.
120. Bermek, H.; Li, K.; Eriksson, K.-E. L. *Tappi J.* **2000**, *83*, 69.
121. Oksanen, T.; Amann, M.; Candussio, A.; Buchert, J.; Viikari, L. *9th International Symposium on Wood and Pulping Chemistry* **1997**, Quebec, 83-81-83-84.
122. Call, H. P.; Mucke, I. *J. Biotechnol.* **1997**, *53*, 163-202.
123. Leonowicz, A.; Cho, N–S.; Luterek, J.; Wilkolazka, A.; Wojtas–Wasilewska, M.; Matuszweska, A.; Hofrichter, M.; Wesenberg, D.; Rogalski, J. *J. Basic Microbiol.* **2001**, *41*, 185-227.
124. Yu, H. S.; Wang, C. H.; Wen, X. H.; Fu, S. Y.; Huang, X. Y. *9th International Symposium on Wood and Pulping Chemistry* **1997**, Quebec, 130-131-130-134.
125. Diamantidis, G.; Effosse, A.; Potier, P.; Bally, R. *Soil Biol. Biochem.* **2000**, *32*, 919-927.
126. Hullo, M.-F.; Moszer, I.; Danchin, A.; Martin-verstraete, I. *J. Bacteriol.* **2001**, *183*, 5426-5430.
127. Bourbonnais, R.; Paice, M. G.; Leech, D.; Freiermuth, B. *Tappi Biological Sciences Symposium* **1997**, Tappi Press, Atlanta, San Francisco, 335-338.
128. Li, K.; Bermek, H.; Eriksson, K.-E. L. *Tappi Biological Sciences Symposium* **1997**, Tappi Press, Atlanta, San Francisco, 349-353.
129. Bourbonnais, R.; Rochefort, D.; Paice, M. G.; Renaud, S.; Leech, D. *Tappi J.* **2000**, 83, 68.
130. DeCarvalho, M. E. A.; Monteiro, M. C.; Sant'Anna, G. L. *Appl. Biochem. Biotechnol.* **1999**, *77-79*, 723-733.
131. Wong, K. K. Y.; Anderson, K. B.; Kibblewhite, R. P. *Enzyme Microb. Technol.* **1999** *25*, 125-131.

132. Bourbonnais, R.; Paice, M. G. *Tappi J.* **1996**, *79*, 199-204.
133. Sealey, J. E.; Runge, T. M.; Ragauskas, A. J. *Tappi J.* **2000**, *82, (9)*. 66.
134. Sealey, J. E.; Runge, T. M.; Ragauskas, A. J. *Tappi Biological Sciences Symposium* **1997**, Tappi Press, Atlanta, San Francisco, 339-342.
135. Vares, T.; Dorado, J.; Almendros, G.; Bocchini, P.; Galletti, G. C.; Martinez, A. T.; Hatakka, A. *Tappi Biolgical Sciences Symposium* **1997**, Tappi Press, Atlanta, San Francisco, 405-412.
136. Poppius-Levin, K.; Wang, W.; Ranua, M.; Niku-Paavola, M. L.; Viikari, L. *Tappi Biological Sciences Symposium* **1997**, Tappi Press, Atlanta, San Francisco, 329-333.
137. Bourbonnais, R.; Paice, M. G.; Freiermuth, B.; Bodie, E.; Borneman, S. *Appl. Environ. Microbiol.* **1997**, *63*, 4627-4632.
138. Li, K.; Xu, F.; Eriksson, K.-E. L. *Appl. Environ. Microbiol.* **1999**, *65*, 2654-2660.
139. Chakar, F. S.; Ragauskas, A. J. *Tappi Pulping Conference* **1998**, Tappi Press, Atlanta, 109-118.
140. Sealey, J.; Ragauskas, A. J. *Enzyme Microb. Technol.* **1998**, *23*, 422-426.
141. Balakshin, M.; Chen, C. L.; Gratzl, J. S.; Kirkman, A. G.; Jakob, H. *Holzforschung* **1999**, *54*, 390-396.
142. Pala, H.; Lemos, M. A.; Mota, M.; Gama, F. M. *Enzyme Microb. Technol.* **2001**, *29*, 274-279.
143. Nagarajan, R.; Sarkar, J. M. *Tappi Papermakers Conference* **1997**, Tappi Press, Atlanta, Philadelphia, 475-480.
144. Bhardwaj, N. K.; Bajpai, P.; Bajpai, P. K. *Appita J.* **1997**, *50*, 230-232.
145. Moran, B. *PIMA magazine* **1996**, pp 44-45.
146. Moran, B. *Pulp and Paper* **1996**, pp 119-121.
147. Mooney, C. A.; Mansfield, S. D.; Beatson, R. P.; Saddler, J. N. *Enzyme Microb. Technol.* **1999**, *25*, 644-650.
148. Stork, G.; Pereira, H.; Wood, T. M.; Dusterhoft, E. M.; Toft, A.; Puls, J. *Tappi J.* **1995**, *78*, 79-88.
149. Gruber, E.; Gelbrich, M. *Papier* **1997**, *51*, 166-174.
150. Pere, J.; Siikaaho, M.; Buchert, J.; Viikari, L. *Tappi J.* **1995**, *78*, 71-78.
151. Oksanen, T.; Pere, J.; Paavilainen, L.; Buchert, J.; Viikari, L. *J. Biotechnol.* **2000**, *78*, 39-48.
152. Pala, H.; Lemos, M. A.; Mota, M.; Gama, F. M. *Enzyme Microb. Technol.* **2001**, *29*, 274-279.
153. Mansfield, S. D.; Wong, K. K. Y.; Dejong, E.; Saddler, J. N. *Tappi J.* **1996**, *79*, 125-132.
154. Sreenath, H. K.; Yang, V. W.; Burdsall, H. H.; Jeffries, T. W. In *Enzymes for Pulp and Paper Processing* **1996**, Jeffries, T. W.; Viikari, L., Eds.; American Chemical Society: Washington, DC, pp 267-279.
155. Lascaris, E.; Forbes, L.; Lonergan, G. *Tappi Biological Sciences Symposium* **1997**, Tappi Press, Atlanta, San Francisco, 271-277.

156. Lascaris, E.; Mew, L.; Forbes, L.; Mainwaring, D.; Lonergan, G. *Appita J.* **1997**, *50*, 51-54.
157. Fitzhenry, J. W.; Hoekstra, P. M.; Glover, D. *Tappi Pulping Conference* **2000**,
158. Caram, F. C.; Sarkar, J. M.; Didwania, H. P.; Espinoza, E.; Benavides, J. C. *Tappi Papermakers Conference* **1996**, Tappi Press, Atlanta, Philadelphia, 481-498.
159. Kaya, F.; Heitmann, J. A.; Joyce, T. W. *Tappi J.* **1995**, *78*, 150-157.
160. Dash, B.; Patel, M. *IPPTA* **1997**, *9*, 61-70.
161. Dinus, R. J.; Welt, T. *Wochenbl. Papierfabr.* **1998**, *126*, 396-407.
162. Heise, O. U.; Unwin, J. P.; Klungness, J. H.; Fineran, W. G.; Sykes, M.; AbuBakr, S. *Tappi J.* **1996**, *79*, 207-212.
163. Jeffries, T. W.; Klungness, J. H.; Sykes, M. S.; Rutledgecropsey, K. R. *Tappi J.* **1994**, *77*, 173-179.
164. Sarkar, J. M.; Cosper, D. R.; Hartig, E. J. *Tappi J.* **1995**, *78*, 89-95.
165. Park, J.; Park, K. *Bioresour. Technol.* **2001**, *79*, 91-94.
166. Park, J. W.; Park, K. N. *Biotechnol. Tech.* **1999**, *13*, 49-53.
167. Treimanis, A.; Leite, M.; Eisimonte, M.; Viesturs, U. *Chemical and Biochemical Engineering Quarterly* **1999**, *13*, 53-57.
168. Viesturs, U.; Leite, M.; Eisimonte, M.; Eremeeva, T.; Treimanis, A. *Bioresour. Technol.* **1999**, *67*, 255-265.
169. Zollner, H. K.; Schroeder, L. R. *Tappi J.* **1998**, *81*, 166-170.
170. Blanco, M. A.; Negro, C.; Gaspar, I.; Tijero, J. *Appl. Microbiol. Biotechnol.* **1996**, *46*, 203-208.
171. Schwingel, W. R. *TAPPI Biological Sciences Symposium* **1997**, Tappi Press, Altlanta, San Francisco, 138-143.
172. Grant, R. *Pulp Paper Internat.* **1998**, 35-37.
173. Dykstra, G. M.; Stoner, M. T. *Tappi Biological Sciences Symposium* **1997**, Tappi Press, Atlanta, San Francisco, 117-123.
174. Wolfanger, D. *PaperAge* **2001**, 34-36.
175. Ratto, M.; Mustranta, A.; Siika-aho, M. *Appl. Microbiol. Biotechnol.* **2001**, *57*, 182-185.
176. Zhang, X.; Cai, Y.; Stebbings, D.; Beatson, R. P.; Saddler, J. N. *7th International Conference on Biotechnology in the Pulp and Paper Industry* **1998**, Vancouver, C151-C154.
177. Zhang, X. *Pulp Pap. Can.* **2000**, *101*, 59-62.
178. Stebbing, D. *Pulp Pap. Can.* **1999**, *100*, 46-49.
179. Francis, D. W.; Ouchi, M. D. *J. Pulp Pap. Sci.* **2001**, *27*, 289-295.
180. Zhang, X.; Beatson, R. P.; Cai, Y. J.; Saddler, J. N. *J. Pulp Pap. Sci.* **1999**, *25*, 206-210.
181. Zhang, X.; Stebbing, D. W.; Saddler, J. N.; Beatson, R. P.; Kruus, K. *J. Wood Chem. Technol.* **2000**, *20*, 321-335.
182. Mustranta, A.; Buchert, J.; Spetz, P.; Holmbom, B. *Nordic Pulp & Paper Res. J.* **2001**, *16*, 125-129.

183.Fischer, K.; Messner, K. *Enzyme Microb. Technol.* **1992**, *14*, 470-473.

184.Fischer, K.; Messner, K. *Tappi J.* **1992**, *75*, 130-134.

185.Fischer, K.; Puchinger, L.; Schloffer, K.; Kreiner, W.; Messner, K. *J. Biotechnol.* **1993**, *27*, 341-348.

186.Fujita, Y.; Awaji, H.; Taneda, H.; Matsukura, M.; Hata, K.; Shimoto, H.; Sharyo, M.; Sakaguchi, H.; Gibson, K. *Tappi J.* **1992**, *75*, 117-121.

187.Thornton, J.; Ekman, R.; Holmbom, B.; Pettersson, C. *J. Wood Chem. Technol.* **1994**, *14*, 177-194.

188.Huttermann, A.; Mai, C.; Kharazipour, A. *Appl. Microbiol. Biotechnol.* **2001**, *55*, 387-394.

189.Lund, M.; Ragauskas, A. J. *Appl. Microbiol. Biotechnol.* **2001**, *55*, 699-703.

190.Chandra, R. P.; Ragauskas, A. J. *Tappi Pulping Conference* **2001**, Tappi Press, Atlanta,

191.Mai, C.; Majcherczyk, A.; Huttermann, A. *Enzyme Microb. Technol.* **2000**, *27*, 167-175.

192.Mai, C.; Milstein, O.; Huttermann, A. *Appl. Microbiol. Biotechnol.* **1999**, *51*, 527-531.

193.Huttermann, A.; Majcherczyk, A.; Braun-Lullemann, A.; Mai, C.; Fastenrath, M.; Kharazipour, A.; Huttermann, J.; Huttermann, A. H. *Naturwissenschaften* **2000**, *87*, 539-541.

194.Felby, C.; Pedersen, L. S.; Nielsen, B. R. *Holzforschung* **1997**, *51*, 281-286.

195.Lund, M.; Felby, C. *Enzyme Microb. Technol.* **2001**, *28*, 760-765.

196.Vachoud, L.; Chen, T.; Payne, G. F.; Vazquez--Duhalt, R. *Enzyme Microb. Technol.* **2001**, *29*, 380-385.

197.Guerra, A.; Ferraz, A. *Enzyme Microb. Technol.* **2001**, *28*, 308-313.

198.Iwahara, K.; Honda, Y.; Watanabe, T.; Kuwahara, M. *Appl. Microbiol. Biotechnol.* **2000**, *54*, 104-111.

199.Akkara, J. A.; Ayyagari, M. S. R.; Bruno, F. F. *Trends Biotechnol.* **1999**, *17*, 67-73.

200.Mansfield, S. D.; Dejong, E.; Stephens, R. S.; Saddler, J. N. *J. Biotechnol.* **1997**, *57*, 205-216.

201.Mansfield, S. D.; Dickson, A. R. *Appita J.* **2001**, *54*, 239-244.

202.Mansfield, S. D.; Swanson, D. A.; Roberts, N.; Olson, J. A.; Saddler, J. N. *Tappi J.* **1999**, *82*, 152-158.

203.Mansfield, S. D.; Saddler, J. L. *J. Pulp Pap. Sci.* **1999**, *25*, 84-89.

204.Lumme, M. S.; Mansfield, S. D.; Saddler, J. N. *Wood Fiber Sci.* **1999**, *31*, 385-393.

205.Linder, M.; Teeri, T. T. *J. Biotechnol.* **1997**, *57*, 15-28.

206.Tomme, P.; Driver, D. P.; Amandoron, E. A.; Miller, R. C.; Antony, R.; Warren, J.; Kilburn, D. G. *J. Bacteriol.* **1995**, *177*, 4356-4363.

207.Simpson, P. J.; Xie, H. F.; Bolam, D. N.; Gilbert, H. J.; Williamson, M. P. *J. Biol. Chem.* **2000**, *275*, 41137-41142.

208.Suurnakki, A.; Tenkanen, M.; Siika-Aho, M.; Niku-Paavola, M. L.; Viikari, L.; Buchert, J. *Cellulose* **2000**, *7*, 189-209.

209.Swanson, B. A.; Ward, M.; Pentilla, M.; Pere, J.; Saloheimo, M. In *International Publication Number WO 99/02693*; Genencor International **1999**.

210.Esteghlalian, A. R.; Srivastava, V.; Gilkes, N. R.; Kilburn, D. G.; Warren, R. A. J.; Saddler, J. N. *Appl. Biochem. Biotechnol.* **2001**, *91-3*, 575-592.

211.Rixon, J. E.; Clarke, J. H.; Hazlewood, G. P.; Hoyland, R. W.; McCarthy, A. J.; Gilbert, H. J. *Appl. Microbiol. Biotechnol.* **1996**, *46*, 514-520.

212.Ali, M. K.; Hayashi, H.; Karita, S.; Goto, M.; Kimura, T.; Sakka, K.; Ohmiya, K. *Biosci. Biotechnol. Biochem.* **2001**, *65*, 41-47.

213.Ali, M. K.; Kimura, T.; Sakka, K.; Ohmiya, K. *FEMS Microbiol. Lett.* **2001**, *198*, 79-83.

214.Henriksson, G.; Salumets, A.; Divine, C.; Pettersson, G. *Biochem. J.* **1997**, *324*, 833-838.

215.Berlin, A. G.; Gusakov, A. V.; Sinitsyna, O. A.; Sinitsyn, A. P. *Appl. Biochem. Biotechnol.* **2000**, *88*, 345-352.

216.Rincon, M. T.; McCrae, S. I.; Kirby, J.; Scott, K. P.; Flint, H. J. *Appl. Environ. Microbiol.* **2001**, *67*, 4426-4431.

217.Kitaoka, T.; Tanaka, H. *J. Wood Sci.* **2001**, *47*, 322-324.

218.Ashan, M. M.; Kaneko, S.; Wang, Q.; Yura, K.; Go, M.; Hayash, K. *Enzyme Microb. Technol.* **2001**, *28,*, 8-15.

219.Bergquist, P. L.; Gibbs, M. D.; Saul, D. J.; Reeves, R. A.; Morris, D. D.; Te'o, V. S. J.; Walsh, D. J. *Tappi Biological Symposium* **1997**, Tappi Press, Atlanta, San Francisco, 169-179.

220.Dalrymple, B. P.; Swadling, Y.; Layton, I.; Gobius, K. S.; Xue, G. P. *Appl. Environ. Microbiol.* **1999**, *65*, 3660-3667.

221.Ebanks, R.; Dupont, M.; Shareck, F.; Morosoli, R.; Kluepfel, D.; Dupont, C. *J. Ind. Microbiol. Biotechnol.* **2000**, *25*, 310-314.

222.Gupta, N.; Reddy, V. S.; Maiti, S.; Ghosh, A. *Appl. Environ. Microbiol.* **2000**, *66*, 2631-2635.

223.Kim, H.; Jung, K. H.; Pack, M. Y. *Appl. Microbiol. Biotechnol.* **2000**, *54*, 521-527.

224.Kimura, T.; Ito, J.; Kawano, A.; Makino, T.; Kondo, H.; Karita, S.; Sakka, K.; Ohmiya, K. *Biosci. Biotechnol. Biochem.* **2000**, *64*, 1230-1237.

225.Kulkarni, N.; Chauthaiwale, J.; Rao, M. *Enzyme Microb. Technol.* **1995**, *17*, 972-976.

226.Mande, S. S.; Gupta, N.; Ghosh, A.; Mande, S. C. *J. Biomol. Struct. Dyn.* **2000**, *18*, 137-144.

227.McLean, B. W.; Bray, M. R.; Boraston, A. B.; Gilkes, N. R.; Haynes, C. A.; Kilburn, D. G. *Protein Eng.* **2000**, *13*, 801-809.

228.Mesta, L.; Rascle, C.; Durand, R.; Fevre, M. *Enzyme Microb. Technol.* **2001**, *29*, 456-463.

229.Moreau, A.; Shareck, F.; Kluepfel, D.; Morosoli, R. *Enzyme Microb. Technol.* **1994**, *16*, 420-424.

230. Reeves, R. A.; Gibbs, M. D.; Morris, D. D.; Griffiths, K. R.; Saul, D. J.; Bergquist, P. L. *Appl. Environ. Microbiol.* **2000**, *66*, 1532-1537.
231. Sunna, A.; Gibbs, M. D.; Bergquist, P. L. *Microbiology-(UK)* **2000**, *146*, 2947-2955.
232. Tabernero, C.; Sanchez-Torres, J.; Perez, P.; Santamaria, R. I. *Appl. Environ. Microbiol.* **1995**, *61*, 2420-2424.
233. Ito, S. *Extremophiles* **1997**, *1*, 61-66.
234. Sanchez-Torres, J.; Perez, P.; Santamaria, R. I. *Appl. Microbiol. Biotechnol.* **1996**, *46*, 149-155.
235. Gibbs, M. D.; Reeves, R. A.; Sunna, A.; Bergquist, P. L. *Curr. Microbiol.* **1999**, *39*, 351-357.
236. Morris, D. D.; Reeves, R. A.; Gibbs, M. D.; Saul, D. J.; Bergquist, P. L. *Appl. Environ. Microbiol.* **1995**, *61*, 2262-2269.
237. Eschenfeldt, W. H.; Stols, L.; Rosenbaum, H.; Khambatta, Z. S.; Quaite-Randall, E.; Wu, S.; Kilgore, D. C.; Trent, J. D.; Donnelly, M. I. *Appl. Environ. Microbiol.* **2001**, *67*, 4206-4214.
238. Wakarchuk, W. W.; Sung, W. L.; Campbell, R. L.; Cunningham, A.; Watson, D. C.; Yaguchi, M. *Protein Eng.* **1994**, *7*, 1379-1386.
239. Miyazaki, C.; Takahashi, H. *FEBS Lett.* **2001**, *509*, 111-114.
240. Fontes, C.; Hazlewood, G. P.; Morag, E.; Hall, J.; Hirst, B. H.; Gilbert, H. J. *Biochem. J.* **1995**, *307*, 151-158.
241. Shibuya, H.; Kaneko, S.; Hayashi, K. *Biochem. J.* **2000**, *349*, 651-656.
242. White, T.; Thibault, L.; Watkinson, J.; Sung, W.; Yaguchi, M.; Wood, M.; Huang, F. *7th International Conference on Biotechnology in the Pulp and Paper Industry* **1998**, Vancouver, A21-A25.
243. Sung, W. L.; Yaguchi, M.; Ishikawa, K.; Huang, F.; Wood, M.; Zahab, D. M. *7th International Conference on Biotechnology in the Pulp and Paper Industry* **1998**, Vancouver, C61-C63.
244. Thibault, L.; Tolan, J.; White, T.; Yee, E.; April, R.; Sung, W. *85th Annual Meeting of the Pulp and Paper Technical Association of Canada* **1999**, B263-B266.
245. Kaneko, S.; Iwamatsu, S.; Kuno, A.; Fujimoto, Z.; Sato, Y.; Yura, K.; Go, M.; Mizuno, H.; Taira, K.; Hasegawa, T.; Kusakabe, I.; Hayashi, K. *Protein Eng.* **2000**, *13*, 873-879.
246. Marrone, L.; McAllister, K. A.; Clarke, A. J. *Protein Eng.* **2000**, *13*, 593-601.
247. Olsen, O.; Thomsen, K. K.; Weber, J.; Duus, J. O.; Svendsen, I.; Wegener, C.; von Wettstein, D. *BioTechnology* **1996**, *14*, 71-76.
248. Cherry, J. R.; Lamsa, M. H.; Schneider, P.; Vind, J.; Svendsen, A.; Jones, A.; Pedersen, A. H. *Nat. Biotechnol.* **1999**, *17*, 379-384.
249. Kuchner, O.; Arnold, F. H. *Trends Biotechnol.* **1997**, *15*, 523-530.
250. Chen, Y.-L.; Tang, T.-Y.; Cheng, K.-J. *Can. J. Microbiol.* **2001**, *47*, 1088-1094.

251. DeGrado, W. F. *Science* **1997**, *278*, 80-81.
252. Dahiyat, B. I.; Mayo, S. L. *Science* **1997**, *278*, 82-87.
253. Lazar, G. A.; Handel, T. M. *Cur. Opin. Chem. Biol.* **1998**, *2*, 675-679.
254. Gordon, D. B.; Marshall, S. A.; Mayo, S. L. *Curr. Opin. Struct. Biol.* **1999**, *9*, 509-513.
255. Malakauskas, S. M.; Mayo, S. L. *Nat. Struct. Biol.* **1998**, *5*, 470-475.
256. Turner, N. A.; Vulfson, E. N. *Enzyme Microb. Technol.* **2000**, *27*, 108-113.
257. Li, M. *Nat. Biotechnol.* **2000**, *18*, 1251-1256.

Wood Deterioration Processes

Termites and Marine Borers

Chapter 13

Termite Physiology in Relation to Wood Degradation and Termite Control

Thomas G. Shelton and J. Kenneth Grace

Department of Plant and Environmental Protection Sciences, University of Hawaii at Manoa, 3050 Maile Way, Gilmore 310, Honolulu, HI 96822

The importance of termites (order Isoptera) in the degradation of wood (cellulose, hemicellulose, and lignin collectively) is discussed, and the relative contributions of termite enzymes and intestinal microfauna (protozoa and bacteria) are presented. We also provide an overview of the areas of cellulose degradation, and physiological (enzymatic and pheromonal) means of termite control. Discussion includes the currently known hormones and pheromones with application in control measures, and some reasons for their current use (or lack of use) in termite control. Termites are social insects, and hormonal/pheromonal control measures often do not have the same results as are expected with solitary pest species. Finally, a short discussion of the current trends in research on feeding and foraging behavior of subterranean termites is presented.

The order Isoptera and the importance of cellulose

The Isoptera, or termites, is an order of relatively primitive insects and is often lumped into a greater taxonomic group, the Dictyoptera, that includes the cockroaches (Blattodea) and the Mantises (Mantodea). The Dictyoptera itself is part of the greater collection of orthopteroid insects, such as grasshoppers, crickets and katydids, all of which have biting and chewing mouthparts (among other shared characteristics). The relationships between and among the various groups within the Dictyoptera and its place among the other orthopteroid taxa are handled well by Boudreaux (*1*). The phylogeny of the order Isoptera is presented in the works of Krishna (*2*) and Thorne and Carpenter (*3*). These works are recommended to readers interested in the relationships between these taxa.

The entire order Isoptera is eusocial (*4*). Thus, termites are not independent or solitary insects, as are other wood destroying organisms such as buprestid beetles, bark beetles, or siricid wasps. Due to their social structure, different termites are assigned different tasks within their colonies (division of labor). These divisions of labor divide termites into physically modified specialists, or castes. Labor is divided in a number of ways; for example, reproduction (queens, kings, supplementary reproductives, and winged alates), defense (soldiers in most species), and other activities such as tunnel construction, egg/brood care, limited defense, and (most important to humans) feeding. This final group of responsibilities belongs to the most versatile of the termite castes, the workers. They also represent the largest group of termites within a colony, as most individuals are workers, with much fewer soldiers, and even fewer still reproductive caste members (*5*). In the lower termites, such as the subterranean termites that are responsible for much damage to buildings, the workers are sometimes referred to as "pseudergates", or false workers, to indicate their prolonged ability to molt into other caste forms if required by the colony. It is generally thought that colonies are founded by a single pair of reproductives, but more recent genetic evidence suggests that this is not necessarily the case in some species (*6*). Kofoid (*7*) provides a still-relevant overview of basic termite biology, colony initiation and structure.

Most termite species are xylophagous, feeding on the lignocellulosic xylem of woody plants. For many years, termites were thought to be unable to produce any of their own cellulases for feeding. For this reason, it was argued that termites would always require the symbiotic association of microorganisms (protozoa, bacteria, or fungi) to produce these enzymes. Otherwise they could not obtain any nutrients from their food. Although this has now been demonstrated to be not entirely the case, it is still clear that a strong symbiotic relationship exists between termites and specific microorganisms. A requirement of symbionts for feeding is thought to have laid the groundwork for the eventual development of eusociality in the termites. More primitive ancestors of the termites, cockroaches of the family Cryptocercidae, share some of the same species of microfauna with the remaining member of the most primitive extant termite family Mastotermitidae (*8*). This information, along with comparative data from mitochondrial cytochrome oxidase subunit II, endo-β-1,4-glucanase, and ribosomal RNA, has led Lo *et al.* (9) to connect these two groups phylogenetically, a relationship also supported by Thorne and Carpenter (*3*), and McKittrick (*8*). Since microfauna are present in the hindguts of termites, they must be replenished at each molt (hindgut and foregut integuments are shed with the exuviae). This can be done by proctodeal feeding (direct anal feeding from one individual to the next) or through trophallaxis (regurgitated food passage between individuals). It is expected that proctodeal feeding is more important for transfer of microfauna, since the feces will have already passed through the donor's hindgut when they are expelled for the recipient. Note that although the hindgut is ectodermal in origin, absorption of nutrients does occur across the hindgut integument in termites (*10*).

Much of the interest in the Isoptera in the United States has been based on their economic importance. However, globally, termites are known for ecological contributions in breaking down cellulosic litter, and improvement of soils through aeration and nutrient cycling (*11-13*). While there are species in North America representing four of the five to seven termite families, only two families are considered economically important in terms of wood destruction. These are the drywood termites (family Kalotermitidae) and the subterranean termites (family Rhinotermitidae). The activities of termites are estimated to cost homeowners in the USA over $1 billion annually (*14*). These costs are due mainly to the activities of subterranean termites, especially those belonging to the native North American genus *Reticulitermes* Holmgren, and the introduced species *Coptotermes formosanus* Shiraki. Due to their destructive potential, these species have been the subjects of much research on termite control technology (*15-18*). The following sections provide an overview of the current state of knowledge in using pheromones, physiological modifications, foraging behavior and nutritional ecology in the prevention of termite-based cellulose degradation. In discussing various species, it will soon become evident that the major effort rests on the Rhinotermitidae, due to its economic importance and subsequent representation in the literature.

The overall biology of the Isoptera is well beyond the scope of this article, but the reader is referred to a second chapter in this book (*19*), as well as to Kofoid (*7*), Krishna and Weesner (*20*), and Grassé (*21*) for further information. For information on the evolution of eusociality in the Isoptera, the reader should consult Bartz (*22*), and Myles and Nutting (*23*), contrasting their work with Hamilton (*24, 25*).

Pheromones

While hormones and pheromones are both chemical messengers in biological systems, designed to produce specific cell responses, they differ mainly in their target organisms. Chemicals that cause altered cell chemistry in the same individual that produced the chemicals are called hormones. Chemicals produced by an individual that alter the cell chemistry and behavior of another individual are pheromones. These definitions are nothing new, and certainly further subdivisions of each category exist. In our discussion, however, we will treat them separately, with hormones discussed in the Physiology section below.

Another reminder of the fact that termites differ from most other wood destroying insects in that they are eusocial is the type of pheromones that they produce. The following paragraphs illustrate the importance of trail-following pheromones in termites, due to their potential value in control efforts. For detailed descriptions of the glands involved in pheromone production by termites see Grassé (*21*).

Trail-following pheromones are perhaps the best studied pheromones aiding in the foraging and wood destroying behavior of termites. These pheromones are deposited from the sternal gland on the ventral abdomen (*26*). Many subterranean termites appear to use the same compound, cis-cis-trans-3,6,8-dodecatrien-1-ol, to mark trails to indicate food availability. Termites can be induced to follow artificial trails of this compound, either as extracts from the sternal gland (*27*), or extracts from decay fungi (*28*), and possible pheromone-based applications to termite control have been discussed since the 1960s. However, Grace (*29*) found that this compound may present a problem for long-term use in misguiding termites, since they acclimate to its presence over time. Other researchers (*26, 30, 31*) have presented evidence for multiple components in the trail-following pheromone blend, but no minor components have as yet been isolated or chemically identified. If identified, such minor components could encourage further development of control methods employing synthetic trails or of baits with more attractive matrices.

In addition to trail-following pheromones, other pheromones have promise for manipulation in termite control. As discussed below, soldier and nymph inhibition pheromones produced and distributed through the colony to maintain caste proportions (*32-34*) may have potential uses in combination with juvenile hormone applications to termite colonies. Inhibition of these pheromones, thought to be produced by termites, could derail the caste proportion systems of termite colonies. However, such pheromones must first be isolated and identified.

Physiology

Certain agricultural insect pests can be controlled using mimics of a certain insect hormone called juvenile hormone. Juvenile hormone is essentially an anti-adult hormone (*35*). By using analogues (such as methoprene and hydroprene) of this compound, advancement to the adult stage can be suppressed and the juveniles induced to form supernumerary larval forms. While this may not directly kill the larvae, it affects the population as a whole by preventing reproduction (only adult insects have functional reproductive organs).

As social insects, termites use a slightly modified hormone chemistry. The majority of termites in a colony are immature forms rather than true adults, and juvenile hormone is used in a different manner from solitary species. With termites, caste may depend on the juvenile hormone concentration in the termite's environment. Lenz (*32*) presents evidence that suggests the role of nutrition in the development of different castes among termites. His studies used the juvenile hormone analogues (JHAs) altozar (ethyl-3,7,11-trimethyldodeca-2,4-dienoate) and altosid (isopropyl-11-methoxy-3,7,11-trimethyldodeca-2,4-dienoate), and various levels of nutrients from plain filter paper to wood decayed by brown rot fungi. Essentially, his work showed a positive influence of nutrition on subsequent molting of lower termite workers that had been exposed to JHA compounds into nymphs and reproductives. In contrast, those workers on relatively low nutritional diets tended to molt into soldiers when exposed to JHA compounds. Different termite species were able to control extraneous soldier development to different degrees; those with higher natural soldier proportions were better able to deter the formation of new soldiers, while those with naturally low soldier proportions became inundated with soldiers. Evidence from *Nasutitermes nigriceps* (Haldeman) suggested that higher termites may also be affected by JHAs in this manner (*32*). Lenz (*32*) suggests that the soldiers are essentially juvenile hormone sinks, perhaps by producing a compound that inhibits its production from the corpora allata of the workers. Henderson (*36*) further expanded upon the concept of termite soldiers as juvenile hormone sponges. Work by Lüscher (*33*) has indicated that higher termites have less plasticity in their developmental 'programming' than lower termites, such that adjustments to caste can only be made at relatively few points along their developmental pathways. There are five known juvenile hormones (0-III, plus 4-methyl juvenile hormone I; *35*), and *R. flavipes* has been shown to use the juvenile hormone III in induction of soldier development (*34*). Prestwich (*34*) was also able to identify the juvenile hormone binding proteins within *R. flavipes*, and suggested that the mechanism for inhibiting soldier development by lower termite soldiers is the release of a pheromone that inhibits these proteins.

Okot-Kober *et al.* (*37*) examined the effects of *n*-sulfenylcarbamate pro-insecticides (precursors of fenoxycarb analogues) as JHAs in several lower termite species. While the compounds did induce increased soldier proportions, this increase could be inhibited by the presence of natural soldiers in the laboratory arenas. Such natural adjustment of soldier proportions argues against the successful use of JHAs for purposes of termite control.

Of the control methods currently available for subterranean termites, the most recent has been the development of various baits (*16-18, 38*). Bait development was plagued for quite a few years by the need for a nonrepellent and slow-acting toxicant that would induce delayed mortality at a distance from the bait itself, thereby removing concerns about learned avoidance (*38*). The first commercially available bait for subterranean termite control was the Sentricon Colony Elimination System™ (Dow AgroSciences), which uses a unique physiological approach to address this issue. The active ingredient for this bait is hexaflumuron, a benzoylphenyl urea compound. Lenz *et al.* (*39*) showed that both hexaflumuron and triflumuron (another benzoylphenyl urea) affected subterranean termites (*C. acinaciformis* [Froggatt]) more quickly than higher termites (*Nasutitermes exitiosus* [Hill]). Their study also demonstrated concentration-dependent differences among species in susceptibility, with the rhinotermitids (subterranean termites) dying more rapidly and from lower concentrations of both materials.

Benzoylphenyl ureas, such as diflubenzuron and hexaflumuron, interrupt the formation of an intact integument during molting, leading to death of the insect during ecdysis. However, the exact mechanism of this inhibition is not yet understood. It is known that these compounds affect chitin synthesis such that chitin is not properly laid down in the new procuticle of the insect during apolysis. Many of the theoretical considerations of how these compounds function relates to chitin synthetase, the enzyme responsible for the conversion of UDP-*n*-acetylglucosamine residues into chains of *n*-acetylglucosamine residues or chitin. Theories for the actual mode of action vary from the potential for a metabolite of benzoylphenyl ureas inhibiting chitin synthetase to the inhibition of UDP-*n*-acetylglucosamine passage across cell membranes. While the exact nature of the mode of action for hexaflumuron may still be unknown, the inhibition of chitin synthetase is thought to be a very good candidate. A full discussion of the theories of benzoylphenyl urea mode of action is provided by Retnakaran (*40*).

While hexaflumuron has provided a very useful bait system (*17, 18*), there will always be room for improvement (*39*). In the next section, we discuss some new information on the feeding and foraging behaviors of termites, and how this may improve bait applications.

Feeding

Since we are ultimately concerned with the destruction of cellulose, we must consider all the active participants in its breakdown. In the Rhinotermitidae (subterranean termites), these participants are the termites, through chewing and mechanical (and some enzymatic) breakdown of wood (cellulose), and the symbiotic protozoa and bacteria found in the termite hindgut. The protozoa participate by the further enzymatic degradation of cellulose in the hindgut, producing acetate which is absorbed by the termite as a nutrient (*41-43*). Since wood contains a low proportion of nitrogen, the termites must also find a way of accumulating nitrogen (*11*). This is accomplished in part by symbiotic bacteria present in the termite alimentary canal (*44-46*). While the bacteria are considered accomplices in the destruction of cellulose, in the Rhinotermitidae they are secondary to the more important cellulase-producing protozoa of the hindgut.

Termite hindgut microfauna species and number vary with termite species. Perhaps the most recent and elaborate work on the various species of microfauna in a single species of termite has been performed with *C. formosanus*, the Formosan subterranean termite (*47-52*). This species is a member of the Rhinotermitidae, is very aggressive in its wood foraging (*53*), and even uses less energy than some other subterranean termites (*R. flavipes*, *54*). Most importantly, as a model for protozoan cellulose degradation, it is quite simple, having only three species in its hindgut (*55*). These species, from largest to smallest, are *Pseudotrichonympha grassii* Koidzumi, *Holomastigotoides hartmanni* Koidzumi, and *Spirotrichonympha leidyi* Koidzumi. The largest protozoan, *P. grassii*, resides in the anterior end of the hindgut, while the smallest, *S. leidyi*, resides mostly in the posterior of the hindgut, with *H. hartmanni* found throughout the hindgut (*48*). Each of the three protozoans is thought to metabolize only certain types of cellulose within the *C. formosanus* hindgut (*50-52*). Defaunation followed by an examination of termite survivorship on various types of cellulose indicated that *P. grassii* could utilize only cellulose with high degrees of polymerization, similar to natural wood (*50*).

At one time, microfauna were thought to be responsible for all cellulase activity in termites, but this is not the case. Termites from both the Termitidae (*Trinervitermes trinervoides* [Sjöstedt], *56*; *Macrotermes natalensis* [Haviland], *57*; *Nasutitermes takasagoensis* [Shiraki], *58*) and the Rhinotermitidae (*C. formosanus*, *49*; *R. speratus* [Kolbe], *59*) have been shown to produce cellulases of their own. In both families, these endogenous cellulases are endo-β-1,4-glucanases and β-D-glucanases (*49, 58*), and their activities extend to areas beyond the hindgut of the termite. It is thought that the termite-produced cellulases are regurgitated from the midgut and are active on wood (or cellulose) fragments throughout the alimentary canal (*49, 51*). It should be pointed out that although they also produce cellulases from their midgut epithelial cells, termites of the subfamily Macrotermitinae do not rely upon protozoa for the remainder of their cellulases (*57*). Instead they cultivate a symbiotic fungus in their colonies, which contains cellulases that are transferred to the termite after feeding upon the fungus.

Yoshimura (*48, 52*) developed an overview of cellulose degradation in *C. formosanus*. Following mechanical degradation of the wood with the mouthparts into smaller pieces, the endo-β-1,4-glucanases and β-D-glucanases produced in the termite midgut break down the cellulose matrix into smaller, although still indigestible, sections. From there the fragments travel to the hindgut and are encountered first by *P. grassii*, which produces a cellulase that operates on highly polymerized cellulose. The less polymerized pieces will be encountered by *H. hartmanni* throughout the hindgut, and are broken down, thus completing the breakdown of cellulose itself into acetate. In the hindgut, *S. leidyi* is thought not to degrade cellulose directly, but instead acts as a host for methanogenic and nitrogen fixing bacteria (*48*).

Current interest in termite breakdown of cellulose has also encouraged new research on the means by which termites locate their food in the first place. Subterranean termite foraging behavior and food-finding capabilities are currently being studied by a number of researchers (*60-62*), using two-dimensional foraging arenas (similar to ant farms) to observe tunneling. Both *R. flavipes* and *R. virginicus* (Banks) will use pre-existing tunnels in their movement through two-dimensional space; however, they show less preference for moving along guide wires placed in the sand (*60*). Re-use of existing foraging tunnels has also been observed by Grace and Su (*18*) in field studies of termite reinvasion of baited sites. *R. flavipes* and *C. formosanus* were found to show an increase in the number of tunnels produced prior to (compared with after) reaching a food source, suggesting a greater searching investment by the termites during the initial phases of food acquisition (*61*). Puche and Su (*61*) argue that the tunnel formation by these species can be adequately modeled using fractal analysis. Robson *et al.* (*63*) discovered that subterranean termites have patterns to their search behavior that, in the absence of other cues, evenly divide the foraging area so as to maximize the likelihood of food discovery. Campora and Grace (*62*) noted that a pattern of successive radial foraging tunnels (similar in appearance to spokes of a wheel) was created as termites encountered either wood in the soil or substrate anomalies such as a hollow space. Overall tunneling efforts shifted towards those parts of the arenas where food (or substrate anomalies) were first encountered. In nature, it may be rare for subterranean termites to be food-limited (*11*), since wood is a resource that few other organisms utilize. Thus, development of radial search patterns triggered by multiple anomalies in the soil rather than solely by discovery of wood (i.e., less selective patterns) would seem to make sense so long as such anomalies are in fact associated with food a majority of the time. Such improved understandings of termite foraging strategies hold promise for improving termite control, particularily with respect to bait placement and monitoring efficiency.

Conclusions

Understanding the feeding behavior, protozoa, and hormone/pheromone physiology of economically important termites will expedite development and improvement of new technologies for controlling pest species. Behavioral examinations will lead to improved bait implementation, particularly better bait placement and improved monitoring methods. Knowledge of the relationships between symbiotic microfauna and their termite hosts could well lead to new control methods targeting metabolic pathways in the degradation of cellulose within the termite gut. The current successful use of chitin synthesis inhibitors in subterranean termite baits provides a good example of a physiologically-based control strategy. Certainly, additional strategies will be spawned by current research efforts.

References

1. Boudreaux, H.B. *Arthropod phylogeny with special reference to insects*; John Wiley & Sons, Inc.: New York, **1979**.
2. Krishna, K. In *Biology of termites*; Krishna, K., F. M. Weesner, Eds. Academic Press: New York, **1970**, Vol. 1, pp 127-153.
3. Thorne, B.L.; Carpenter, J.M. *Sys. Entomol.* **1992**, *17*, 253-268.
4. Wilson, E.O. *The insect societies*; Belknap Press: Cambridge, MA, **1971**.
5. Haverty, M.I. *Sociobiology* **1977**, *2*, 199-216.
6. Hussenender, C. Ph.D. dissertation, University of Bayreuth, Bayreuth, GERMANY, **1998**.
7. Kofoid, C.A. *Termites and termite control;* University of California Press: Berkeley, CA, **1934**.
8. McKittrick, F.A. *Ann. Entomol. Soc. Am.* **1965**, *58*, 18-22.
9. Lo, N.; Tokuda, G.; Watanabe, H.; Rose, H.; Slaytor, M.; Maekawa, K.; Bandi, C.; Noda, H. *Curr. Biol.* **2000**, *10*, 801-804.
10. Chapman, R.F. *The insects: structure and function*, 4th ed., Cambridge University Press: Cambridge U.K., **1998**.
11. Waller, D.A.; LaFage, J.P. In *Nutritional ecology of insects, mites, spiders, and related invertebrates;* Slansky, F.; J. G. Rodriguez, Eds.; Wiley and Sons Publishing: New York, **1987**, pp 487-532.

12. Kooyman, C.; Onck, R.F.M. *The interactions between termite activity, agricultural practices and soil characteristics in Kisii district, Kenya;* Agricultural University Wageningen Papers, Drukkerij Veenman: Wageningen, GERMANY, Vol 87-3, **1987**.
13. Curtis, A.D. Ph.D. dissertation, Old Dominion University, Norfolk, VA, **1996**.
14. Su, N.-Y.; Scheffrahn, R.H. *Sociobiology* **1990**, *17*, 77-94.
15. Grace, J.K.; Yates, J.R.; Tamashiro, M.; Yamamoto, R.T. *J. Econ. Entomol.* **1993**, *86*, 761-766.
16. Su, N.-Y. *J. Econ. Entomol.* **1994**, *87*, 389-397.
17. Grace, J.K.; Tome, C.H.M.; Shelton, T.G.; Oshiro, R.J.; Yates, J.R. *Sociobiology* **1996**, *28*, 511-520.
18. Grace, J.K.; Su, N.-Y. *Sociobiology* **2000**, *37*, 310-310.
19. Husseneder, C.; Vargo, E.L.; Grace, J.K. *this volume* **2003**.
20. Krishna, K.; Weesner, F.M. *Biology of termites*; Academic Press: New York, Vol. 1, 2, **1970**.
21. Grassé, P.P. *Termitologia*; Masson Publishing: Paris, Vol. 1, **1984**.
22. Bartz, S.H. *PNAS* **1979**, *76*, 5764-5768.
23. Myles, T.G.; Nutting, W.L. *Quart. Rev. Biol.* **1988**, *63*, 1-23.
24. Hamilton, W.D. *J. Theor. Biol.* **1964**, *7*, 1-16.
25. Hamilton, W.D. *J. Theor. Biol.* **1964**, *7*, 17-52.
26. Kaib, M.; Bruinsma, O.; Leuthold, R.H. *J. Chem. Ecol.* **1982**, *8*, 1193-1205.
27. Grace, J.K.; Wood, D.L.; Frankie, G.W. *J. Chem. Ecol.* **1988**, *14*, 653-667.
28. Grace, J.K.; Wilcox, W.W. *Pan-Pac. Entomol.* **1988**, *64*, 243-249.
29. Grace, J.K *Sociobiology* **1989**, *16*, 175-182.
30. Runcie, C.D. *J. Chem. Ecol.* **1987**, *13*, 1967-1978.
31. Cornelius, M.C.; Bland, J.M. *Environ. Entomol.* **2001**, *30*, 457-465.
32. Lenz, M. In: *Phase and caste determination in insects: endocrine aspects*, Lüscher, M., Ed. Pergamon Press: New York, **1976**, pp. 73-90.
33. Lüscher, M. In: *Phase and caste determination in insects: endocrine aspects*, Lüscher, M., Ed. Pergamon Press: New York, **1976**, pp. 91-104.
34. Prestwich, G.D. *Sociobiology* **1988**, *14*, 175-191.
35. Nijhout, H.F. *Insect hormones*; Princeton University Press: Princeton, NJ, **1994**.
36. Henderson, G. In: *Pheromone communication in social insects*, VanderMeer, R.K.; Breed, M.D.; Winston, M.L.; Espelie, K.E., Eds. Westview Press: Boulder, CO, **1998**, pp. 314-330.
37. Okot-Kober, B.M.; Ujváry, I.; Mollaaghababa, R.; Szurdoki, F.; Matolcsy, G.; Prestwich, G.D. *Sociobiology* **1991**, *19*, 77-90.

38. French, J.R.J. *Sociobiology* **1994**, *24*, 77-91.
39. Lenz, M.; Gleeson, P.V.; Miller, L.R.; Abbey, H.M. *IRG/WP* **1996**, *96-10143*, 1-11.
40. Retnakaran, A.; In: *Chitin in nature and technology*, Muzzarelli, R.; Jeuniaux, C.; Gooday, G.W., Eds., Plenum Press, New York, **1986**, pp. 147-164.
41. Hungate, R.E. *Ann. Entomol. Soc. Am.* **1943**, *36*, 730-739.
42. Brauman, A.; Kane, M.D.; Labat, M.; Breznak, J.A. *Science* **1992**, *257*, 1384-1387.
43. Breznak, J.A.; Brune, A. *Annu. Rev. Entomol.* **1994**, *39*, 453-487.
44. Potrikus, C.J.; Breznak, J.A. *Insect Biochem.* **1980**, *10*, 19-27.
45. Potrikus, C.J.; Breznak, J.A. *PNAS* **1981**, *78*, 4601-4605.
46. Noda, S.; Ohkuma, M.; Usami, R.; Horikoshi, K.; Kudo, T. *Appl. Environ. Microbiol.* **1999**, *65*, 4935-4942.
47. Tsunoda, K.; Ohmura, W.; Yoshimura, T.; *Jpn. J. Environ. Entomol. Zool.* **1993**, *5*, 166-174.
48. Yoshimura, T. Ph.D. dissertation, Kyoto University, Kyoto, JAPAN, **1995**.
49. Yoshimura, T.; Watanabe, T.; Tsunoda, K.; Takahashi, M. *Mat. und Org.* **1992**, *27*, 273-284.
50. Yoshimura, T.; Azuma, J.; Tsunoda, K.; Takahashi, M. *Mokuzai Gakkaishi* **1993**, *39*, 221-226.
51. Yoshimura, T.; Azuma, J.; Tsunoda, K.; Takahashi, M. *Mokuzai Gakkaishi* **1993**, *39*, 227-230.
52. Yoshimura, T.; Azuma, J.; Tsunoda, K.; Takahashi, M. *Mokuzai Gakkaishi* **1993**, *39*, 1322-1326.
53. Delaplane, K.S.; LaFage, J.P. *Sociobiology* **1989**, *16*, 183-189.
54. Shelton, T.G.; Appel, A.G. *J. Ins. Phys.* **2001**, *47*, 213-224.
55. Lai, P.-Y.; Tamashiro, M.; Fujii, J.K. *Proc. Hawaiian. Ent. Soc.* **1983**, *24*, 271-276.
56. Potts, R.C.; Hewitt, P.H. *Comp. Biochem. Phys.* **1974**, *47B*, 317-326.
57. Martin, M.M.; Martin, J.S. *Science* **1978**, *199*, 1453-1455.
58. Yoshimura, T.; Tsunoda, K.; Takahashi, M. *Mokuzai Gakkaishi* **1996**, *42*, 1250-1257.
59. Tokuda, G.; Lo, N.; Watanabe, H.; Slaytor, M.; Matsumoto, T.; Noda, H. *Biochem. Biophys. Acta* **1999**, *1447*, 146-159.
60. Pitts-Singer, T.L.; Forschler, B.T. *J. Insect Behav.* **2000**, *13*, 273-290.
61. Puche, H.; Su, N.-Y. *Environ. Entomol.* **2001**, *30*, 545-549.
62. Campora, C.E.; Grace, J.K. *J. Econ. Entomol.* **2001**, *94*, 1193-1199.
63. Robson, S.K.; Lesniak, M.G.; Kothandapani, R.V.; Traniello, J.F.A.; Thorne, B.L.; Fourcassie, V. *Naturwissenschaften* **1995**, *82*, 526-528.

Chapter 14

The Biology of Marine Wood Boring Bivalves and Their Bacterial Endosymbionts

Daniel L. Distel

Department of Biochemistry, Microbiology, and Molecular Biology and The School of Marine Sciences, University of Maine, Orono, ME 04469

Marine bivalves of the superfamily Pholadacea include species that consume wood as a primary nutrient source (teredinid shipworms and deep-sea xylophagainid clams) with the aid of intracellular bacterial endosymbionts, a fact that has been recognized only in the last quarter century. In addition this superfamily includes a number of species that bore in wood and other substrates for shelter only. Though little is known about the mechanism of symbiotic xylotrophy (wood-eating) in bivalves, morphological and molecular evidence demonstrate that these marine systems are quite different from symbiotic xylotrophy and cellulotrophy observed in terrestrial animals (termites, ruminants, etc.). This review focuses on our current state of knowledge regarding the systematics and biology of wood boring bivalves and their endosymbiotic bacteria, the role of these associations in wood degradation, and current methods for the control of damage caused by wood boring bivalves.

Introduction

Wood boring bivalves have presented a challenge to mankind since man first plied the sea in wooden vessels, fished its waters with wooden gear, and built wooden structures on its shores (*1*). The choice of wood for marine construction is a natural one. Its desirable features include high strength to weight ratio, resilience when subjected to sudden loads, durability, resistance to abrasion, and of course, buoyancy. Wood is also widely available, inexpensive, renewable and easy to fabricate into structures (*2*). Despite these favorable characteristics, a principal deterrent to the use of wood in marine construction today, and throughout history, has been its susceptibility to destructive attack by wood boring bivalves.

The Biology of Wood Boring Bivalves

Systematics

Wood boring bivalves are members of the superfamily Pholadacea. This superfamily contains two related families, Teredinidae (shipworms) and Pholadidae (piddocks), which together include 20 genera and approximately 175 wood boring species (*3*). Teredinids,with the exception of the seagrass borer *Zachsia* and the mud boring *Kuphus,* are obligate woodborers that utilize wood both as a means of shelter and as a source of nutrient (*4*). Pholadidae is less highly specialized, containing species that burrow in a variety of hard substrates including wood, mud, stone, shells, clay, sand and peat (*5*). The wood boring pholadids are placed in two subfamilies, Martesiianae and Xylophagainae. Members of Martesiinae lack morphological features associated with wood feeding (*5*). They apparently burrow for protection but do not ingest wood. Members of Xylophagainae, on the other hand, ingest wood particles and wood likely comprises a significant portion of their diet (*6, 7*).

Distribution

Wood boring bivalves occur in all oceans (*3*). The teredinids are widely distributed in shallow waters, from tropical to temperate seas, where they are the dominant wood consumers. Although teredinids have been observed to occur at considerable depths, it is not clear if they invade wood or reproduce in deep waters. For example, living specimens of *Bankia carinata* have been reported in wood and *Pandanus* fruit dredged from depths as great as 7488 m (*5*). However, these materials may have been invaded before sinking from shallower water.

Teredinids have not been observed to invade test wood at depths greater than about 200m (*8*).

Xylophagainid bivalves occupy a similar trophic position in deep water. They are distributed throughout the world's oceans, primarily at depths from 150m to greater than 7000m. At higher latitudes some xylophagainid species may be found in shallow sublitoral waters, though (*3, 8*) none are found in the intertidal zone or in driftwood. These observations suggest that cold adaptation, intolerance to variation in temperature and/or salinity, and limited resistance to exposure and/or desiccation may limit their distribution.

Salinity, Temperature and Oxygen

Many teredinid species exhibit broad tolerance to fluctuations in salinity, temperature and oxygen concentration. For example, *Teredo navalis* reproduces over a salinity range from normal seawater (~ 35 parts per thousand) to as low as 9 parts per thousand, and survives for extended periods at 4 parts per thousand (*9*). Some Teredinid species may be adapted to brackish or even freshwater; e.g. *Nausitora hedleyi* which reproduces optimally in a salinity range from 11-14 parts per thousand (*10*), and *Psiloteredo healdi,* which lives and reproduces in freshwater lakes and streams in Central and South America. Though adults may be resistant, changes in salinity in rivers and bays may result in wide fluctuations in larval settlement by brackish water species (*10*). Similarly, many teredinids, including tropical species, survive wide temperature variation as adults, though spawning and survival of young is often limited to relatively narrow species-specific temperature optima (*5*).

Many shipworm species also display considerable tolerance to anoxia. For example, *T. navalis* was reported to survive and maintain normal respiratory currents for 23 days in oxygen free seawater (*11*). Many species can survive for days or weeks in wood that has been removed from seawater; a condition that does not allow normal respiration. This ability, which aids shipworms in surviving tidal exposure and other unfavorable periodic changes in their environment, further demonstrates their considerable tolerance to extended periods of anoxia.

Shipworms

By far the best known, best understood, and most destructive of the wood boring bivalves are the teredinids, commonly referred to as shipworms, pileworms, augerworms or shellworms (*12*). Though these common names are taxonomically misleading, they accurately reflect the peculiar appearance, characteristics, and life habits of these clams. Shipworms are highly modified in body plan compared to more typical bivalves (*5*) (Figure 1). Their elongated worm-like shape and habit of boring deep, shell-lined burrows in wooden ships

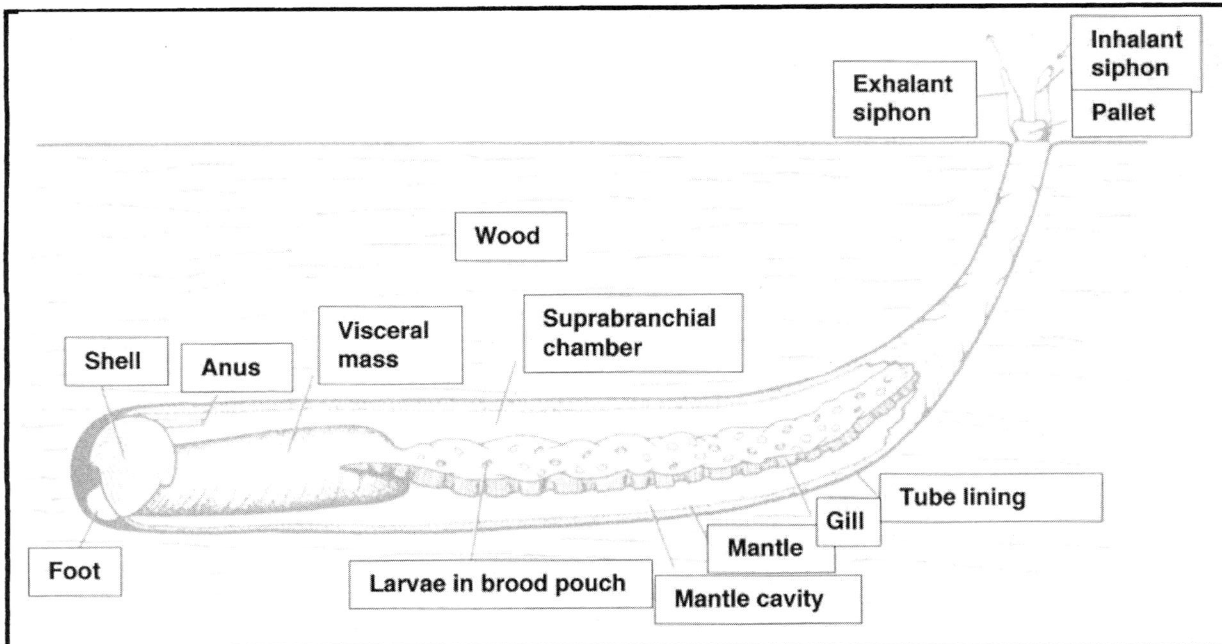

Figure 1. Diagram of a shipworm in its burrow. (adapted from Barnes, R.D., 1974, *Invertebrate Zoology*, W.B. Saunders Co., PA).

and piers explains their unusual nicknames. Appearance notwithstanding, shipworms are true bivalves. They are quite closely related to the soft-shell or "steamer clam", *Mya arenarea*, a common staple of New England cuisine (*5*). However, unlike their more highly-prized relatives, shipworms have long maintained an adversarial relationship with mankind.

History of Shipworm Damage

Historical accounts detailing shipworm damage to manmade wooden structures date at least to the third century BC in naval records and in the classical writings of Theophrastus (371-287 BC), Pliny and Ovid (*1, 13*). Among the earliest scholarly writings on shipworms are those of the Dutch naturalists. This is not surprising given that much of the Netherlands was once protected from the sea by wooden dykes. It was the Dutch naturalist Godfrey Sellius who in 1733 first demonstrated that shipworms are indeed bivalve mollusks (*1*). The importance of shipworms to mankind is evident in the accounts of many famous European voyages of exploration from the Viking voyages to those of Cook, Drake, and Columbus (*1*). In fact the famously disastrous fourth voyage of Columbus saw all four of his ships destroyed by shipworms (*14*), leaving Columbus and his crew marooned on the then remote island of Jamaica.

Costs of Shipworm Damage

Though the use of wooden vessels has declined in recent history, shipworms remain an important cause of economic loss (*15*). Wood is still widely used in marine construction, often without protective treatment in areas where shipworm damage is not known to be severe. However, for reasons that are not fully understood, the range and distribution of shipworm species may fluctuate widely, leading to dramatic and sometimes devastating episodes of destruction. For example, the first recorded occurrence of the shipworm *T. navalis* in San Francisco bay was in 1913. Populations of this species rapidly increased over the next decade, resulting in an estimated $3.1 billion (current US dollars) in damage to ships and waterfront structures in the three-year period between 1919 and 1921 (*16*). Similarly, areas of the northeastern coast of the United States and the Baltic coast of Germany have recently been subjected to rapid, costly and unexplained increases in damage due to shipworm activity. Though these increases in shipworm activity are anecdotally attributed to improvements in water quality in these regions (*17*), no scientific evidence exists to support this conclusion.

Accurate estimates of the current annual worldwide dollar cost of shipworm damage are lacking (*15*). However, the combined damage to marine structures, including ships, boats, docks, piers, groins, lock gates, fishing equipment and aquaculture enclosures, due to wood boring bivalves and crustaceans, was estimated at $1 billion in 1986 (*2*). In 1993 it was estimated that more than $100 million dollars had been spent to remediate shipworm damage by New York City alone (*17*).

Wood Boring Habits of Shipworms

The damage caused by shipworms is primarily due to their destructive boring habits. As shipworms grow, they excavate deep burrows that may extend from a few millimeters to more than two meters into wooden substrates (*12*). Infestation densities can typically exceed 5000 individuals per square foot with the closely spaced burrows following tortuous and intertwined, but non-intersecting paths within the wood (*18*) (Figure 2). Though these burrows often exceed one centimeter in diameter, each has only a single tiny opening to the external environment, usually less than 1-2 mm across (*19*). Such burrowing activity may eventually excavate most of the internal volume of a wooden timber while leaving the outer surface relatively unblemished. Therefore shipworm infestations often go undetected until revealed by the structural failure of the wood.

The mechanism of wood boring used by shipworms is primarily mechanical. Shipworms bore by abrasion of the substrate using their highly modified shells. While typical bivalves have large paired shells that encase and protect the entire animal, the shells of shipworms are very small, covering only the anterior tip of the shipworm's elongated body. Because the wood provides protection to the adult shipworm and the burrow lining prevents dessication, shells are not needed for these functions and instead have become adapted for the specialized function of grinding wood. Tiny tooth-like projections cover the surface of the shells, forming a rasp-like surface (Figure 3). The shipworm bores into wood, using the shells as a drill bit; rotating them against the wood with powerful actions of its foot and adductor muscles (*18,20*). This mechanism of boring is surprisingly efficient. For example, boring rates as high as 12cm per month have been recorded for *B. setacea* (*21*).

As the shipworm bores it lays down a thin calcareous lining on the burrow walls, forming a tube that opens to the external environment only at one end. The opposite end opens to the wood at the boring face of the burrow. The shipworm is permanently attached to its tube by a ligament near the burrow opening. Therefore the shipworm must remain in the same burrow throughout its adult life and must grow at a rate that matches the elongation of the burrow.

Shipworms can close their burrows using a pair of calcified plates, called pallets, that can be forced like a stopper into the opening of the tube. The pallets form a watertight seal that help the shipworm to survive exposure of the wood during tidal fluctuations or other unfavorable changes in their environment.

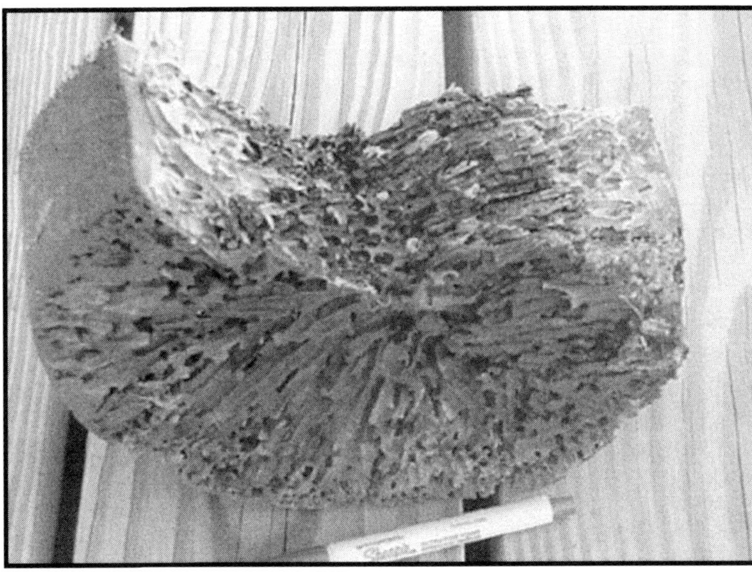

Figure 2. Damage caused by shipworm activity to an eight-inch diameter red oak pile in service for less than one year in Belfast, Maine, USA.

(Reproduced with permission from Invertebrate Zoology, Sixth Edition, *by Ruppert, ISBN 0–0302–6688–8. Copyright 1994 Thompson Learning.)*

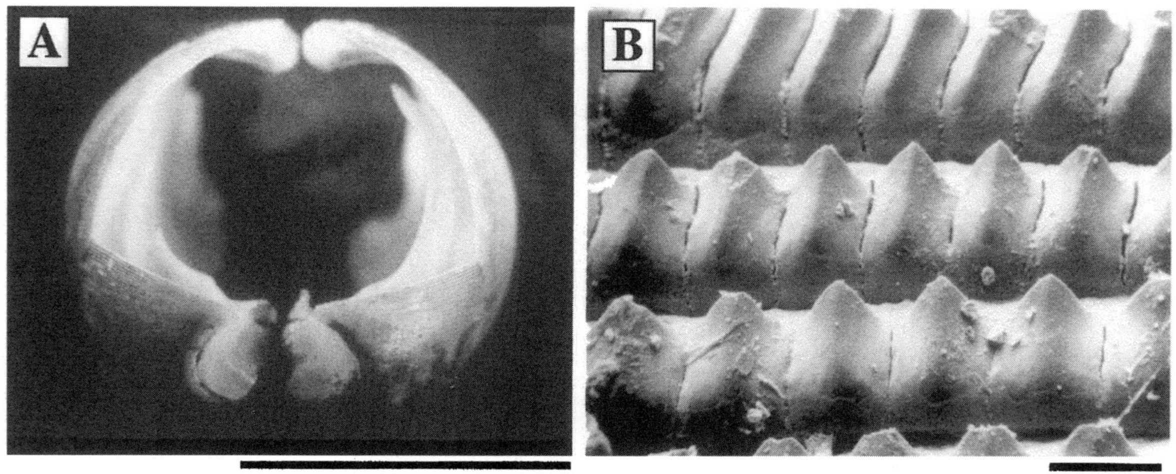

Figure 3. **(A)** Shipworm shell (scalebar = 0.5 cm). **(B)** Scanning electron micrograph showing shellteeth of *Xylophaga washingtona*. (scalebar = 5.0μm).

(Reproduced with permission from Invertebrate Zoology, Sixth Edition, *by Ruppert, ISBN 0–0302–6688–8. Copyright 1994 Thompson Learning.)*

Life History and Reproduction

Life histories and reproductive strategies vary widely among wood boring bivalves, even within individual families. For example, among members of the family Teredinidae, reproductive strategies range from ovipary with planktotrophic development in the subfamily Bankiinae to larvipary with live brooding in Teredininae. Within the Teredininae fertilization occurs internally and larvae are retained within brood sacs located on the ventral side of the gills. In some species, e.g., *Lyrodus pedicellatus*, larvae are retained until the pediveliger stage and are capable of settlement and metamorphosis within hours of release. Other species, e.g., *T. navalis*, release larvae as straight hinged veligers that must spend several weeks feeding in the plankton before they are ready for settlement and metamorphosis (*22*).

So far as is known, all teredinids and xylophagainids are hermaphrodites (*23*). While most appear to be protandrous, at least some may become simultaneous hermaphrodites at some point in development and are capable of self-fertilization (*24*). This may be an advantage to organisms that live and feed on ephemeral and unpredictable substrates like driftwood, as a single larva can produce a new breeding population when suitable wood is discovered.

All teredinids appear to be highly fecund, with planktotrophic species releasing as many as one million eggs per spawning event (*25*). Larvae are efficiently dispersed by currents and may travel thousands of kilometers, possibly across ocean basins, before metamorphosis (*26*). Teredinid larvae are common both in coastal waters and in the open sea. One study found mature teredinid larvae in 19% of 742 surface samples taken throughout the temperate and tropical North Atlantic Ocean, including waters of the North Atlantic gyre (*26*).

Feeding

As a shipworm bores in wood it ingests the excavated wood particles. These particles are passed first to the stomach and then to a finger-like sac called the appendix or caecum. The microvillar brush border that lines the caecum indicates that this is a primary site of wood digestion and nutrient absorption (*27*). In some species the anal canal is also enlarged and lined with brush border, suggesting that this may serve as a secondary site of wood digestion (28). Although all shipworms are probably capable of filter feeding, wood serves as a primary nutrient source in most species (3). In fact at least one species has been shown to be capable of normal growth and reproduction with wood as its sole nutrient source (4). To date no other marine organism has been shown to have this capability.

Wood as a Food

Relatively few animal species consume wood as a primary food source. These include certain termites and related insects in the terrestrial world, and teredinid and xylophagainid bivalves and possibly certain isopod crustaceans (*Limnoria*) in the marine environment. The reasons for this limited distribution are not fully understood, however, the inability of animals to synthesize cellulase (endo-β-1-4-glucanase, EC 3.2.1.4) has frequently been offered as an explanation (*29*). It has been argued that symbiotic microorganisms in the digestive system are the sole source of cellulolytic enzymes in cellulose-consuming higher animals (*30*). Recently, however, this view has been challenged by the discovery of a nuclear encoded cellulase (endo-β-1-4-glucanase, EC 3.2.1.4) gene in the termite *Reticulitermes speratus* (*31*). More recently, endogenous endoglucanase genes have been found in a small number of additional insect and arthropod species (*32*). These results indicate that these animals, and possibly others, may contribute cellulolytic enzymes to cellulose digestion in addition to (or possibly in lieu of) those provided by symbiotic microbes.

Another barrier to the use of wood as a food is its nutritional imbalance. Wood is rich in carbohydrates, but contains little protein, amino acid, or other sources of combined nitrogen (*33*). Certain prokaryotic microorganisms can fix atmospheric nitrogen; however, the enzymes required for this pathway are absent from the eukaryotic genome (*34*). Therefore, higher organisms that obtain a substantial part of their nutriment from wood must typically supplement their diet with alternative nitrogen sources. This may include nitrogen fixed by symbiotic microorganisms (*35*).

Cellulotrophic Symbioses in Terrestrial Animals

Although it now appears that certain animals have at least some native ability to degrade cellulose, conspicuous symbiotic microbial populations are frequently, if not universally, observed in cellulotrophic animals (*35-37*). In fact, functionally similar symbiotic systems have been described in many terrestrial cellulotrophs including insects (e.g. termites and roaches) (*35*), ruminants (e.g. cattle and sheep) (*36*), and other animals that utilize cellulose from woody or leafy plant materials (*37*). In these animals dense and highly complex populations of symbiotic protists and/or bacteria are observed in the digestive tract. Enzymes produced by these microbes have been shown to aid in the fermentation of cellulose in the anaerobic environment of the gut. The host animals subsequently utilize the waste products of this fermentation, mainly volatile fatty acids, as their primary nutritional carbon source. Some members of these symbiotic communities also fix nitrogen and provide a necessary supplement to the host's nitrogen deficient diet (*35*).

Shipworm Symbionts

Cellulotrophic Symbioses in Wood Boring Bivalves

Teredinid and xylophagainid bivalves also harbor microbial symbionts (*7, 38, 39*), but these symbioses are quite different from those observed in terrestrial cellulotrophs (*7, 40*). The digestive systems of these bivalves lack the dense and conspicuous microbial populations observed in the gut of terrestrial cellulotrophs [(*40*) and unpublished observations of the author]. Instead, dense populations of endosymbiotic bacteria are found in specialized cells (bacteriocytes) within the gills (Figure 4). These bacteriocytes constitute the tissue formerly but erroneously referred to as the Gland of Deshayes (*41*). Rather than glandular tissue, these cells can now be reinterpreted as bacteriocytes in the basal region of the interlammelar extensions in the gills (*22, 40*). These structures are homologous to those observed to contain chemoautotrophic and methanotrophic symbionts in lucinid, solemyid, vesicomyid, and mytilid bivalves (*42*).

Unlike terrestrial cellulotrophic symbioses, the mechanism of the shipworm symbiosis and the role of the shipworm symbionts in facilitating a wood-based diet is not well understood. Initial investigations have indicated that gill symbionts in the shipworm *B. setacea* synthesize essential amino acids lacking in the host's diet (*43*). However, the subsequent isolation of a dinitrogen fixing, cellulolytic bacterium from the gill tissues of 24 species of shipworm representing 9 of 14 genera of the bivalve family Teredinidae (*44, 45*) suggested additional functions for the shipworm symbionts. This rare combination of cellulolytic and nitrogen fixing capabilities suggests that this bacterium may function to aid the host in the digestion of wood and to supplement the host's protein-deficient diet (*44*). Indeed, nitrogen fixation has been observed experimentally in intact shipworms (*33*).

Bacterial Symbionts of Shipworms

Considerable progress has been made to characterize the shipworm symbiont isolates. Sequence analysis has shown that the small subunit ribosomal RNA (16S rRNA) genes from four bacterial isolates taken from four different species of shipworms are identical, suggesting that these isolates represent a single symbiont species (*40*). Extensive physiological characterization of 14 isolates also supports this contention (*44, 45*). Phylogenetic analysis has further shown that these isolates are members of the gamma subdivision of the proteobacteria, a group that includes many common pathogenic, symbiotic and free-living bacteria (*40*). Within this group, however, the shipworm bacterial

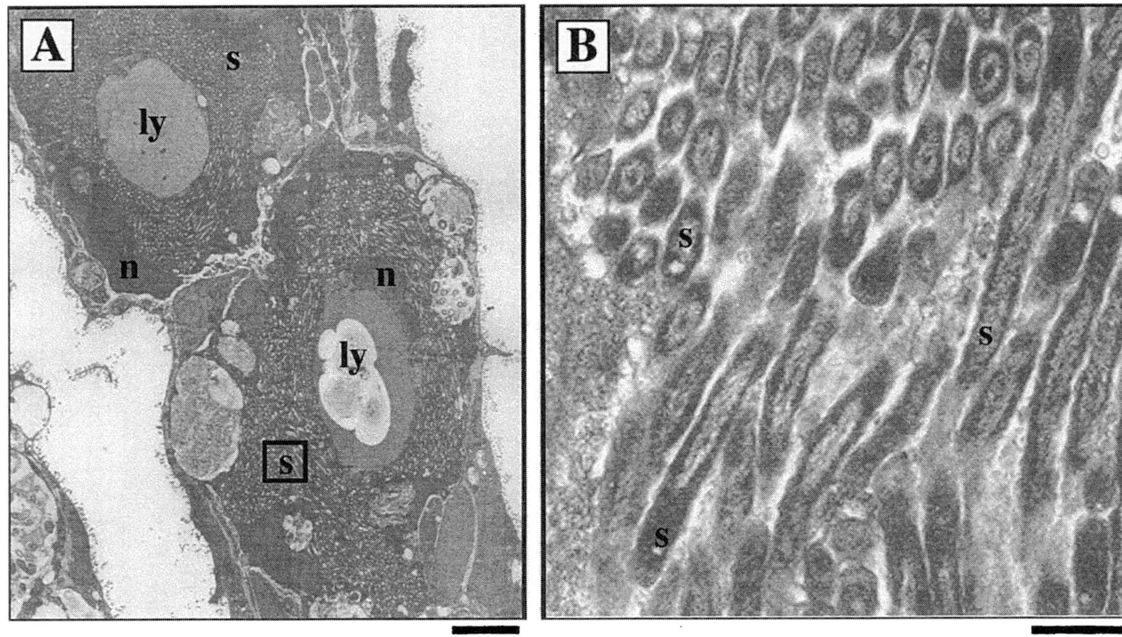

Figure 4. (A) Transmission electron micrographs showing portions of two bacteriocytes within a single gill filament: lysosomes (ly) , nuclei (n), symbionts (s) (scale bar 5.0μm). (B) Detail from box in A; Symbionts within bacteriocytes (scale bar = 0.5μm).
(Reproduced with permission from Invertebrate Zoology, Sixth Edition, *by Ruppert, ISBN 0–0302–6688–8. Copyright 1994 Thompson Learning.)*

isolates form a unique lineage, without compelling similarity to named and characterized species. On the basis of their unique properties, phylogenetic position, and widespread symbiotic association with teredinid bivalves, it was proposed that these isolates represent a new genus and species, *Teredinibacter turnerae* (type strain T7902^T, ATCC 39867) (*45*).

Teredinibacter turnerae

All strains of *T. turnerae* share certain unique features. The cells are Gram-negative, rigid, rods (0.4-0.6 x 3- 6μm) that bear a single polar flagellum. All isolates are capable of chemoheterotrophic growth in a simple mineral medium supplemented with cellulose as the sole carbon and energy source, but will also grow on a variety of carbon sources. All isolates fix dinitrogen under microaerobic conditions. Doubling time for growth under nitrogen fixing conditions is slow (1-2 days), however, with addition of combined nitrogen growth accelerates and doubling times decrease to 8-15 hours. The pH, temperature, and salinity optima for growth are approximately 8.5, 30-35°C and 0.3 M NaCl respectively. The isolates are marine, i.e. in addition to NaCl they require elevated concentrations of Ca++ and Mg++ that reflect the chemistry of seawater. The DNA mol% G+C ranged from 49-51 for six isolates tested. Isolates examined by 16S rRNA analysis fall within a unique clade most closely affiliated with the genera *Pseudomonas* (sensu stricto) and *Oceanospirillum* (*45*).

Enzymes of *Teredinibacter*

When grown in pure culture, cells of *T. turnerae* secrete cellulolytic and proteolytic enzymes into the medium (*46*). These enzyme activities have been described for strain T8201, isolated from *Psiloteredo healdi.* A cellulase (endo-β-1-4-glucanase) (*47*) and a serine protease (*48*) secreted by *T. turnerae* were purified to homogeneity and characterized. Both can be considered hardy enzymes, with unusual characteristics that may make them candidates for commercialization (*46*). The endoglucanase has high specific activity, good thermal stability, long half-life in solution, and functions over a broad range of salt concentrations (0-4M NaCl). The protease has similar characteristics but in addition has increased activity under alkaline conditions and in the presence of strong oxidants; qualities that are unusual and potentially valuable in commercial applications. Both enzymes are produced constituitively, but activity of the protease is increased when cells are grown in the presence of combined nitrogen (*46*).

In a separate investigation, two cellulases secreted by *T. turnerae* have been cloned, expressed, and partially characterized (*49*). One is a typical modular endoglucanase (63 kD) with a single catalytic domain (CD, family V) and cellulose-binding domain (CBD, family II). The second is a nonmodular endoglucanase (40kD) consisting of a single catalytic domain (CD family V) but lacking a CBD.

The function of cellulolytic enzymes in shipworm symbioses seems self evident, however, it is less clear why secretion of proteases would be beneficial for an intracellular endosymbiont. Proteases are secreted by some wood degrading fungi and have been shown to enhance hydrolysis of woody plant materials (*47*). Alternatively, the protease may be involved in processing other secreted proteins, or may play a role in defense against the host's immune response.

Additional Symbionts

Though the presence of *T. turnerae* has been confirmed by molecular methods (fluorescent *in situ* hybridization) in the gill bacteriocytes of the shipworm *Lyrodus pedicellatus* (*40*), evidence has been reported to suggest that other phylogenetically distinct symbiont types (phylotypes) may be present in some shipworm species. Using 16S rRNA based PCR amplification and cloning methods Sipe et al. (*50*) demonstrated the presence of a symbiont phylotype that is closely related to but phylogenetically distinct from *T. turnerae* in the gills of the shipworm *B. setacea. In situ* hybridization and PCR-based experiments provided evidence that this bacterium is passed vertically from parent to offspring through the egg.

Recent evidence indicates that the shipworm *L. pedicellatus* contains additional symbiont phylotypes that coexist with *T. turnerae* in the gill tissue (*51*). Each new phylotype is distinct from but closely related to *T. turnerae.* Multiple coexisting intracellular endosymbionts within a single tissue have been observed in only a few symbiotic bivalves (*52, 53*). In each case the existence of multiple symbiont types is explained by the need to utilize multiple substrates (sulfide and methane) in the environment. The role of multiple symbionts in the shipworm associations, however, remains unclear. In natural terrestrial systems, wood is typically degraded by complex microbial consortia, with each species contributing a different complement of cellulolytic enzymes (*54*). This consortial strategy likely reflects the need to accommodate the structural and chemical diversity of wood. Perhaps multiple symbionts in shipworms also provide flexibility in adapting to and optimizing digestion of various wood types.

Although little is yet known about the function of the shipworm symbioses, it is clear from morphology that these symbioses must function differently from those observed in terrestrial cellulotrophs. Since the symbionts are located intracellularly in the gills, they are never in direct contact with wood or with the contents of the digestive system. Therefore, any cellulolytic enzymes contributed by the symbionts must be transported to the digestive system by an as yet unknown mechanism. Unlike other bacterial systems, contact of these symbionts with the wood is apparently unnecessary for efficient degradation. Also, the symbiont's location in the gill suggests that the symbionts are aerobic and that fermentation is not a central feature of the symbiotic strategy, as is the case in terrestrial cellulotrophs. Research is needed to elucidate the mechanism of this unusual symbiosis.

Control of Shipworm Damage

Control of damage to marine wood is made difficult by the diversity of organisms that may contribute to wood degradation in the sea. In addition to the molluscan borers already mentioned, isopod crustacean borers (*Limnoria* and *Sphaeroma*) and fungal and bacterial microorganisms may also contribute significantly to wood decay (*2*). In combination these organisms present complex control problems, as measures effective against one organism may be ineffective or may even aggravate damage caused by another. In addition to the potential diversity of boring species in a given location, successful treatment methods must take into account temperature, salinity, oxygen, and current regimes.

In many cases, shipworm damage can be readily controlled by traditional treatments. Application at 20-25 lbs./cubic ft. of marine grade creosote meeting American Wood Preservers Association (AWPA) Standard P1/P13-91, or treatment with chromated copper arsenate (CCA) at 2.5-lbs/cubic ft. is typically effective in controlling shipworm damage (54). However, either treatment alone may be ineffective when pholadid borers like *Martesia* are present. In this case combined CCA/creosote treatment may be required. The crustacean borer *L. tripunctata* also resists creosote treatment, and various *Sphaeroma* species, found mostly in brackish water, may resist high loadings of CCA (*2*).

Although creosote and CCA treatments may be very effective against borer damage, there is increasing concern over their environmental impacts. Many states in North America, Europe and Asia regulate the use of these compounds in marine environments. The combination of environmental concerns, legal uncertainties, and a complex regulatory environment often proves to be an effective deterrent to the use of CCA and creosote for marine wood preservation.

Other factors also limit applications for creosote and CCA. Creosote treatment may add considerable weight to structures and can produce objectionable odors and initial oil slicks that discourage its use in many applications (*57*). Also, CCA treatment has been perceived to increase the brittleness of wood and this has sometimes discouraged its use in applications where impact resistance is an important criterion (*57*).

A number of alternatives to traditional chemical treatments have been used with various degrees of success. One of these has been replacement of domestic wood with exotic hardwoods that show greater resistance to borer damage. For example the heartwood of greenheart (*Ocotea rodiaei*) often shows excellent resistance to borer attack (*2*). However, this wood is largely harvested in environmentally sensitive tropical rain forests and there is concern that its use exports, rather than resolves, environmental problems. There is also concern that use of imported exotic hardwoods may have negative effects on local economies in terms of increased costs and loss of revenue by local lumber producers.

Various types of wraps and physical barriers have also been used to discourage or remediate borer attack. For example, polyethylene sheaths may be added to pilings before or after shipworm attack. This method is proposed to protect wood by limiting oxygen supplies to adult shipworms and preventing settlement of larvae. Reinforced wraps may also be designed to restore structural integrity to damaged piles. While wraps hold considerable promise, they present significant design challenges. They are effective only when intact and their susceptibility to damage by tearing, abrasion, ultraviolet light and chemical corrosion may limit their application. Wraps may also result in increased fungal and bacterial decay and may complicate design and fabrication of wood structures.

Various chemical modifications to wood have been tried with limited success; e.g. acetylation, treatment with propylene oxide, butylene oxide, or butyl isocyanate (*58*). These modifications are intended to decrease the susceptibility of wood to enzymatic degradation. Such treatments have shown effectiveness against teredinids and *Limnoria* spp. in trials, but have little effect on *Sphaeroma* spp., presumably because these borers do not ingest wood. By the same reasoning these treatments would not be expected to be effective against shallow water pholadid borers.

Some success has been observed with wood treatments that mimic the natural resistance of certain tropical timbers. For example, organo-silicate formulations have been used to simulate the natural resistance of siliceous timbers such as *Syncarpia glomulifera*. These treatments have shown efficacy against some molluscan borers but have less effect on crustacean borers (*59*). Similarly, treatments with extractives from other resistant species, including *Chlorophora excelsa*, *Nauclea diderichii* and *Dalbergia retusa*, have shown various degrees of effectiveness against a variety of boring species (*2*).

Though many of the described methods show promise for effective control of wood borers, current practice indicates a continued reliance on the use of broad-spectrum biocides. Increasing environmental concerns will likely shift the emphasis of control measures toward more targeted strategies that focus on the unique biology of individual wood boring taxa. Surprisingly, little is known regarding borer nutrition, mechanisms of wood digestion, symbiotic associations with microbes, and mechanisms of recognition and selection of wood substrates by boring animals. Research in these areas has great potential to foster the development of alternative wood treatments that are effective, inexpensive, and environmentally acceptable.

Control in Relation to the Ecological Position of Wood Boring Bivalves

While control of shipworm damage is a highly desirable goal, shipworm eradication would be disastrous. Shipworms hold a critical position in marine carbon cycles. As the primary consumers of wood and woody plant materials in the marine environment, shipworms provide an important link between terrestrial and marine food chains. Although accurate estimates are not available, vast quantities of wood and woody plant materials enter the seas each day as a result of riparian and coastal erosion and through *in situ* production in coastal mangrove forests and sea grass beds. Shipworms convert these biologically recalcitrant cellulosic materials, which include wood, bark, fibers, fruits, nuts, seeds, roots, rhizomes, leaves and stems, into shipworm tissue, larvae, and fecal material. This "upgraded", nitrogen-supplemented, biomass provides a ready food source for many marine organisms. In this way shipworms play an important role in remineralization of plant carbon in the sea. Unlike terrestrial cellulose degraders that release significant quantities of greenhouse gasses into the atmosphere, organic carbon remineralized by shipworms is released in sequestered form as dissolved bicarbonate. Finally, shipworms provide a valuable service to mankind by preventing the fouling of bays and waterways with waterlogged wood. Clearly, efforts to control shipworm damage must take into account the important environmental position of these unique marine organisms.

References

1. Turner, R. D. In *Marine boring and fouling organisms*; Ray, D. L.; Ed.; 1959; University of Washingtom Press: Seattle, 1959; Vol., pp 124-136.
2. Cragg, S. M.; Pitman, A. J.; Henderson, S. M. *Int. Biodet. Biodeg.* **1999**, *43*, 197-205.
3. Hoagland, K. E.; Turner, R. D. *Malacologia* **1981**, 21, 111-148.
4. Gallager, S. M.; Turner, R. D.; Berg, C. J. *J. Exp. Mar. Biol. Ecol.* **1981**, *52*, 63-77.
5. Turner, R. D. In *Treatise on Invertebrate Paleontology*; Moore, R. C.; Ed.; 1969; The Geol. Soc. Am. and the University of Kansas, 1969; Vol. Part N, Volume 2, Mollusca 6, Bivalvia, pp 703-741.
6. Turner, R. D. *Science* **1973**, *180*, 1377-1379.
7. Distel, D. L.; Roberts, S. J. *Biol. Bull.* **1997**, *192*, 253-261.
8. Turner, R. D. *Basteria* **1972**, 36, 98-104.
9. Miller, R. C. *Ecology* **1926**, *7*, 247-254.
10. Saraswathy, M.; Nair, N. B. *Hydrobiologia* **1974**, *44*, 397-411.
11. Lane, C. E.; Sims, R. W.; Clancey, E. J. *Am. J. Physiol.* **1955**, *183*, 3.
12. Turner, R. D. *Aust. Nat. Hist.* **1971**, *17*, 139-145.
13. Walton Smith, F. G. *Nat.Geog. Magazine* **1956**, *CX*, 563.

14. Scofield, J. *Nat. Geog. Magazine* ; **1975**., *615*,148.
15. Cohen, A. N.; Carlton, J. T., *Nonindigenous aquatic species in a United States estuary: A case study of the biological invasions of the San Francisco Bay and Delta.* 1995, A report for the United States Fish and Wildlife Service, Washington, D.C. and the National Sea Grant College Program, Connecticutt Sea Grant.
16. Thompson, J.; Parchaso, F.; Alpine, A.; Cloern, J.; Cole, B.; Mace, O.; Edmunds, J.; Baylosis, J.; Luoma, S.; Nichols, F., *The history and effects of exotic species in San Francisco Bay,San Francisco Bay Project, Water Resources Division.* 1999, U.S. Geological Survey.
17. Gruson, L. *New York Times* **1993**, *1*,32.
18. Lane, C. E. *Sci. Amer.* **1961**, *204*, 132-142.
19. Turner, R. D. In *Marine borers, fungi and fouling organisms of wood*; Jones, E. G. B.; Eltringham, S. K.; Eds.; Organization for economic co-operation and development: Paris; **1971**, pp 17-64.
20. Morton, B. Oceanography and Marine Biology, Annual Review **1978**, 16, 107-144.
21. Quayle, D. B. In *Marine boring and fouling organisms*; Ray, D. L.; Ed.; 1959; University of Washingtom Press: Seattle, **1959**; pp 175-185.
22. Turner, R. D. In; Eds.; The Museum of Comparative Zoology: Harvard University, Cambridge, MA, 1966; pp Pages.
23. Turner, R. D. American Malacological Union **1967**, Annual Reports for 1967, 46-48.
24. Eckelbarger, K. J.; Reish, D. J. Bull. S. Cal. Acad. Sci. **1972**, *71*, 48-50.
25. Hoagland, K. E. *Malacologia* **1986**, *27*, 323-329.
26. Scheltema, R. S. *Int. J. Life in Oceans and Coastal Wat.* **1971**, *11*, 5-11.
27. Bazylinski, D. A.; Rosenberg, F. A. *Veliger* **1983**, *25*, 251-255.
28. Lopes, S. G. B. C.; Domaneschi, O.; De Moraes, D. T.; Morita, M.; Meserani, G. In *The evolutionary biology of the Bivalvia*; Harper, E. M.; Taylor, J. D.; Crame, J. A.; Eds.; The Geological Society of London, Special Publications: London, 2000; Vol. 177, pp 257-271.
29. Breznak, J. A.; Brune, A. *Ann. Rev. Entomol.* **1994**, *39*, 453-487.
30. Martin, M. M. Phil. *Trans. R. Soc. Lond. B* **1991**, 333.
31. Watanabe, H., Hiroaki, N; Tokada, G.; Lo, N. *Nature* **1998,** *394*, 330-331.
32. Watanabe, H.; Tokuda, G. *Cellular mol. life sci.* **2001**, *58*, 1167-1178.
33. Carpenter, E. J.; Culliney, J. L. *Science* **1975**, *187*, 551-552.
34. *Brock Biology of Microorganisms*; Madigan, M. T.; Martinko, J. M.; Parker, J. ed, Eds.; Prentice Hall: Upper Saddle River, NJ, 2000, 9th ed..
35. Kane, M. D. In *Gastrointestinal Fermentation and Ecosystems*; Mackie, R. I.; White, B. A.; Eds.; Chapman and Hall, USA, 1996; Vol. 1, pp 38.
36. Dehority, B. A. In *Biosynthesis and biodegradation of cellulose*; Haigler, C. A.; Weimer, P. J.; Eds.; Marcel Dekker, Inc.: New York, 1991.
37. Troyer, B. A. In *Biosynthesis and biodegradation of cellulose*; Haigler, C. A.; Weimer, P. J.; Eds.; Marcel Dekker, Inc.: New York, 1991.
38. Popham, J. D.; Dickson, M. R. *Mar. Biol.* **1973**, *19*, 338-340.

39. Popham, J. D. *Veliger* **1974**, *18*, 55-59.
40. Distel, D. L.; Delong, E. F.; Waterbury, J. B. *Appl. Environ. Microbiol.* **1991**, *57*, 2376-2382.
41. Sigerfoos, C. P. *Bull. Bureau Fish.* **1908**, *37*, 191-231.
42. Distel, D. L. *BioScience* **1998**, *48*, 277-286.
43. Trytek, R. E.; Allen, W. V. *Comp. Biochem. Physiol.* **1980**, *67A*, 419-427.
44. Waterbury, J. B.; Calloway, C. B.; Turner, R. D. *Science* **1983**, *221*, 1401-1403.
45. Distel, D. L.; Morrill, W.; MacLaren-Toussaint, N.; Franks, D.; Waterbury, J. **submitted 2002**.
46. Greene, R. V. *Soc. Industrial Microbiol. News* **1994**, 44, 51-59.
47. Greene, R. V.; Griffin, H. L.; Freer, S. N. *Arch. Biochem. Biophys.* **1988**, 1, 334-341.
48. Greene, R. V. *Curr. Microbiol.* **1989**, 19, 353-356.
49. Li, L., Ph.D thesis, University of Maine, Orono,ME, 2000.
50. Sipe, A. R.; Wilbur, A. E.; Cary, S. C. *Appl. Environ. Microbiol.* **2000**, *66*, 1685-1691.
51. Beaudoin, D. J. Ph.D thesis, University of Maine, Orono,ME, **1998**.
52. Distel, D. L.; Cavanaugh, C. M. *Proc. Natl. Acad. Sci. USA* **1995**, *92*, 9598-9602.
53. Fisher, C. R.; Brooks, J. M.; Vodenicher, J. S.; Zande, J. M.; Childress, J. J.; Burke Jr, *R. A. P. S. Z. N. I. : Mar. Ecol.* **1993**, *14*, 277-289.
54. Coughlan, M. P.; Mayer, F. In *The Prokaryotes*; Ballows, A.; Truper, H.; Harder, W.; Schleifer, K. H.; Eds.; Springer-Verlag: Berlin, 1992; Vol. 4, pp 451-516.
55. Cassens, D. L.; Johnson, B.; Feist, W. C.; DeGroot, R., *Selection and use of preservative treated wood.*, Forest Products Society: Madison, WI. 1995, p. 36-40.
56. U.S. Department of Agriculture, Forest Service, Forest Products Laboratory: Madison, WI, 1999: Vol. Gen. Tech. Rep. FPL-GTR-113.; pp Pages.
57. Gardner, D.; Goodell, B.; Lopez-Anido, R.; Eckelbarger, K. *Bangor Daily News*, February 16 2001: Bangor, ME.
58. Johnson, B. R.; Rowell, R. M. Mat. u Org. **1988**, 23, 147-156.
59. Scown, D. K.; Cookson, L. J. a.; McCarthy, K. *International research group on wood preservation* **2001** 01-30270.

Chapter 15

Marine Wood Boring Arthropods: Ecology, Functional Anatomy, and Control Measures

Simon M. Cragg

Institute of Marine Science, School of Biological Sciences, University of Portsmouth, Ferry Road, Portsmouth PO4 9LY, United Kingdom

Wood boring arthropods cause damage of major economic significance to coastal structures. The curculionid *Pselactus* is restricted to the upper intertidal in decayed wood. Limnoriids occur in full salinity subpolar to tropical waters, from the intertidal zone to over 1000m. Sphaeromatid borers are euryhaline temperate and tropical, intertidal species. These isopods have low fecundity, iteroparity and extended parental care, and thus compete effectively once established in a substrate. They migrate to new substrata as young adults. Limnoriids usually lack a resident microbial flora, but ingest wood particles containing wood-degrading microorganisms. Cellulose is degraded during gut passage, but the source and functioning of a full suite of lignocellulose degrading enzymes remains to be demonstrated. Sphaeromatids may not ingest wood. Neither creosote nor CCA treatments provide complete protection from these organisms. Protocols for evaluating anti-borer measures that take account of migratory and substrate seeking behaviour are needed.

Introduction

In temperate waters, wood boring isopod crustaceans from the family Limnoriidae limit the service life of timber structures. In tropical waters, members of the Sphaeromatidae are also a serious threat to such structures (Figure 1.). The relative importance of damage caused by these crustaceans and by the wood boring bivalves (shipworms and *Martesia*) depends on environmental conditions. Unlike the shipworms, attack by these crustaceans is readily detectable even at the early stages. Species from the two crustacean families are capable of attacking timber treated to control shipworms. Other isopods and the amphipod crustacean *Chelura* also burrow into wood. Insects can cause damage in the high intertidal zone.

Reviews of the biology of wood boring crustaceans (1,2,3) will be updated herein, focusing on unresolved questions about their functioning and ecology.

Identification and Systematics of Wood Boring Arthropods

The type of borer can usually be determined from features of the tunnels that they excavate. Recognition to family level may not be sufficient to enable appropriate decisions about anti-borer measures: some species of the Limnoriidae have greater resistance to chemical treatments than others (3).

The main wood-boring Crustacea and their diagnostic features are illustrated by Kühne (4). Studies of comparative anatomy of limnoriids (5,6,7) and of sphaeromatids (8) have facilitated identification by providing keys to species and genera, detailed line drawings and SEM images (Figs. 2 and 3). Intensive collection in tropical and subtropical waters of Central and North America (5) and of Australasia (6) has increased the number of known limnoriid species considerably. Collections from deeper waters have also added to the list of species (9,10). More additions to the species list can be expected when other waters are examined more closely (11).

The genera *Phycolimnoria* and *Limnoria* were separated on the basis of having clearly distinct mandible types (5), but it is now known that there is a gradation in mandible anatomy so the genera have been merged into *Limnoria*, which can be distinguished from *Paralimnoria* on the basis of uropod anatomy (6). There are two wood-boring species of *Paralimnoria* and forty nine species of *Limnoria*, encompassing algal burrowers, seagrass borers and many wood borers. *Hadromastax*, is no longer considered a member of the Limnoriidae, but *Lynseia*, a seagrass borer, has been placed within the family and a revised key to the limnoriid genera is available (7).

Figure 1. Port Moresby, Papua New Guinea: CCA-treated hoop pine pilings destroyed by infestation in the intertidal zone by *Sphaeroma triste*

(Reproduced with permission from an unpublished source. Copyright IRG Secretariat, Stockholm.)

Figure 2. SEM image of ventral view partially curled specimen of *Limnoria cristata* from Singapore. Features of the dished pleotelson (to right) are used to identify the species. Note partially folded peraeopods (walking limbs) with dactyls (distal claw-shaped articles) extending away from body.

Figure 3. SEM image of posterior view of *Sphaeroma triste* showing its characteristic broad, toothed exopods of the uropods (limbs attached to the final body component – the telson).

(Reproduced with permission from an unpublished source. Copyright IRG Secretariat, Stockholm.)

Some members of the genus *Sphaeroma* have been separated, on the basis of mouth part anatomy, into a new genus, *Lekanesphaera* (12). The borers *S. annandalei*, *S. quoyanum* and *S. terebrans* remain within *Sphaeroma*.

Other isopods have been reported from tunnels in wood in tropical intertidal sites: the cirolanids, *Ceratolana*, a borer of decayed mangrove wood, and *Anopsilana willeyi*, probably a predator (13,14), and a number of species of *Corallana* (Corallanidae), some from CCA-treated piling (15).

The amphipod borer family Cheluridae has not had any revisions or additions since the appearance of Kühne's key (4), which describes three species within the genus *Chelura*, though previously Barnard allocated each of the three species to a separate genus (16).

The curculionid *Pselactus* is included in a key of the Cossoninae (17).

Ecology of Arthropod Borers

Biogeography

Wood boring insects are generally assumed to be terrestrial organisms. However, wood-boring weevils occur in decayed wood in the intertidal zone around the coast of southern Great Britain (18). Attack by subterranean termites has also been observed in the intertidal zone on wood attached to a concrete wharf in Papua New Guinea. The earth runways of the termites extended about 100m under the concrete deck (unpub. obs.).

Some wood boring members of the Limnoriidae are only known from single sites. Others are distributed in waters around the globe within a certain latitudinal range: *L. borealis* - boreal; *L. lignorum* - boreal and temperate northern; *L. quadripunctata* - temperate and warm temperate North and South of the equator; *L. tripunctata* - temperate and tropical; *Paralimnoria andrewsi* – subtropical and tropical (4,5,6).

Wood boring sphaeromatids are restricted to warmer waters: *S. quoyanum* occurs in New Zealand, warm temperate Australia and, probably as an introduced species, in California, *S. triste* has been reported from India, northern Australia and Papua New Guinea while *S. terebrans* may be circumtropical (8).

Chelura terebrans occurs in temperate and subtropical waters North and South of the Equator, while *C. insulae* has only been reported from islands in the Caribbean and the tropical Pacific, and *C. brevicauda* from Japan (19).

Environmental Factors Determining Distribution

The geographical ranges of crustacean borers are principally determined by water temperature (1). Studies of fluctuations in populations of three sympatric species of *Limnoria* from the temperate waters off southern Great Britain support this notion (20), with *L. lignorum* being most tolerant of unusually low temperatures, *L. quadripunctata* being stenothermal and *L. tripunctata,* eurythermal. Local variations in temperature regime due to a warm water plume from a power station affected species dominance. Temperature probably determines the latitudinal range of *Sphaeroma quoyanum* on either side of the equator and the circumtropical distribution of *S. terebrans*.

Local distribution is strongly influenced by salinity. *Sphaeroma* spp. in Australasia and India are markedly euryhaline (14, 21). Laboratory studies indicate that the *Limnoria* species studied to date are much more stenohaline (1), which probably accounts for their otherwise surprisingly low profile in major intertidal wood resources - mangrove ecosystems, as these are most profuse where there is a freshwater input. Limnoriids have, however, been reported as root borers in mangroves in full salinity seawater (22).

Pselactus ranges from the high intertidal into the splash zone above (23). *Sphaeroma* burrowing is virtually restricted to the intertidal zone (24). In the first case the distribution may be reflect the range of conditions tolerated by basidiomycete decay fungi; in the second, it may reflect adaptation to their natural habitat - the less heavily corticated, intertidal prop roots of the mangrove *Rhizophora* (22). Limnoriids occur in the intertidal zone, but have also have been found in waterlogged wood at depths of over 1000m and over 400km from the nearest forest (10).

Population Dynamics and Dispersal

Adult *Limnoria* occur as pairs in burrows and may remain paired for over 10 months. Broods of up to 30 are produced by the boreal *L. lignorum*, but no more than 6 in tropical *P. andrewsi* (25). Eggs are released into a ventral brood pouch and are retained there throughout embryonic development. Juveniles are released into the adult burrow and frequently form side burrows from the parental burrow. Parental care is also reported for *Sphaeroma*, sometimes with juveniles of one species being found with the brood of another (26,27). Mean brood size varies with female size between 20 and 48 in *S. terebrans* and 19 and 101 in *S. annandalei* (28).

The life cycle strategy of wood boring isopods - low fecundity, iteroparity, extended parental care – is characteristic of *K*-selected species. As such, they would be expected to be strong competitors capable of maximising resource share once established. On the other hand, there are many typical *r*-selected species among the wood boring molluscs (teredinids and pholads), which produce hundreds of thousands of minute planktotrophic larvae with no period of parental care. These are well adapted to colonising new environments due to high fecundity and high dispersal potential due to transport on ocean currents during the planktonic larval phase. Other wood boring bivalves brood their larvae during some or all of larval development, so that they display more *K*-selected characteristics (29), but even in brooding species, fecundity is much higher than in crustacean borers.

Juvenile *Sphaeroma* establish new tunnels close to the parental tunnels, though young adults have been reported to migrate in the water column (30). Late juveniles and young adults of *Limnoria* have also been reported to disperse during specific periods of the year (1,31), perhaps triggered to move by overcrowding (5). The dispersal potential of crustacean borers is likely to be less than that of bivalve borers, due to much smaller numbers of offspring and the inability to feed, unlike the planktotrophic bivalve veligers, when dispersing, which limits in the dispersal period. Nonetheless, the colonisation of wood in the deep sea argues for an effective mechanism for locating such a patchily distributed food resource as wood. Evidence for contact chemical cues has been provided by laboratory experiments (32), but distance chemical reception, of value during migration, is harder to demonstrate (33). Wood boring species of *Sphaeroma* are not necessarily dependent on wood as a substrate. Large colonies occur in inert intertidal substrates (34,35).

Interactions with other Organisms

Competition may occur between *Sphaeroma* and molluscan borers (36). Crustacean borers may also either be outcompeted or removed by predators (37). They play host to a range of epizooites (1) and bacteria (38). No resident gut microflora has been found in either *Limnoria* or *Chelura* (39), except when living on creosote treated wood (40), but there is a commensal relationship between the non-borer *S. serratum* and a hindgut-resident trychomycete fungus (41). The lignolitic ascomycete fungus *Lulworthia* readily colonises wood with hyphae, but appears to only form peritheca after *Limnoria* forms tunnels (42). A range of cellulose-degrading microorganisms has been reported to be associated with wood-boring isopods (43).

The co-occurrence of *Limnoria* and *Chelura* has been interpreted as a loose form of symbiosis. *Chelura* is reported to consume the faecal pellets of *Limnoria* (1, 19), but may not depend on this, as it is capable of forming its own tunnels and produces faecal pellets consisting of particles of wood (44 and Fig.4).

Functional Anatomy

Appendages and Feeding Mechanisms of Isopods

Wood boring isopods have a sequence of paired, jointed appendages: antennae, mouth parts, thoracic limbs (peraeopods), abdominal limbs (pleopods) and uropods. A clearer idea of complex inter-appendage interactions is provided in illustrations of mouth parts *in situ* and with the attached muscles (6). The innermost mouth parts, the mandibles, have heavily scleratosed cutting processes responsible, together with head movements, for tunnel forming. The outermost mouthparts, the maxillipeds, can reach back to clean the peraeopods during grooming. The peraeopods are walking limbs and pleopods serve as fans for creating respiratory, locomotory and feeding currents. *S. quoyanum* filter feeds within its burrow, by beating its pleopods to draw seawater between the long setae on its peraeopods (45). In *S. terebrans*, these setae have fine side branches that form an effective fine net (46 and Fig. 5).

Gut Structure and Function

The gut of the wood boring isopods consists of a simple oesophagus leading into a complex midgut region of filters and spines, which directs particles into the cuticle lined hindgut. Opening off the midgut are elongated glandular side pockets which form the hepatopancreas. Liquid in the hindgut can be exchanged with the midgut and hepatopancreas following contraction of muscles surrounding the hepatopancreas pockets or the hindgut. The hepatopancreas has been a focus for the search for endogenous cellulase production (47), but though gut resident microorganisms may not be involved in the production of enzymes (38), ingested wood degrading microorganisms could be an enzyme source (48). Cellulase activity has been detected in hepatopancreas extracts from *Sphaeroma terebrans* (49).

The lignin-to-cellulose ratio in faecal pellets of *Limnoria* is markedly higher than that of the source wood (50,51), indicating cellulose breakdown during gut passage. However, the presence and source(s) of the full suite of enzymes

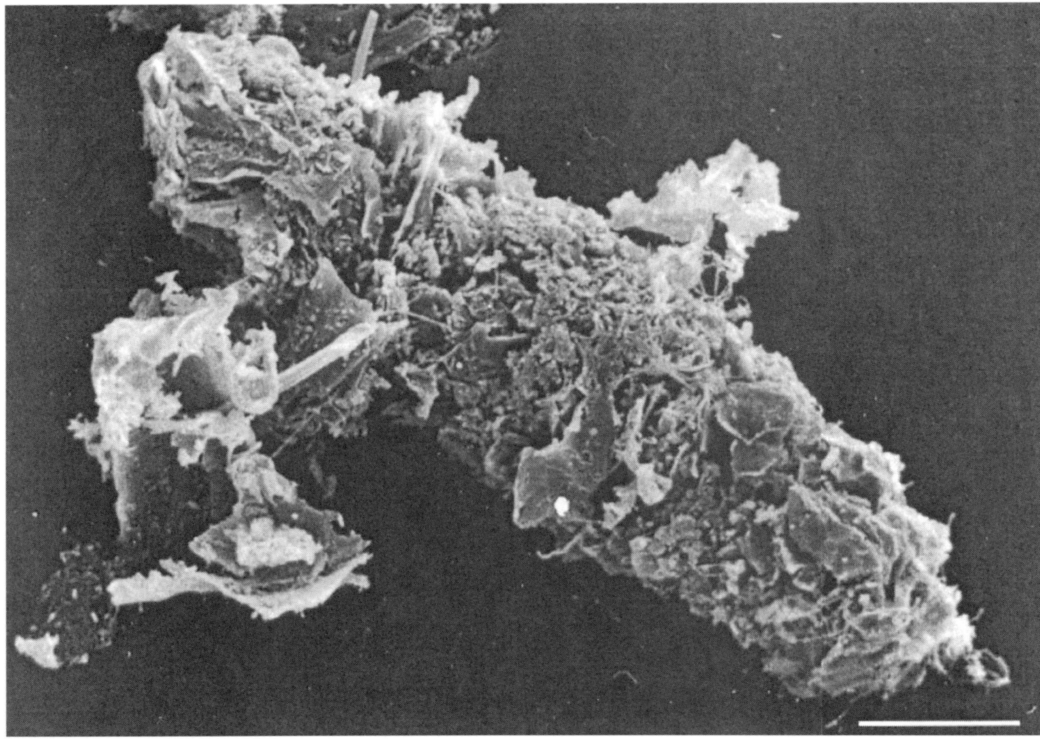

Figure 4. SEM image of faecal pellet from *Chelura terebrans* reared separate from *Limnoria* on balsa wood (*Ochroma lagopus*).

(Reproduced with permission from an unpublished source. Copyright IRG Secretariat, Stockholm.)

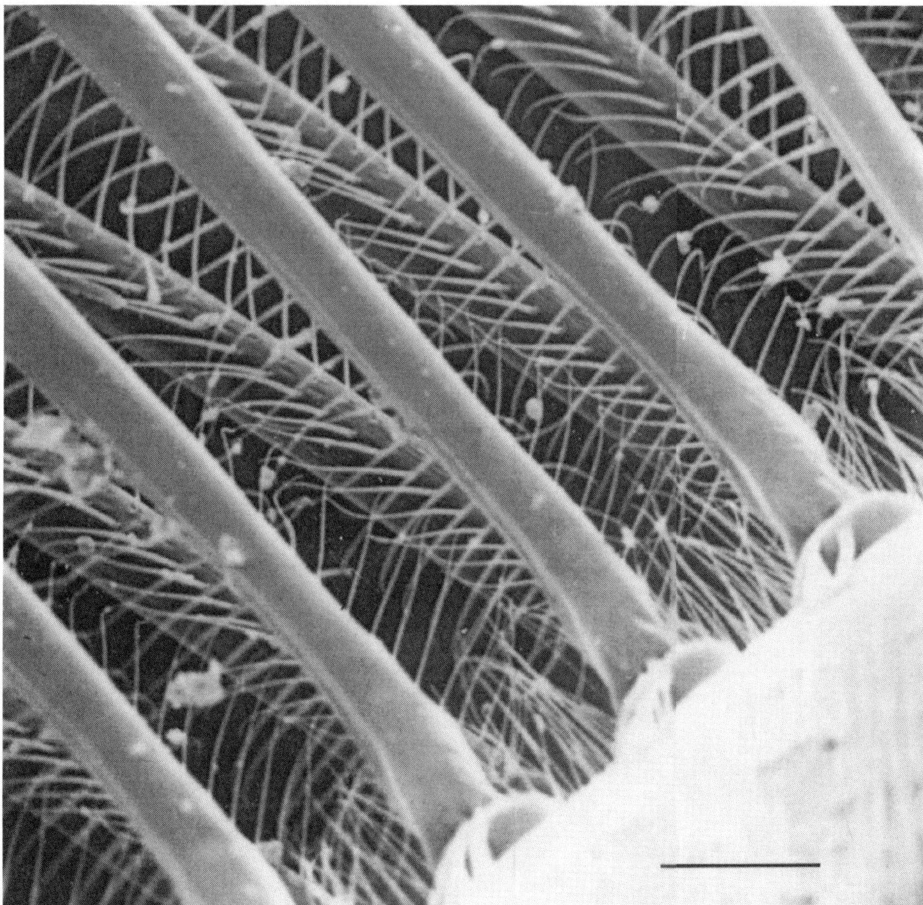

Figure 5. SEM image of detail of the plumose setae on peraeopod of *Sphaeroma terebrans* (source: reference no. 46)

(Reproduced with permission from an unpublished source. Copyright IRG Secretariat, Stockholm.)

required to break down crystalline cellulose within the lignocellulose complex remains to be demonstrated by critical experimentation.

Wood alone is a poor source of dietary nitrogen. Extra nitrogen may be gleaned during the grooming process (38), during which bacteria are removed from the exoskeleton and may be transferred via the setae of the peraeopods (Fig. 6) and maxillipeds to the mouth. Bacteria and fungi colonising the surfaces of tunnels may also be a source of additional nitrogen (48).

Within the hepatopancreas cells of *Limnoria* and *Sphaeroma* are granules containing copper. These are more numerous and have a more varied morphology in animals burrowing into CCA-treated wood, but no evidence of chromium or arsenic accumulation has been detected (46,52). Such granules occur in other isopods living in contaminated environments and appear to enable the organisms to tolerate high environmental levels of metals. Tolerance to creosote appears to be achieved in a different manner. Individuals of *L. tripunctata* growing on creosote treated wood have larger numbers of bacteria within the food mass in the gut, but also bacteria associated with the hind gut cuticle. It is suggested that these bacteria may break down the toxic components of the creosote and may provide a supplementary source of nitrogen when they are lysed in the gut (39). The report of crystals of hydrated protein complexed with iron in *Limnoria* hepatopancreas cells (50) deserves further investigation in the light of findings regarding other metals.

Protection of Wooden Structures Against Arthropod Borers

Tropical hardwoods with heartwood containing bioactive extractives are used in maritime construction because of their ability to resist borer attack. Field and lab observations have identified timbers resistant to limnoriid attack, though resistance may vary with latitude and species of limnoriid (3, 53). Furthermore, microbial degradation may reduce the protective value of these extractives (3,54). Laboratory tests could provide more precision in identifying specifically arthropod-resistant species. Choice-type or force feeding tests may yield different findings (55), with the former approach being relevant to initial settlement while the latter may evaluate the capacity to minimise subsequent colony growth.

Standard timber treatments offer good protection against teredinids, but species of *Limnoria* and *Sphaeroma* are capable of attacking timber treated with CCA or creosote and many novel treatments have been tested without identifying one that provides demonstrable efficacy against all types of arthropod (3). Even physical barriers are vulnerable to *Sphaeroma* (56). However, at least in the case of wood ingesters like limnoriids, modification of wood chemistry to reduce access for cellulases may prove effective (57). Biological control, using

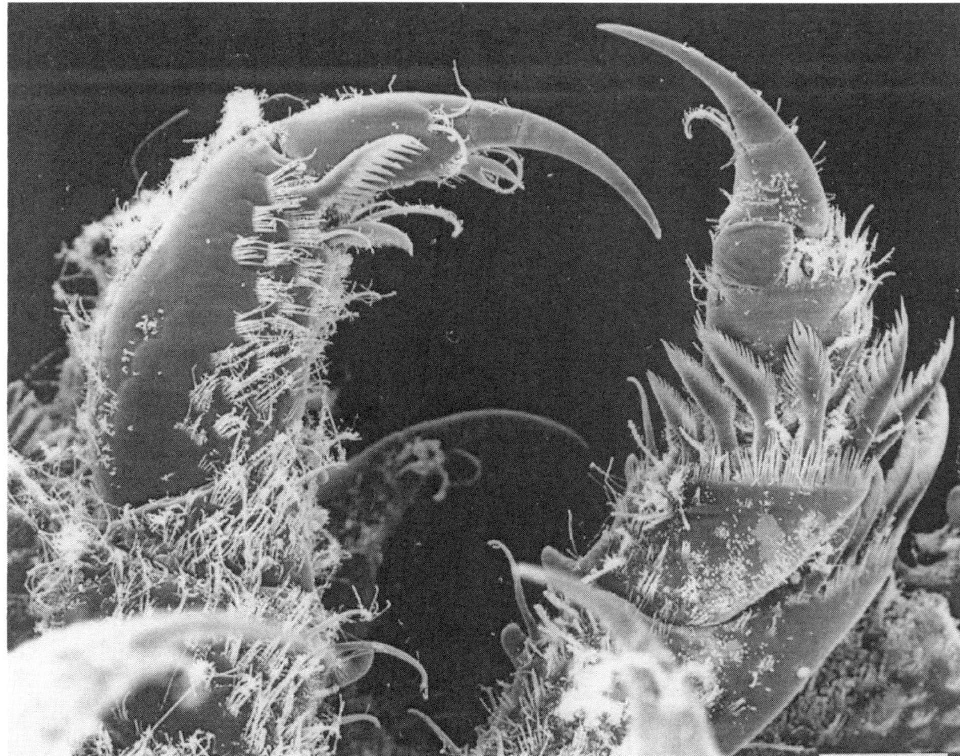

Figure 6. SEM image of distal articles of peraeopods 1 (left of image) and 7 (right) of *Limnoria cristata* showing comb setae used in grooming.

(Reproduced with permission from an unpublished source. Copyright IRG Secretariat, Stockholm.)

competing or predatory organisms has not yet been explored (36). The need for innovation is becoming more pressing as environmental concerns call into question the use of existing treatments.

Field trials, even adjacent to natural populations, suffer from unpredictable recruitment and are difficult to interpret as the intensity of attack is due to success of initial settlement and to subsequent colony formation. In the laboratory, testing relies on animals changing many of their sensory responses from their burrow-dwelling mode to migration mode. Laboratory assays tailored to borer behaviour promise to assist the development of new approaches to the problem of arthropod attack on wood below the high tide mark.

References

1. Becker, G. In *Marine Borers, Fungi and Fouling Organisms* Jones E.B.G.; Eltringham S.K. Eds., OECD: Paris, 1971.
2. Boyle, P.J. In *Marine Biodeterioration; Advanced Techniques Applicable to the Indian Ocean* Thompson M.-F.; Sarojini R.; Nagabhushanam R. Eds. Oxford IBH Publishing Co., New Delhi, 1988.
3. Cragg, S. M.; Pitman, A.; Henderson, S. M. *Int. Biodet. Biodeg.* **1999**, *43*, 197-205.
4. Kühne, H. In *Marine Borers, Fungi and Fouling Organisms* Jones E.B.G.; Eltringham S.K. Eds., OECD: Paris, 1971.
5. Menzies, R. J. *Bull. Mar. Sci. Gulf Caribb.* **1957**, *7*, 103-200.
6. Cookson, L.J. *Mem. Mus. Vic.* **1991**, *52*, 137-262.
7. Cookson, L.J.; Poore, G.C.B. *Mem. Mus. Vic.***1994**, *54*, 174-189.
8. Harrison, K.; Holdich, D.M. *Zool. J. Linn. Soc.***1984**, *81*, 275-387.
9. Schotte, M., *Proc. Biol. Soc. Wash.***1989**, *102*, 716-725.
10. Kussakin, O. G.; Malyutina, M. V. *Crustaceana* **1989**, *56*, 8-13.
11. Cookson, L. J.; Lorenti, M. *Crustaceana* **2001**, *74*, 339-346.
12. Jacobs, B. J. M. *Zool. Verhandelingen* **1987** *238*, 1-71.
13. Bowman, T.E. *Proc. Biol. Soc. Wash.* **1977** *90*, 819-825.
14. Cragg, S.M.; Aruga, J., **1987** In *Advances in Aquatic Biology and Fisheries* Natarajan, P. Ed, St. Josephs Press, Trivandrum, India.
15. Jones, D.A.; Icely, J.D.; Cragg, S.M. *J. Nat. Hist.* **1983** *17*, 837-847.
16. Barnard, J. L. *Pacif. Nat.* **1959** *1*, 3-10.
17. Folwaczny, B. *Entomolog. Blätter* **1973** *67*, 157-187.
18. Oevering, P.; Matthews, B. J.; Cragg, S. M.; Pitman, A. J.; *Int. Biodet. Biodeg.* 2001 **47**, 175-181.
19. Kühne, H; Becker, G. *Z. angew. Zool.* **1964** *Beiheft 1*, 1-141.
20. Hockley, A. *Mat. Org.* 1965, *Suppl. 1*, 457-4

21. Cragg, S.M., **1988**. In *Biodeterioration 7* Houghton, D.; Smith, R.W.; Eggins, H.O.W., Eds, Elsevier, London.
22. Ellison, A. M.; Farnsworth, E. *J. Exp. Mar. Biol. Ecol.* **1990** *142*, 91-104.
23. Sawyer, G. S.; Cragg, S. M. *Mat. Org.* **1995**, *29*, 67-79.
24. Cragg, S.M.; Levy, C.R., *Int. J. Wood Pres.* **1979** *1*, 161-168.
25. Menzies, R. J. *Bull. Mus. Comp. Zool., Harvard* **1954** *112*, 363-388.
26. Messana, G.; Bartolucci, V.; Mwaluma, J.; Osore, M. *Ethology Ecology & Evolution* **1994** *Spec. Issue 3*, 125-129.
27. Thiel, M. *J. Nat. Hist.* **2000** *34*, 737-745.
28. Krishnan, R. V.; Nair, N. B. *Zool. Anz.* **1973** *191*, 119-122.
29. Hoagland , K. E.; Turner, R. D. *Malacologia* **1981** *21*, 111-148
30. Cheriyan, P.V. *Forma et Functio* **1973** *6*, 1-68.
31. Eltringham, S. K.; Hockley, A. R. *Limnol. Oceanogr.* **1961** *6*, 467-482.
32. Geyer, H. *Int. J. Wood Pres.* **1981** *2*, 77-89.
33. Henderson, S. M.; Cragg, S. M.; Pitman, A. J. *Int. Res. Grp. Wood Pres.* **1995** *96-10157*, 18pp.
34. Krishnan, R. V.; Nair, N. B. *Zool. Anz.* **1973** *191*, 119-122.
35. Talley, T.; Crooks, J. A.; Levin, L. A. *Marine Biology* **2001** *138*, 561-573.
36. Si, A.;Alexander, C. G.; Bellwood, O. *J. Mar. Biol. Ass. UK* **2000** *80*, 1131-1132.
37. Kirkegaard, J. B.; Santhakumaran, L. N. *Vid. Med.Dansk Nat.Foren.* **1967**, *130*, 213-216.
38. Boyle, P.J. & Mitchell, R. *Appl. Environ. Microbiol* **1981** *42*, 720-729.
39. Boyle, P.J. & Mitchell, R. *Science*, **1978** *200*, 1157-1158.
40. Zachary, A.; Parrish, K.K.; Bultman, J.D. *Mar. Biol.* **1983** *75*, 1-8.
41. Charmantier, G.; Manier, J.-F. *Vie Milieu* **1981** *31*, 101-111.
42. Corte, A. M. Giorn. Bot. Ital. 1975 109, 227-237.
43. El Shanshoury, A.R.; Mona, M.A.; Shoukr, F.A.; El Bossery, A.M. *Mar. Biol.* **1994** *119*, 321-326.
44. Cragg, S.M.; Daniel, G., *Int. Res. Grp. Wood Pres.* **1992** *4180-92,* 1-10.
45. Rotramel, G. *Crustaceana* **1975** *28*, 7-10.
46. Cragg, S.M.; Icely, J.D. *Int. Res. Grp. Wood Pres.* **1982** *491*, 1-26.
47. Ray, D. L. In *Marine Wood Boring and Fouling Organisms* Ray D. L. Ed. Washington University Press, 1959, pp 46-59.
48. Daniel, G.; Nilsson, T.; Cragg, S. *Holz Roh- Werkstoff* **1991** *49*, 488-490.
49. Benson, L.; Rice, S.; Johnson, B. *Florida Scientist* **1999**, *62*, 128-144.
50. Fahrenbach, W.H. In *Marine Wood Boring and Fouling Organisms* Ray D. L. Ed. Washington University Press, 1959.
51. Seifert, K. *Holz Roh- und Werkstoff* **1964**, *6*, 209-215.
52. Tupper, C.; Pitman, A. J.; Cragg, S. M. *Holzforschung* **2000** *54*, 570-576.
53. Kalnins, M. *Proc. Amer. Wood Pres. Assoc.* **1976** 250-262.
54. Pitman, A. J.; Cragg, S. M.; Daniel, G. *Mat. Org.* **1997** *31*, 281-291.
55. Kühne, H. Mat. Org. **1964** *3*, 107-118.
56. Eaton, R. A. *Int. Res. Grp. Wood Pres.* **1996** *Doc. No. IRG/WP/96-10170.*
57. Johnson, B.R.; Rowell, R.M. *Mat. Org.* **1988** *23*, 147-156.

Detection and Test Methodology

Chapter 16

Concepts in the Development of New Accelerated Test Methods for Wood Decay

Darrel D. Nicholas[1] and Douglas Crawford[2]

[1]Department of Forest Products, Mississippi State University, Mississippi State, MS 39762–9820
[2]Forest Products Technologist, Forest Service, Forest Products Laboratory, U.S. Department of Agriculture, Madison, WI 53705–2398

Abstract

Efforts to develop new environmentally friendly wood preservatives are seriously handicapped by the extended time period required to carry out the evaluation needed to establish confidence in the long term performance of new preservative systems. Studies in our laboratory have shown that using strength loss as a measure of the extent of wood decay makes it possible to detect the early stages of decay that results from non-enzymatic reactions. We have developed specialized equipment and techniques that have applications for both above ground and soil contact preservative systems. By coupling these evaluation techniques with a better understanding of moisture control, microbial succession, soil chemistry and soil microbial dynamics, it may be possible to develop improved test methods that can greatly reduce the time required to evaluate wood preservative systems.

Introduction

The development of new or modified wood preservative systems is severely hampered because of inadequate test methods. Although current methods for evaluating the efficacy of wood preservatives provide valid data, the time required to obtain definitive results is too long because reliable decisions on efficacy against wood decay organisms must rely on long-term field test data. In order to reduce the test time, it will be necessary to develop reliable laboratory and field tests that optimize the wood decay process. However, this goal will not be realized until we develop a better understanding of the many variables that influence microbial decay rates, develop improved methods for detecting and quantifying the extent of wood decay and couple these developments with improved designs for test specimens and methods. The factors which affect decay rates are reviewed in this chapter. The variables affecting decay rates are briefly reviewed in this chapter in order to provide the necessary background needed before decay test methodology can be addressed. Following this, detailed suggestions on research needed to develop reliable accelerated wood decay test methods that are critical to the future of the wood preserving industry are presented.

Variables Influencing Microbial Decay Rates

A basic understanding of the many variables associated with microbial decay rates is needed before serious consideration can be given to development of accelerated decay tests. Accordingly, the material presented in this section is presented in order to provide this background.

Water and Oxygen

Water is essential for the growth of wood decay fungi. It serves as a solvent for metabolic processes, for the transport of metabolites, enzymes, and organelles, is a vital skeletal ingredient and is the driving force behind extension growth (*1*). It is generally accepted that a minimal wood moisture content (MC) at or above the fiber saturation point is needed to support the wood decay processes. At lower levels there is insufficient water to meet metabolic requirements. Furthermore, the MC of wood plays a major role in determining which fungal species colonize wood and the extent to which they utilize this substrate.

The optimal MC for growth of many of the wood decaying basidiomycetes is unknown, but for those that have been studied it ranges between 40 and 80% (*2*). The reason that optimum decay rates are associated with this MC range is mainly due to the oxygen requirements for these fungi. At higher MC levels, aeration is reduced and this impacts fungal growth. Rypacek (*3*) studied wood decay in sticks that had MC gradients from one end to the other and found that decay was inhibited when the amount of air in the wood was less than about 10-20% of the void volume. Other studies (*2*) support this conclusion, thus it is generally considered that 20% residual air volume in wood is close to the minimum level needed for basidiomycete wood decay fungi. Since the void volume of wood varies inversely with specific gravity, the MC levels that inhibit fungal growth are considerably higher in lower density wood.

When wood is subjected to exterior exposure it undergoes alternate wetting and drying which can impact the survival and growth rate of fungi inhabiting the wood, but very little information is available on this subject. With the exception of fully saturated wood it does not appear that high MC levels affect the survival of most basidiomycetes (*2*). At MC levels below fiber saturation, prolonged periods of low MC levels appear to have a significant effect on the survival of some fungi. For example, Scheffer and Chidester (*4*) showed that when wood was stored at a MC of 12% seven of eight sapstain fungi present died within 7 months. On the other hand, most of the wood decay fungi such as *Postia placenta*, *Gloeophyllum trabeum,* and *Gloeophyllum sepiarium* were still alive after 3 years. One exception was *Meruliporia (Poria) incrassata*, a water-conducting fungus that inhabits damp wood structures, which died within 25 days.

In general, soft-rot fungi have a greater tolerance for low oxygen concentration than basidiomycete decay fungi, and as a result they can decay wood that is saturated with water. This explains why these fungi are associated with decay in cooling tower slats and similar products maintained at high MC levels. The influence of low oxygen tolerance by soft-rot fungi is also evident in soils. When soils are maintained at a relatively high MC, basidiomycetes are inhibited and soft-rot fungi become the major wood-degrading microorganisms. This has recently been demonstrated by the junior author when it was shown that only soft-rot fungi were present in a soil bed containing soil with a water holding capacity of 34.1%, determined by AWPA Std. E-10 (*5*), with the soil maintained at a MC of 29% (oven-dry basis) and a corresponding wood MC of approximately 96%. Fungi that were isolated from the soil in this study included species from the genera Fusarium, Penicillium, Phialemonium, Curvularia, Exophiala and Geotrichum. Although all of these genera are common soil-dwellers, many of these species also have soft-rotting capabilities (*6*). This knowledge is useful in conducting soil contact tests by making it possible to screen for either soft-rot or basidiomycete decay based on selection of the appropriate soil MC.

In decay tests that subject wood to soil contact, the MC of the wood and soil are both important because it influences the microbial populations and decay rates. In order to optimize wood decay by basidiomycete fungi the wood MC should be in the range of approximately 40-80%. However, a linear correlation between the soil and wood MC does not exist. As the soil MC is increased, wood shows a corresponding increase up to a certain level after which the wood MC increases very sharply until it levels off around 100%. The significance of this relationship is discussed in more detail in the section on soil contact decay tests.

Temperature

The effects of temperature on the growth rate of fungi have been studied extensively. Most wood decay fungi show optimal growth between 12 to 40°C (*2*). At the lower and higher extremes of this range growth is either inhibited or sharply retarded. The temperature optimum for fungal growth varies widely within species as well as among species, so it is not possible to select a single temperature that will maximize decay rates. For most common wood-decay fungi studied the optimum temperature for growth, which is correlated to decay rate, appears to be in the range of 20 to 36°C. For the fungi generally used in laboratory decay tests, the optimum temperature for decay ranges from 28 to 36°C.

The fact that some fungi are consistently associated with certain exposures of wood products may be at least partially associated with temperature optima (*2*). An example is the predominance of *G. trabeum* and *G. sepiarium* in the exterior wood of various structures which attain fairly high temperatures during the summer months.

Hydrogen Ion Concentration

The pH of wood has an effect on the growth of fungi. In general, wood decay fungi grow best within the pH range of 3 to 6; whereas many bacteria and actinomycetes grow best at a more neutral pH. Brown-rot fungi prefer more acidic conditions than white-rot fungi and have an optimum pH of around 3 (*2*). In general, the pH of wood decreases during the decay process.

Nutrient Requirements of Fungi

In addition to the carbon compounds supplied by wood, fungi require the nutrients and vitamins development and growth (Table 1). Of these nutrients, nitrogen is unquestionably the most important because substantial amounts are required for the synthesis of proteins and other cell constituents (*7*). Phosphorous is also very important because a deficiency of this element has been shown to limit the uptake of nitrogen by fungi (*1*). Wood contains some nitrogen, which originates in the cytoplasm of cambial cells that differentiate and mature into woody plant tissue. The nitrogen is located primarily in the living parenchyma cells (*8*). In trees, the nitrogen content decreases from the cambium to the heartwood and then remains constant.

Table I. Nutrient and vitamin requirements of wood decay fungi (*2, 9*)

Major Elements	*Minor Elements*	*Vitamins*
Nitrogen	Iron	Thiamine (B1)
Phosphorus	Zinc	Biotin
Potassium	Copper	
Magnesium	Manganese	
Sulfur	Molybdenum	
Calcium		

The total amount of nitrogen present in wood is quite variable within as well as among wood species, ranging from 0.038-0.227% w/w in softwoods and from 0.051-0.106% w/w in hardwoods (*8*). It also appears that the nitrogen content varies within individual growth rings. In this regard, it has been shown that for *Pinus sylvestris* the level of nitrogen in the springwood was 1.5 to 2 times greater than in the summerwood. Other studies with both softwoods and hardwoods suggest that higher levels of nitrogen are common in the springwood compared to summerwood (*7*). Nitrogen levels are extremely low in relation to carbon content, with C:N ratios ranging from approximately 250:1 to 1250:1 (*1*).

A number of studies have demonstrated that the nitrogen level influences the rate of wood decay. For example, Merrill and Cowling (*8*) showed in lab tests that the decay rate by *G. trabeum* and *Trametes versicolor* was twice as great in wood from the outer rings that contained 0.149% nitrogen as compared to that from the inner rings that contained 0.044% nitrogen. Levi and Cowling (*10*) obtained similar results for the same fungi with oak wood. In another study, the decay rate by soft-rot fungi could be accelerated 5- to 10-fold by pre-treating wood with 2% casein hydrolysate (*11*). It has also been shown that the addition of nitrogen to agar media, soil, and expanded mica, or use of soils naturally high in nitrogen, resulted in greater weight loss of wood exposed to fungi growing on these media (*8*). Further substantiation of the importance of nitrogen levels in wood on the degradation by fungi is provided by studies that have shown that when nitrogen is extracted from wood it is less susceptible to decay (*8*).

The type of wood decay that occurs in soil contact appears to be related to the nitrogen content of wood. For example, King *et al.* (*12*) showed that soft-rot fungi require higher levels of nitrogen than basidiomycetes to decay wood.

Accordingly, they suggested that a nitrogen threshold level of approximately 0.2% w/w is required before decay of wood in soil contact by soft-rot fungi assumes significant proportions. Furthermore, it was demonstrated that the nitrogen levels of wood increase by up to 600%, based on the original pre-burial weight of the wood, during decay by soft-rot fungi in soil contact. This accumulation of nitrogen during the decay process by soft-rot fungi is in sharp contrast to pure culture basidiomycete decay where the nitrogen levels remain relatively constant during the decay process. King *et al.* (*12*) also found that failure of wood exposed to soils that have high nitrogen levels is invariably caused by soft-rot fungi.

Decay fungi are able to effectively degrade wood despite the relatively low levels of nitrogen in wood. This is somewhat surprising but can be explained by a number of unique mechanisms that wood decay fungi employ to overcome the nitrogen deficiency. Fungi are able to conserve nitrogen by hyphal autolysis and recycling nitrogen (*7*). Wood decay fungi can also parasitize the hyphae of some ascomycetes and Fungi Imperfecti as well as bacterial cells (*10, 13*). This may explain why there is often a close relationship between wood stain fungi, bacteria, and wood destroying basidiomycetes. Wood decay fungi are also able to tranlocate nitrogen and other nutrients from external substrates such as soil (*14*). Some species of bacteria that colonize wood may exhibit nitrogenase activity and are therefore able to fix atmospheric nitrogen (*15,16*). Consequently, invasion of wood by these bacteria would result in elevated nitrogen levels by fixation as well as by the presence of the bacterial cells. When wood is placed in contact with soil, the movement of water into the wood transports nitrogen as well as other nutrients (*17*).

Microbial Colonization of Wood

Because of its cellular structure, wood can be readily colonized by a variety of microorganisms. The overall process is a dynamic one in which the nature of the microenvironment continually changes. A number of studies suggest that the general sequence of wood colonization by microorganisms, either in above-ground or soil contact exposure, is: bacteria → stain-fungi – soft-rot fungi → basidiomycetes (*18-20*). The time required to complete this colonization sequence is variable and depends on a number of factors such as wood species (*21*) and treatment with wood preservatives. Carey (*22*) found that the colonization sequence in untreated wooden joinery was completed in 71 days. Basidiomycetes were not detected in wood treated with either tributyltin oxide or pentachlorophenol until after 221 and 730 days, respectively. However, colonization and detection of visible decay are two different parameters. Our observations indicate that even in high decay hazard climates decay in coniferous sapwood test samples exposed above ground is not evident before six months of exposure and in some pieces the lag period may extend to 24 months. Obviously, some inhibitory factors are involved since in a pure culture test decay is apparent within a few days.

The delay in colonization of untreated wood by basidiomycetes may be partially due to antagonism expressed by some of the microorganisms. For example, a number of studies have shown that actinomycetes and species of Eubacteria are highly antagonistic toward wood degrading basidiomycetes (*23*). These bacteria also degrade wood extractives that affect colonization by basidiomycetes. Removal or alteration of these extractives eventually permits rapid colonization by wood decay fungi. However, Hulme and Shields (*24*) suggested that the major factor involved in repression of the basidiomycetes is the rapid depletion of nutrients that are required for rapid growth and successful colonization. Although some of the microorganisms that colonize wood in the early stages are antagonistic to basidiomycetes, others have the opposite effect (*25, 28*). For example, some bacteria develop a synergistic relationship with wood decay fungi (*23*). This synergism stimulates fungal growth and accelerates the rate of wood decay. In this relationship, bacteria are provided with a supply of carbon sources as a result of the extracellular enzyme reactions of the fungus that would not normally be available. Growth of the bacteria depletes concentrations of cellobiose, simple sugars, etc., around the hyphae so that cellulase production is not repressed. In turn, bacteria may supply essential vitamins and other growth promoting substances such as nitrogen to the fungus.

Another aspect of early colonization by actinomycetes, ascomycetes, and Fungi Imperfecti is the ability of some of those microorganisms to detoxify wood preservatives. Merrill and French (*26*) showed that ascomycetes and Fungi Imperfecti detoxified a number of wood preservatives. Carey (*18*) has shown that a *Phiolophora species* isolated from TBTO-treated wood effectively detoxified this preservative when it was incorporated into agar tests. In another study by King *et al.* (*27*), it was noted that actinomycetes isolated from treated wood have the ability to biodegrade pesticides in soil. Hence, a number of microorganisms that colonize wood have the potential to detoxify wood preservative chemicals. Such detoxification could then set the stage for wood degradation by basidiomycetes.

Spore Germination

It is generally assumed that fungal spores are the principal infection mode of wood exposed above ground under conditions conducive to decay. Although the presence of spores produced by microorganisms capable of degrading wood are ubiquitous, their germination on wood is problematic (*29, 30*). In order for fungal spores to develop and colonize wood, there must be a sufficient supply of readily assimilable nutrients, a favorable microclimate, including adequate moisture, and an absence of competitors. Much of the time these conditions do not exist simultaneously so the mere presence of spores does not necessarily ensure that the wood decay process will be initiated. The failure of spores to germinate when they are in contact with wood may be attributed to nutrient and water availability, pH, temperature, wood extractives, and microbial metabolites, etc. (*1*). Unfortunately, only a limited number of studies have been conducted in this area so our understanding of this complex subject is limited.

Morton and French (*31*) studied spore germination with *G. trabeum*, *G. sepiarium*, and *Fomitopsis roseus* on wood and found that sterilizing wood significantly increased the percentage germination of *G. trabeum* basidiospores, with a greater percentage germinating on wood that was autoclaved dry rather than wet. Furthermore, when wood was subjected to either hot or cold water extraction, the percentage of *F. rosea* spores that germinated was significantly less than that for non-extracted wood. This difference suggests that some water-soluble extractives in Douglas-fir sapwood stimulate germination. In additional experiments, they compared the effect of sodium pentachlorophenate-treated wood on germination of *G. trabeum* basidiospores to the growth of mycelium from the same fungus. From this experiment it was concluded that the basidiospores were about 10-fold more sensitive to the biocide than was the mycelium.

Schmidt and French (*30*) studied the effect of sterilization methods on spore germination of *G. trabeum*, *Trametes hispida*, and *Poria tenuis* and found that some methods severely retarded germination. They concluded that the best way to minimize variation in spore germination was to briefly dip the wood in boiling water.

One of the problems with spore germination studies is that a given fungus can produce several different types of spores. For example there are basidiospores, chlamydospores, and secondary spores (conidia). These different types of spores may have varying requirements for germination, so experiments with only one type of spore may not apply to the others. A good example of this are the thick-walled resting spores of the white-rot fungus *Ganoderma lucidum* which must pass through a fly larvae gut before they will germinate (*1*).

Test Methodology

Methods of Detecting and Measuring Wood Decay

At the present time, lack of accurate, rapid, non-destructive methods for detecting and quantifying wood decay is a major deterrent to wood preservation research. It will be necessary to develop improved methods for measuring the extent of wood decay in test samples before accelerated test methods can become a reality. A partial review of current methods and new approaches that may have potential are outlined in this section.

Visual Observation

At present, visual observation is the method normally used to evaluate outdoor exposure wood decay test specimens exposed outdoors. This method is subjective, rather than quantitative, does not detect early decay and consequently does not provide the information needed for accelerated testing.

Mass Loss

During the decay process, fungi utilize wood as a carbon source and produce CO_2 and water as by-products. As a result, the wood mass is gradually reduced. This can be measured at specific times or progressively by determining the weight of test specimens and comparing this with the samples original weight. This method is widely used in laboratory soil block decay tests and provides reasonably good results. Shortcomings of this method are associated with difficulty in making adjustments for variation in wood moisture content, loss of wood preservatives and inability to make adjustments for the weight gain due to colonization by the fungus.

Because of these problems, the use of mass loss as a method for detecting decay is essentially limited to laboratory soil block tests. When used for this application, the main weakness of this test is its inability to detect the early stages of decay. It is generally conceded that mass losses of less than 2-3% are statistically insignificant because of the inherent variability of this method. Even small mass losses in the early stages of decay are associated with significant strength losses, especially when the wood is colonized by brown-rot fungi (*32*). For example, it has been shown that the compression strength of wood has been reduced by at least 20% by the time a mass loss of 2-3% has been attained (*33*).

Mechanical Properties

The mechanical properties of wood are adversely affected by wood decay fungi, so that the progressive measurement of selected strength properties could be used to monitor the wood decay process. This supposition has been verified in a number of studies and appears to have considerable potential (*33*). Although there are a number of strength properties that can be measured, compression, bending, and torsion strength are perhaps the most suitable for detecting and measuring decay in laboratory and field tests.

Compression Strength

Recent studies demonstrate that the extent of wood decay can be quantified by measuring the compression strength of wood wafers in the radial direction (*33, 34*). This method is based on the comparative compression strength at the 5% level of end matched wafers, one of which is exposed to the fungus in a soil block test and the other serving as a non-exposed control. The use of thin wafers (5 mm in the longitudinal direction) and compression strength analysis has made it possible to carry out tests in approximately 6 weeks, compared to approximately 16 weeks necessary for the standard soil block test utilizing mass loss.

Janzen's study (*33*) also indicates that it may be desirable to use modulus of elasticity (MOE) rather than 5% compression strength as a measure of wood decay. Use of MOE results in less variability among samples and appears to provide a better estimate of toxic threshold values.

Bending Stiffness

MOE of wood decreases during the wood decay process (*32*). As a result, progressive measurement of bending stiffness is a potential method for following the decay process by determining the initial MOE at strain levels below the proportional limit and then making subsequent measurements. Crawford (*35*) used laboratory soft-rot tests to show that there is a good correlation between the reduction of MOE in thin wood sticks and the number of soft-rot cavities. Hence, this is an excellent method for evaluating soft-rot decay test samples. Although detailed studies have not been carried out, it appears that changes in bending stiffness could be used to measure the extent of decay in test specimens that have been attacked by basidiomycetes. This concept has been evaluated in preliminary studies in our laboratory and the results obtained to date are encouraging.

Torsion Strength

The torsional strength of wood is often used to measure the shear strength. As such, it is conceivable that torsional shear might be used to measure the extent of decay in test specimens exposed to wood decay fungi. We have initiated some studies in our lab evaluating the use of torsion strength as a measure of wood decay. In this process, the sticks (14 mm square by 200 mm long) are torqued at a level below the proportional limit to establish the initial torsion strength. The samples are then exposed to wet, unsterile soil in special tube containers that leave 50 mm of the stick ends exposed to air. These sticks are then removed periodically and re-tested, with reductions in torsion strength serving as a measure of decay. Additional work needs to be done, but this method shows promise for accelerated soil bed tests.

Permeability

Carey (*36*) showed that the permeability of wood exposed above-ground increased during the early stages of decay. This increase is due to the colonization of wood by bacteria and various stain and soft-rot fungi. As a result of this observation it was suggested that measurement of permeability could be used to detect incipient decay. One method of doing this is to measure the uptake of decalin when the samples are dipped in this low viscosity liquid. However, due to the extreme variability of wood permeability, this method has not been very successful. Nevertheless, if better methods of measuring the permeability of wood could be developed then this method might have some potential. In this regard, one possibility is the use of the air pressure method developed by Wallace and Schmidt (*37*).

Immunodiagnosis

In this method, polyclonal and monoclonal antibodies are used to detect the presence of wood decay fungi (See chapter by Clausen). This method is very sensitive and has the potential to detect incipient stages of decay in above-ground test samples. One limitation of this method is the inability to quantify the progressive development of wood decay. As a consequence, it appears necessary to couple this procedure with some other method, such as bending stiffness, to provide adequate evaluation of test samples.

NIR Analysis

In recent years near infrared (NIR) spectroscopy has been used for many applications that involve rapid analysis of various substrates. Preliminary studies with decayed wood indicate that NIR may be useful in detecting and quantifying biodeterioration of wood. By coupling NIR analysis with multivariate regression or principal components analysis, it has been demonstrated that there is a good correlation between the NIR analysis and mass loss of wood decayed by brown-rot fungi (*38*). On the basis of additional preliminary studies, there also appears to be a good relationship between NIR analysis and changes in compression strength of brown-rotted wood.

Additional studies will need to be conducted before the potential of this spectroscopic method can be discerned. So far, only wood degraded by brown-rot fungi has been investigated and additional studies with wood decayed by white-rot and soft-rot fungi are needed since these microorganisms are also associated with the wood degradation process. Furthermore, it is not known whether this method can detect the early stages of decay, so additional work is also needed in this area.

The availability of lightweight, portable NIR instruments makes this potential method of decay detection particularly attractive. Furthermore, the fact that these instruments can be fitted with fiber optic probes provides the versatility needed for on-site evaluation of various test sample configurations.

Soil Block Test

In the evaluation of new wood preservatives, the soil block test is typically used to provide initial information on the efficacy against wood decay fungi. Both white- and brown-rot fungi are generally included in this evaluation. The toxic threshold value, i.e., the minimum amount of preservative that inhibits wood decay, is determined in this test by using test specimens that are treated with varying levels of the preservative in question.

The standard soil block test (AWPA E 10) specifies that either 14 mm or 19 mm wood cubes can be used and the presence of decay is determined by mass loss (*5*). In order to have a valid test the untreated control blocks must attain a minimum of 50% mass loss, which requires about 12 weeks exposure for brown-rot fungi. By using 14 mm cubes the rate of decay is increased and the incubation time can be reduced from 12 to 8 weeks. However, when softwoods are tested against white-rot fungi the exposure time needs to be increased in order to achieve sufficient decay in the untreated blocks. Using this standard method, the total time required to run the test is in the range of three to four months.

Another approach to measuring the extent of decay in the soil block test is to use compression strength, which decreases in proportion to mass loss during wood decay (*39*). The advantage of using compression strength is that it permits the use of smaller samples without sacrificing accuracy. By using wood samples measuring 5mm thick it is possible to reduce the incubation time to approximately four weeks. Furthermore, the conditioning period required to equilibrate the test samples to a specified moisture content is eliminated (*34*). As a consequence, the total time required to carry out the test can be reduced to approximately six weeks.

Although the soil block test is fairly rapid, it has a number of limitations. Because it uses a pure culture of a given fungus growing under near ideal conditions, the test is more severe than natural (outdoor exposure) conditions would be, where colonization of the wood is dependent on spore germination and decay fungi must compete against each other as well as other microorganisms. There are also numerous fungi that can decay wood and these can exhibit a wide range of tolerance to different biocides. The virulence of fungi are also dependent on which strain is being used and this can vary with the extent of subculturing, the type of wood, the type of soil, etc. (*40- 42*). Finally, most organic biocides are susceptible to microbial or chemical degradation and this effect is less apparent in the artificial short term lab decay test. As a result of these weaknesses, the soil block test results are best used as an initial screen to compare a given candidate preservative to known preservatives in order to determine whether or not further evaluation in field trials is justified.

When wood is used in exterior above-ground applications, the decay process is primarily initiated by spore germination. This suggests that it might be possible to use spore germination to establish toxic threshold values of candidate preservatives. Morton and French (*31*) carried out a series of experiments to evaluate this concept and concluded that such a test method might be feasible. This concept was subjected to further evaluation by Savory and Cary (*43*) using basidiospores of *G. trabeum*. The toxic threshold values for tributyltin oxide, pentachlorophenol and chromated copper arsenate, determined by inoculation with mycelium and also by spores, were found to be essentially the same. Furthermore, it was found that spores would germinate at some preservative concentrations but failed to colonize the wood. Hence, the concept of using spore germination as a rapid method for establishing toxic threshold values for wood preservatives may not be possible. However, it might be possible to monitor spore germination and further hyphal development for a relatively simple and reasonably rapid test method. Additional research in this area would be desirable.

Above-ground Decay Tests

These tests are carried out to evaluate wood preservative systems that are designed for treated wood used in exterior applications where the product is not subjected to soil contact. Typical applications are millwork, decking, fascia boards, etc. A number of different designs have been used over the years with the most popular being the post-rail, L-joint, and Lap-joint units. In recent years the L-joint and Lap-joint designs have been used for most tests (Figure 1).

A common element in the design of these test units is the inclusion of some type of joint that effectively traps rainwater. Although these units provide a realistic evaluation of the performance of wood preservative treatments, an extended exposure period, with the time being dependent on the rainfall and temperature at the test site, is needed to develop a reliable prediction of preservative performance. Also, the use of visual observation and probing techniques in the evaluation process makes it difficult to detect incipient decay and provide quantitative data, and the ratings are often dependent on the moisture content of the sample at the time of evaluation.

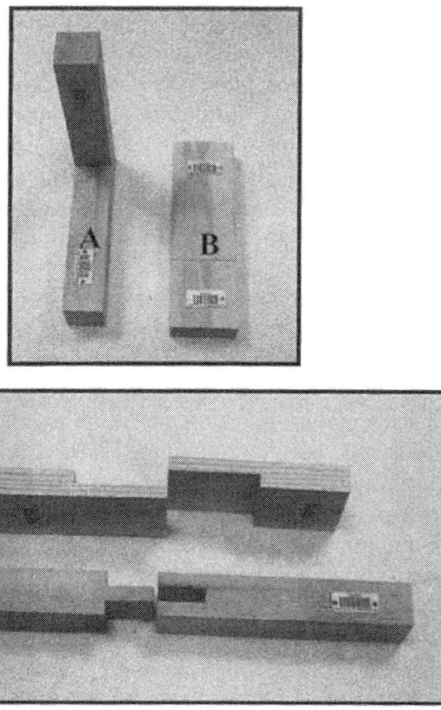

Figure 1. Top and side view illustrations of L-Joint (A) and Lap-Joint (B) above ground test units.

One of the major weaknesses of the current above-ground test unit procedures is the extended time required to evaluate a given preservative. In order to address this problem two concepts need to be examined. These are the development of methods for accelerating the decay process and methods that accurately detect incipient decay and subsequent progressive deterioration of the wood structure.

Accelerating Decay Rates

Optimizing decay rates requires developing a better understanding of the spore germination process as well as providing appropriate nutrient, moisture and temperature levels. Factors influencing spore germination on wood have been discussed in detail in earlier sections of this chapter. From this review, it is apparent that considerably more research is required to permit the development of methods that optimize spore germination on wood. Such information is critical in our quest to develop accelerated test methods.

With regard to nutrients, it seems logical that providing appropriate levels of the essential elements for fungal growth will be necessary to accelerate decay test methods. Nitrogen is of particular interest because wood is inherently deficient in this element. Since autolysis of other microorganisms and possibly airborne particles are the only source of additional nitrogen in wood exposed above-ground, serious thought needs to be given to injecting nitrogen compounds into the wood test units. The inclusion of other elements, such as phosphorus which is essential for nitrogen utilization by fungi, also needs to be considered. Fungi also require the vitamins thiamine and biotin which are not present in wood and are probably provided by bacteria that colonize wood prior to basidiomycete attack (*13*). Hence, it would be of interest to see if the addition of vitamins would accelerate the colonization by basidiomycetes. Although the above suggestions may be of value, it should be emphasized that the addition of nutrients could have a major impact on microbial populations and a considerable amount of research will be required to ensure that artificial condition are not created that result in unrealistic test conditions. Great care will need to be taken to ensure that these modifications do not accelerate decay rates in a manner that is unrealistic for the intended end use applications.

As discussed previously, the optimum wood MC for decay by basidiomycetes is in the range of 40-80%. Consequently, developing a system that would maintain wood at these moisture levels should accelerate the decay process as long as appropriate temperatures were also maintained. One approach to maintaining a high MC of non-soil contact test specimens is to place them on concrete blocks that rest on the ground. Further reduction in moisture loss is achieved by covering the test unit with shade cloth, which allows rainwater to enter but reduces evaporation rates. This test method has been used successfully in Hilo, Hawaii to accelerate rates of biodeterioration (*44*).

By giving consideration to the above factors it should be possible to significantly accelerate the wood decay process in above-ground tests. This approach must be rigorously evaluated to ensure that unrealistic artificial conditions are not created so that the results reflect real world conditions. Accordingly, the accelerated method must mimic the results obtained from actual long-term exposure.

An indication of the validity of this approach is provided by our experience with conducting above-ground decay tests in Hilo, Hawaii and Saucier, Mississippi. The Hilo location has ideal temperature and rainfall conditions to support wood decay. In contrast, the Saucier location has considerably less rainfall and lower temperatures in the winter. Our results from testing the same wood treatments at both sites clearly show that the rate of decay is accelerated 2-3 fold at the Hilo test site. Figure 2 shows decay ratings over a 10-year period for dip-treated Ponderosa pine L-joints. Clearly, this represents a significant acceleration. Nevertheless, it may be possible to further accelerate the decay rate by incorporating appropriate nutrients in the test specimens as long as this can be done without creating unrealistic decay conditions that do not apply to real world applications.

In addition to accelerating decay rates, improved methods for detecting and measuring the extent of decay are needed. Methods that appear to have the best potential for above ground tests are based on mechanical tests, NIR analyses and immunodiagnosis. A conceptual test unit for using a combination of immunodiagnosis and mechanical test is shown in Figure 3. This test utilizes a typical lap-joint with thin sticks incorporated in the joint. The immunoassay will be used to detect the onset of colonization by decay fungi by periodically removing small samples near the ends of the sticks. In addition, the sticks will be removed periodically and subjected to a bending test to measure the MOE, which decreases as decay develops.

If additional research confirms that both brown- and white-rot decay can be detected and quantified by NIR analyses, it would be a valuable tool for use in above ground decay tests. Portable NIR units are available and it may be possible to measure decay in lap-joints as well or other test units by using small fiber optic probes.

Soil Contact Decay Tests

Treated wood products are often used in applications that involve soil contact. This results in a very severe exposure condition for a number of reasons. Soil contains large populations of microorganisms, some of which can biodegrade wood. A number of these microorganisms also have the ability to biodegrade organic wood preservatives while others can modify inorganic preservatives and make them more susceptible to leaching. Soil contains numerous chemical components that provide essential nutrients required by the fungi and also have an impact of the rate of depletion of wood preservatives. The leaching of wood preservatives is generally more severe when the wood is

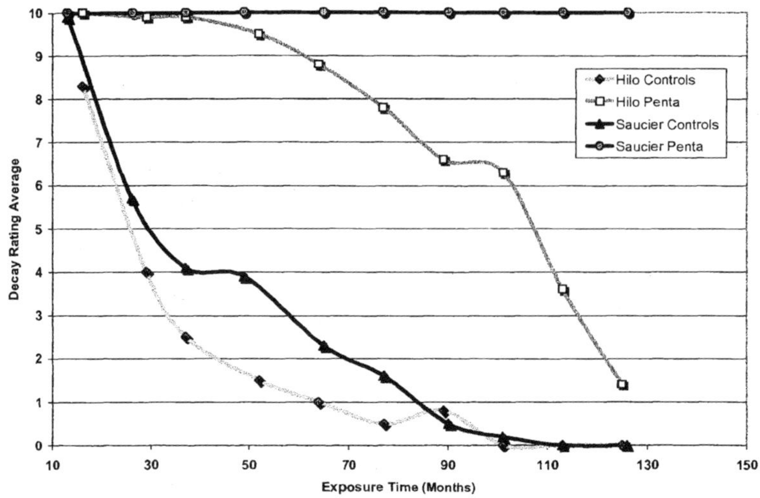

Figure 2. Comparative decay rating for ponderosa pine L-joints after exposure at Hilo, Hawaii and Saucier, Mississippi. The test samples were dip treated with 5% Penta in mineral spirits and the controls were dip treated with escorez resin (5%) in mineral spirits. A 10 rating denotes no decay and a 0 denotes failure.

exposed to water that contains soil than to water alone. Soil also serves as a reservoir for water, which helps maintain wood MC levels in the range that favors wood decay.

In testing wood preservatives for soil contact applications, treated test specimens of various sizes ranging from ¾ inch square by 18 inches long to full size pole stubs are inserted to one-half their length into pre-drilled holes in the ground at selected test sites. These test specimens are inspected periodically and rated for the extent of decay and insect attack. These are generally long-term tests that require a minimum of five years of exposure of smaller test specimens in locations that have relatively high decay hazard conditions. Although the small test specimens decay faster, they do not provide information on potential preservative distribution problems and may exacerbate preservative depletion rates.

One method for accelerating the decay rate when wood is subjected to soil contact is to control the soil moisture content and temperature conditions so that they are optimum for wood decay fungi. This can be done either by selecting test sites that have warm/wet climates or by the use of soil beds that are artificially watered and housed in buildings that have temperature and humidity

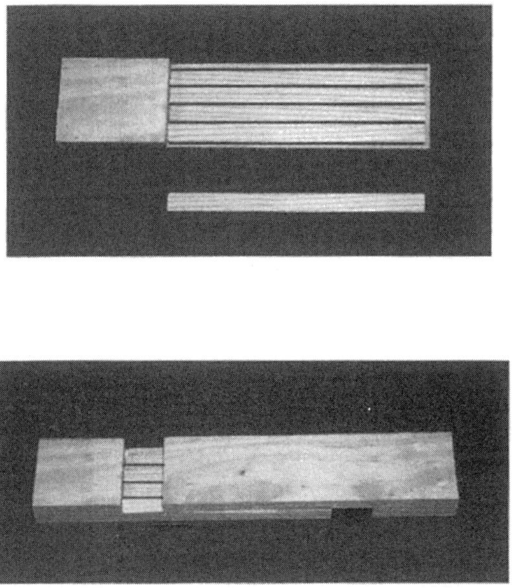

Figure 3. The modified Lap-joint designed to incorporate bending strength and immunoassay to detect and measure decay.

controls. This concept was originally developed by Gersonde and Becker (*45*) and further refined by others (*44,46-48*). On the basis of these studies and additional work by Archer (*49*), a soil bed test procedure was developed (*5*). The use of this accelerated method to predict long-term performance of treated wood exposed to soil has not been widely accepted and long term field tests are still required before one can have confidence in the performance of new wood preservative systems. Some concerns about this method are discussed in recent publications (*49, 50*). The basic problems with this concept are associated with difficulty in maintaining consistent moisture levels and failure to recognize the role that soil physical and chemical properties play in wood preservative depletion rates and microorganism populations. Another major problem is the lack of accurate methods for detecting and quantifying the extent of decay in test specimens.

Moisture Control

Most soil beds are operated on the basis of controlling moisture at some specified MC in relation to the water holding capacity of the soil. Accurate control of soil MC is extremely difficult and wide variations generally exist in the soil beds, which are often overlooked. Nevertheless, even if adequate soil MC control is achieved this approach is often flawed because it fails to take into consideration the soil/wood MC relationships. Wood MC is critical because basidiomycete fungi are most active when the wood MC (oven dry basis) is in the range of 40 to 80%. Recent studies in our laboratory on the MC of wood in contact with soil shows that the relationship between soil MC and wood MC is not linear (Figure 4). From this graph it is apparent that there is a sharp rise in wood MC when the soil MC reaches about 22%. This steep rise then levels off when the soil MC reaches about 25%. The soil used in this study had a water holding capacity of 34.6%. We have observed similar relationships for other soils, but the inflection point in the curve occurs at widely different soil moisture contents. Hence, it is important that soil/wood MC relationships be developed for wood samples used in soil bed tests. Although the MC of some areas of the stakes - at ground line and slightly above – may be in the desired range the soil MC is still important because it can influence the type and number of microorganism present.

Physical and Chemical Properties of Soils

Often little consideration has been given to the potential impact of soil types, chemical composition and soil collection methods on the results obtained in soil beds despite their potential impact on biocide depletion, microbial populations and fungal virulence. Biocide depletion rates are an important factor in soil bed tests because of the relatively small test specimens and consistently high MC levels. A recent study has shown that wood preservative depletion rates vary widely in different types of soils (*51*), but the specific components or properties of the soil responsible for leaching differences were not identified. Additional research is needed in this area. With regard to microorganism populations, a study by Nilsson and Daniel (*52*) has shown that some soils favor brown-rot fungi and others favor white-rot fungi and garden compost favors soft-rot fungi. The reason for these differences is not clear, but is undoubtedly related to the soil chemical and physical properties. As discussed previously, several elements and vitamins are critical for the growth and development of fungi. Since the relative amounts and types of these compounds in soil undoubtedly have an influence on the rates of wood decay, analysis of soil and addition of deficient elements would help assure that optimum decay rates are achieved in soil bed tests.

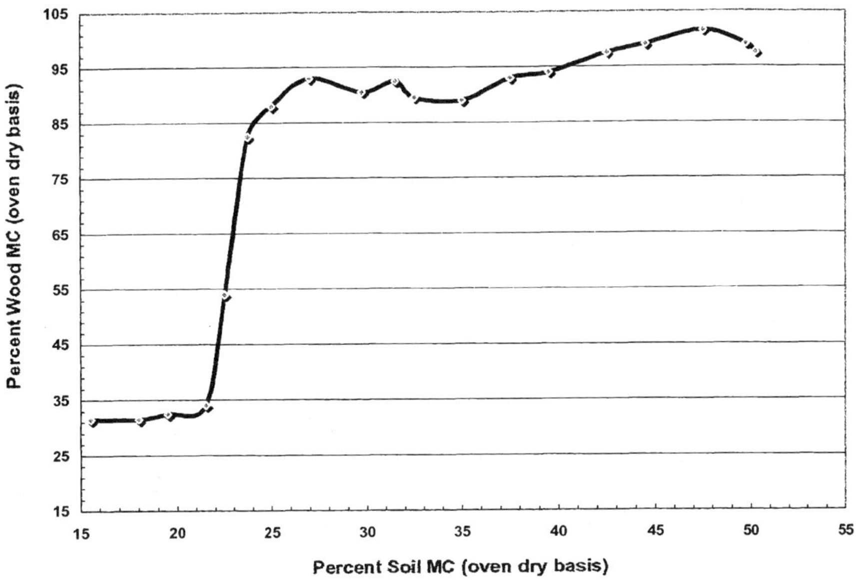

Figure 4. Relationship between soil MC and the EMC of southern yellow pine sapwood sticks buried in the soil. The water holding capacity of the soil determined using AWPA standard E10 (5), was 34.6%.

Oxygen level in soil is also an important factor that has not been given much attention. In the typical soil beds, which have stakes inserted vertically, optimum oxygen levels are undoubtedly available at the ground line areas much the same as they are in field stakes. In the soil bed designs, such as the tube test described below in this paper, the oxygen levels at the interior portion of the stakes may or may not have optimum oxygen levels. Research is needed in this area to determine whether or not aeration of the soil beds would enhance decay rates.

Microbial Populations in Soil

Soils contain large populations of microorganisms—fungi, bacteria, nematodes, protozoa and microarthropods. The types and relative numbers of these microorganisms is highly dependent on the composition, moisture, oxygen levels, and temperature of the soil. These factors need to be considered to optimize wood decay rates in the soil, but very little research has been carried out in this area. Nevertheless, the potential of this approach is illustrated by a study recently carried out in the laboratory of the senior author. In this study, a soil obtained from a forested area was fortified with composted wood that added additional nitrogen, phosphorus and organic matter (Table II). This soil and non-amended soil were used in a soil bed test with untreated pine sapwood stakes. After 8 months of exposure, the amount of decay in the stakes in the amended soil was considerably greater (Table II). These two soil samples were also analyzed for bacteria and fungal populations. It is apparent from Table III that the addition of compost had a significant effect on microbial growth, particularly on the basidiomycete populations. Although limited in scope, this study illustrates the promise of this approach to accelerating decay rates in soil bed tests.

Table II. Chemical Analysis of Soils Used in a Soil Tube Test and Decay Ratings for Untreated Pine Sticks After Eight Months Exposure

Soil Composition	*Total Organic Carbon (μg/g)*	*Total Organic Nitrogen (μg/g)*	*Total Organic Phosphorus (μg/g)*	*Decay Rating*[a]
Soil	27104	988	44	9.8
Soil + Compost (25% w/w)	91543	3603	2585	7.5

[a]A rating of 10 denotes sound and 0 denotes failure.

Table III. Microbial Analysis of Soils Used in a Soil Tube Test

Soil Composition	*Active Bacterial Biomass (μg/g)*	*Total Bacterial Biomass (μg/g)*	*Active Fungal Biomass (μg/g)*	*Total Fungal Biomass (μg/g)*	*Average Hyphael Diameter*[a] *(μm)*
Soil	2.2	214	0.7	174	3
Soil + Compost	2.8	216	3.7	287	3

[a]Hyphael diameter of 3 denotes a community dominated by basidiomycetes.

New Soil Bed Design

Accurate control of soil and wood moisture contents is critical in soil bed tests. Many soil bed facilities use large containers, such as bathtubs or large concrete bins. In general, the larger the container the more difficult it is to control soil and wood moisture levels. As a result, we have been investigating the use of small PVC tubes as soil containers (Figure 5). In this configuration the test specimens (14 x 14 x 250 mm long sticks) are inserted into holes in the side of the tube so that only the center portion is in contact with the soil. The soil MC is controlled by weighing the tubes periodically and adding water through a perforated tube extending along the bottom of the tube. The wood MC is controlled by inserting thin wood sticks into the soil from the top opening in the tube, and then removing them periodically and determining the MC by weight.

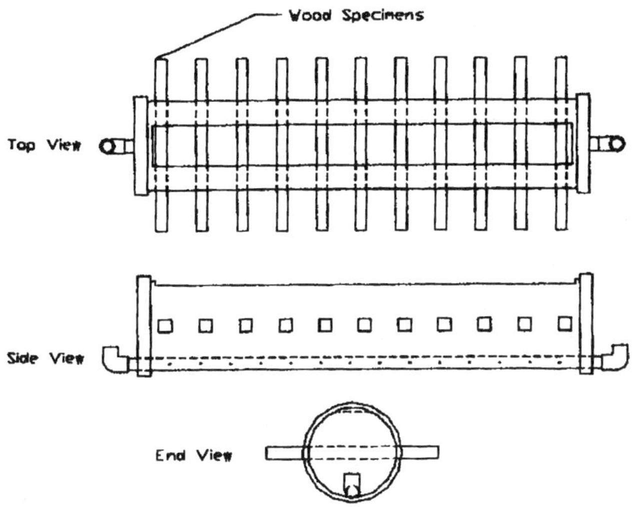

Figure 5. Schematic of soil tube test apparatus.

CONCLUSIONS

There is a need to develop accelerated test methods for evaluating the efficacy of wood preservatives. To achieve this goal it will be necessary to develop improved methods for detecting and quantifying the extent of wood decay. It will also be necessary to develop a better understanding of the many variables that influence microbial wood decay rates. In this regard, we need more knowledge on the process of microbial colonization of wood and the interaction of the various types of microorganisms involved in the process. In addition, our knowledge of the role that essential nutrients play in the decay process and their effect on the efficacy of wood preservatives is limited and needs to be addressed. Once we develop a better understanding of these relationships it should be possible to develop vastly improved wood decay test methods. However, it needs to be emphasized that there is a danger in developing accelerated test methods because the results may not provide a realistic evaluation of commercial treated wood products. Consequently, it is essential that comparative field tests with known preservatives be included in the development process.

REFERENCES

1. Rayner, A.D.M.; Boddy, L. *Fungal decomposition of wood: its biology and ecology*; John Wiley & Sons, New York, 1988.
2. Zabel, R.A.; Morrell, J.J. *Wood Microbiology Decay and Its Prevention.* Academic Press: San Diego, CA, 1992.
3. Rypacek, V. 1966. *Biologic Holzzerstorender*. Pilze, Fischer, Jena.
4. Scheffer, T.C.; Chidester, *MS Southern Lumberman*. **1948**, *177*,110-112.
5. *American Wood Preservers' Association Standards* 2000: Granville, TX.
6. Wang, C.J.K.; Zabel, R.A. Identification manual for fungi from utility poles in the eastern United States. *American Type Culture Collection*, Rockville, MD, 1990.
7. Cowling, E.B. 1970. Nitrogen in Forest trees and its role in wood deterioration. *Abstracts of Uppsala Dissertations in Science*, 164.
8. Cowling, E.B.; Merrill, W. *Can. J. Bot.* **1966**, *44*,1533-1544.
9. Cochrane, V.W. *Physiology of fungi*. John Wiley & Sons, New York, 1958; 525 pp.
10. Levi, M.P.; Cowling, E.B. *Phytopathology* **1968**, *58,* 246-249.
11. Levi, M.P.; Merrill, W.; Cowling, E.B. *Phytopathology*. 1968, 58: 626-634.
12. King, B., Mowe, G., Bruce, A., Smith, G.M.. *Biodeterioration* 5. Oxley, T.A. and Barry, S. (Eds.) John Wiley & Sons, New York, 1983.
13. Levy, J.F.; Millbank, J.W.; Dwyer, G.; and Baines, E.F. The role of bacteria in wood decay. *Record Brit. Wood Pres. Ann Conv*. 1974, pp. 3-13.
14. King, B.; Waite, J. *Biodeterioration Bull.* **1979**, *15*(1), 29-35.
15. Levy, J.F. *Colonization of wood by fungi*. News Sheet No. 130. Brit. Wood Pres. Assn., 1973.
16. Baines, E.F.; Millbank, J.W. Nitrogen fixation in wood in ground contact. *Organismen und Holz*: International Symposium Berlin-Dahlem. 1976, 3,167-173.
17. Baines, E.F.; Levy, J.F. *J. Inst. Wood Sci.* **1979**, *8,* 109-113.
18. Carey, J.K. *Holz als Roh-und Werkstoff.* **1982**, *40*, 269-274.
19. Clubbe, C.P. The colonization on succession of fungi in wood. *International Research Group on Wood Preservation*,1980, Document No. IRG/WP/1107.
20. Butcher, J.A. *Biodeterioration of Materials Microbiology and Allied Aspects*; Walters, A.H.; Elphick, J.J., Eds.; Elsevier: Amsterdam, 1968; pp 444-459.
21. Carey, J.K.; Bravery, A.F. Developments in the assessment of joinery preservatives. *Record Brit. Wood Pres. Ann. Conv.* 1985, 3-13.

22. Carey, J.K. *Biodeterioration 5*; Oxley, T.A.; Barry, S.; eds.; Wiley Interscience, 1983; pp 13-25.
23. Erickson, KEL., R.A. Blanchette, and P. Ander. *Microbial and Enzymatic Degradation of Wood and Wood Compounds*. Springer-Verlay: Berlin, 1990.
24. Hulme, M.A.; Shields, J.K.. *Can. J. Bot.* **1972**, *50*, 1421-1427.
25. Greaves, H. *Wood Science Tech.* **1971**, *5*, 6-16.
26. Merrill, W.; French, D.W.. *Phytopathology*. **1966**, *56*, 301-303.
27. King. B.; Eaton, R.A.; and Baecker, A.A.W. A summary of current information on actinomycetes and wood. *International Research Group on Wood Preservation*, 1978, Document No. IRG/WP/177.
28. Blanchette, R.A.; Shaw, C.G. *Phytopathology*. **1978**, *68*, 631-637.
29. Schmidt, E.L.; French, D.W. *Plant Dis. Reptr.* **1979**, *63*, 30-31.
30. Schmidt, E.L.; French, D.W. *Phytopathology*,. **1979**, *69*, 688-689.
31. Morton, H.L.; French, D.W. For. Prod. J. **1966**, *16*, 25-30.
32. Wilcox, W.W. *Wood Fiber* **1978**, *9*, 252-257.
33. Janzen, S. M.S. thesis, Department of Forest Products, Mississippi State University, Mississippi State, MS, 2001.
34. Nicholas, D.D.; Jin, Z. Use of compression strength loss for measuring decay in the soil block test. *International Research Group on Wood Preservation*.1966, Document No. IRG/WP 96-20083.
35. Crawford, D.M. M.S. thesis, Department of Forest Products, Mississippi State University, Mississippi State, MS, 1994.
36. Carey, J.K. Biodeterioration **1983**, *5*, 13-25
37. Wallace, B.L.; Schimdt, E.L. *Forest Prod. J.* **2000**, *50*(7/8), 71-74.
38. Kelley, S. 2001. *Personal communication*.
39. Toole, R.E. *Wood Science* **1971**, *3*,172-178.
40. Leithoff, H.; Peek, R-D; Borck, V.; Gottsche, R.; Kirk, H.; and Grinda, M. Toxic values derived from EN113 tests – are they determined by virulence of a test fungus. *International Research Group on Wood Preservation*, 1994, Document No. IRG/WP/99-20176.
41. Hegerty, B. Virulence investigations on several standard basidiomycete test fungi. *Record Brit. Wood Pres. Assn. Conv.* 1987, pp 77-83.
42. Behr, E.A. In *Decay Test Methods in Wood Deterioration and Its Prevention by Preservative Treatments*; Nicholas, D.D., Ed.; *Degradation and Protection of Wood*; Syracuse Univ. Press: Syracuse, NY, 1973; Vol. 1, pp 217-246.
43. Savroy, J.G.; Carey, J.K. Laboratory assessment of the toxic efficacy of joinery preservative treatments. *Record Brit. Wood Pres. Assn. Conv.* 1976, pp 3-24.
44. Preston, A.; McKig, P.; Walcheski, P. Proceedings: *American Wood Preservers' Association*. 1983, 79,207-212.
45. Gersonde, M.; Becker, *G. Holz als Roh-und Werkstoff* **1958**, *16*, 346-357.

46. Hedley, M.E. Comparison of decay rates of preservative-treated stakes in field and fungus cellar tests. *International Research Group on Wood Preservation.* 1980, Document No. IRG/WP/2135.
47. Johnson, G.C.; J.D. Thornton: Greaves, H. The accelerated field simulator (fungal cellar). *International Research Group on Wood Preservation.* 1982, Document No. IRG/WP/2170.
48. Vinden, P.; Savory, J.G.; Dickson, D.J.; and Levy, J.F. Soil bed studies. *International Research Group on Wood Preservation*, 1982, Document No. IRG/WP/2181.
49. Archer, K. Proceedings: *Canadian Wood Preservers' Association.* 1996, 17-28.
50. Nicholas, D.D.; Archer, K.J. 1994. *Forest Products Society Conference Proceedings No. 7308 "Wood Preservatives in the 90's and Beyond.*" Sept. 26-28, pp. 110-115.
51. Wang, J.H.; Nicholas, D.D.; Sites, L.S.; Pettry, D.E. Effect of soil chemistry and physical properties on wood preservatives leaching. *International Research Group on Wood Preservation.* 1998, Document No. IRG/WP/98-50111.
52. Nilsson, T.; Daniel, G. 1990. Decay types observed in small stakes of pine and Alstonia scholaris inserted in different types of unsterile soil. *International Research Group on Wood Preservation.* 1990, Document No. IRG/WP/1433.

Chapter 17

Use of Fatty Acid Profiles To Identify White-Rot Wood Decay Fungi

Susan V. Diehl, M. Lynn Prewitt, and Fatima Moore Shmulsky

Forest Products Laboratory, Mississippi State University, Mississippi State, MS 39762

Identification libraries have been created for the white-rot fungi *Phanerochaete chrysosporium, P. sordida* , *P. sanguinea, Trametes versicolor, T. hirsuta*, and *T. pubescens*. Libraries were compiled using five to ten isolates of each species with each isolate cultured four times, extracted in duplicate and analyzed by gas chromatography for its fatty acid methyl ester (FAME) profile. Single libraries were created for *T. versicolor* and *T. pubescens*, indicating that all isolates of these species clustered as a single species. More than one library was created for *P. chrysosporium, P. sordida,* and *T. hirsuta,* indicating that not all isolates of these species clustered as a single species. The *P. chrysosporium* libraries provided a good identification of all but one blind sample. Analysis of that sample identified *P. sordida* as first choice, but the second choice of *P. chrysosporium* was too close for a good species match. The genus *Phanerochaete* was correctly identified in all blind samples for both species tested. Blind samples of *T. pubescens* and *T. hirsuta* were not distinguished at the species level. There appears to be considerable variability within *T. hirsuta*. It will be interesting in the future to compare the DNA sequences of these outliers to determine if the genetic information supports the separations seen in the FAME profiles.

Introduction

Classification of wood decay fungi is currently in a state of revision and review. Recent phylogenetic studies of the Homobasidiomycetes based on nuclear and mitochondrial small subunit rDNA tentatively divided this group into eight major clades (*1*). Many of the estimated 1,568 species of wood-decay homobasidiomycetes fall within the polypore and corticoid groups (*2, 3*). Only 6-7% of the wood-decay homobasidiomycetes are brown-rot species, with the remainder being white-rot species (*2*). Recent molecular studies have suggested that many groups are polyphyletic and that the evolution of the brown-rot decay mode has repeatedly evolved from the ancestral white-rot modes (*4*). Most phylogenetic analyses are based on a single isolate of a species and, in many cases, a single species of a given genus. These types of studies do not consider the variation that occurs among isolates of a given species and among species. Tanesaka et al. (*5*) found that variation among species accounted for 63% of the total variation of 17 genera analyzed. Many wood decay isolates show wide ranges of variability in physiological characteristics, appearances and abilities. The identification of wood decay fungi is typically accomplished by isolation and observation of growth characteristics on different culture media and spore formation. Identification keys are used to identify the fungi in question (e.g. *6* and *7*). This process is time consuming and can be difficult due to the similar cultural and morphological characteristics of many different fungi and poor (or lack of) sporulation in culture of many decay isolates. Other methods are being sought to identify unknown fungi more quickly, with higher accuracy, and requiring less expertise. Currently, serological detection, comparisons of cellular fatty acid methyl ester (FAME) profiles, and identification of unique sequences of fungal DNA are being developed for wood decay fungi. Each method has its advantages and disadvantages. The focus of this chapter is the application of FAME profiles for identification of the wood decay fungi. Identification using unique DNA sequences is discussed in a separate chapter of this book (*8*).

Microorganisms synthesize over 200 different fatty acids and the presence of specific fatty acids and their relative amounts are constant for a particular species (*9*). The type and amount of fatty acids produced are used to identify a particular genus, species, or strain in bacteria and yeasts. Since the early 1960's, analysis of fatty acids (derivatized to methyl esters) by gas chromatography has been used for identification of bacteria, and more recently, fungi (*10 - 14*). MIDI, Inc. (Newark, DE) has developed a database of FAME profiles called the Sherlock Microbial

Identification System using over 5,000 strains of microorganisms including sixty common fungi. Although this system does not contain profiles of wood decay fungi, it has the capability for the creation of new libraries of fatty acid profiles as deemed necessary by the user. This system of FAME profiles combined with multivariate discriminate analysis was used to speciate glomalean endomycorrhizal fungi for taxonomic purposes (*10*) and seven species of *Penicillium* were correctly identified with a 98% accuracy rate (*14*). Muller et al. (*13*) compared the fatty acids of 42 different fungi including 11 genera and 16 species and found that not all genera could be separated by fatty acid profiles alone, but identification improved when sterols were added to the profiles of the slow-growing fungi.

A thesis project in the Department of Forest Products, Mississippi State University evaluated if a MIDI Sherlock System library built for select wood decay fungi could be used for taxonomic purposes (*15*). The wood decay fungi, *Gloeophyllum trabeum, G. sepiarium, Trametes hirsuta*, and *T. versicolor*, were extracted and analyzed for their FAME profiles and libraries of these profiles were created. Ten isolates of each species were tested. Each species had its own unique fatty acid profile, however , there was poor extraction of the fatty acids from slow-growing isolates. Poor extraction efficiency led to low peak area counts on the gas chromatogram, and low percent area counts "lost" fatty acids that were present in low concentrations. In comparison, high extraction efficiency led to high percent area counts and numerous trace fatty acids "appeared". Since the library is based on the presence and percent amounts of fatty acid in a given species, differences between the high and low area counts contributed to a weak match for two species. In addition, the MIDI software selected the data to be included in the library, which led to an omission of some of the trace fatty acids of a given species from the library. Given the problems encountered, we felt that the system could distinguish among the different wood decay fungi, but, certain culture, extraction and library generation variables would need to be optimized. The data given in this paper are a result of tightening a number of variables and generating a more accurate identification system.

Research Methods

Preparation of Fungal Cultures

The isolates in this study were purchased from the American Type Culture Collection, or provided by the Forest Products Laboratory, Madison, WI or Dr. Dana Richter, Michigan Technological University, Houghton, MI. Fresh cultures of seven to ten isolates of *T. versicolor, T. hirsuta*, and *T. pubescens*, eight isolates of *Phanerochaete chrysosporium,* five isolates of *P. sordida* and seven isolates of

P.sanguinea were started on Sabouraud dextrose agar (SDA) plates (Difco Laboratories) and incubated at 28 C for three to seven days. One 0.5 cm plug was removed from the leading edge of the culture, placed on a fresh SDA plate, and incubated at 28 C for four days. Four 0.5 cm plugs from the leading edge of the culture were placed into 100 ml of Sabouraud dextrose broth (Difco Laboratories) and incubated for seven days in a Lab-Line environmental shaker at 28 C and 150 revolutions per minute (rpm). This culture procedure was replicated four times for each isolate. Mycelia were filtered through a Whatman 541 (4.7 cm) filter paper, rinsed three times with distilled water, and excess water was blotted from the mycelia. The mycelia were separated into 0.05 g and 0.10 g weights and placed into 16 mm x 125 mm culture tubes. Each weight (g) was duplicated for each isolate replicate. Culture tubes were capped with Teflon-lined caps and frozen at -70 C overnight. *Aspergillus fumigatus* (ATCC 36607) was extracted and run as a positive control and a reagent blank served as a negative control for every set of samples analyzed.

Extraction and Esterification of the Fatty Acids

To each culture, 2 ml of Extraction Solution A (45 g sodium hydroxide, 150 ml methanol and 150 ml deionized water) was added , the mixture vortexed for 10 - 15 s, heated in a boiling water bath for 5 min, re-vortexed for 10 - 15 s, and heated an additonal 25 min in the boiling water bath. After cooling to room temperature, 4 ml of Extraction Solution B (325 ml 6.0 N hydrochloric acid and 275 ml methanol) was added to each tube, and the mixture vortexed for 10 - 15 s, heated in a water bath for 10 min at 80 C, then cooled to room temperature. The fatty acid methyl esters were extracted with 1.2 ml of Extraction Solution C (50:50 solution of methyl tert-butyl ether and hexane) and 10 min of rotation at 20 rpm on a Thermolyne Vari-Mix. The top layer was transferred to a 13 mm x 100 mm culture tube with Teflon-lined cap, 3 ml of Extraction Solution D (5.4 g sodium hydroxide in 450 ml deionized water) was added to this layer, and the mixture rotated for an additional 5 min. A small portion of this top layer was transferred to a 2 ml autosampler vial containing a 100 μl glass insert. Vials were sealed with a Teflon-lined cap for analysis by gas chromatography.

Establishment of the Fatty Acid Profiles by Gas Chromatography

Fatty acid methyl esters were identified and quantified using an Agilent Technologies, Inc., (formerly Hewlett-Packard) 6890 Gas Chromatograph (GC) equipped with an automatic injector and flame ionization detector, and connected to a computer containing ChemStation Data Acquisition and Sherlock Microbial Identification System software packages. GC conditions were: injector 250 C; detector at 300 C; and oven-initial temperature 170 C, 5 C/min to 260 C, 40 C/min to 310 C then hold 1.5 minute. The column used for analysis was the Agilent Ultra-2 capillary column, (25m,0.22mm,0.33μm) with hydrogen as the carrier gas. The fatty acid profile for each isolate was analyzed by GC approximately 16 times. There were a few slow-growing isolates that were analyzed fewer times. Since two mycelial weights (0.05 and 0.1 g) were extracted for each analysis, only isolates that produced area counts between 100,000 and 600,000 were used to generate the library. This provided approximately 8 GC profiles per isolate for generation of a library.

Creation of the Wood Decay Fungal Identification Libraries

The samples were cataloged in each library using the Sample Catalogue feature of the Sherlock Microbial Identification System software package which searches all the data files and creates a master alpha-numeric index. After cataloging the samples, comparisons among the samples were made using the 2-D plot feature of the software. This program determines relationships among a large number of organisms by generating a 2-D plot of Principle Component 1 on the x axis and Principle Component 2 on the y axis. Closely related organisms will form clusters. A line is drawn around the cluster and perpendicular lines are drawn from the minimum and maximum x and y values to the corresponding x and y intercept. The difference in the minimum and maximum x and y values are multiplied to give the Euclidean Distance Squared (ED^2). For bacterial speciation, samples clustered in a group with ED^2 of about 110 or less are assumed to be of the same species; about 60 or less are of the same subspeices; and about 30 or less are of the same strain. There are no suggested guidelines for speciation of fungi. This program is helpful in determining contaminated samples and samples that are outliers. Samples that are outliers or questionable or contain small total area (<40,000) or large single peak area (> 400,000) are flagged either by the software or manually and then eliminated from the database. After each analysis, a report is generated which gives the Similarity Index. The Similarity Index is a numeral value which expresses how closely related the fatty acid composition of an unknown is to the library entry. An exact match would result in a similarity index of 1.000 and will decrease in proportion to the cumulative variance between the composition of the unknown and the library entry. Similarity indexes of 0.500 or higher with a separation of at least 0.100 between the first and second choice are considered to be a good match.

Results and Discussion

The *Phanerochaete chrysosporium* Library

This paper will discuss the creation of the *P. chrysosporium* (PC) library in detail to illustrate how the library selections were made. The libraries of the other species will not be reported in as much detail due to space constraints. A summary of pertinent information will be provided for the other libraries. The PC library was a compilation of 96 data points, with each data point representing a FAME profile. The 2-D plot for all 96 points is given in Figure 1. The ED^2 calculated for Figure 1 is 394 (Table I). This number is greater than 110, indicating that some of the isolates tested may not be the same species, therefore, sub-groups were created. It is obvious from Figure 1 that some isolates were outside of the main cluster. For PC, there were 4 sub-group libraries (A - D), with the largest library (A) composed mainly of 5 isolates, B composed of 2 isolates, C composed of 1 isolate, and D composed of 1 isolate. The 2-D plot for library B is shown in Figure 2. Separate FAME profiles for a given isolate are circled. These same data points are also circled in Figure 1. The ED^2 for PC library B is 75.2, indicating these isolates are the same species but not the same subspecies. Table I lists the ED^2 for all libraries created for all species thus far. When an unknown is analyzed and its identification sought using the wood decay libraries, the identification reported distinguishes among the sub-group libraries of a given species. Thus an organism will be identified as *P. chrysosporium* A or B, not simply as *P. chrysosporium*.

"Blind" samples were run against the PC library to test the accuracy of the identification system. The purpose of the sub-grouping is to strengthen the accuracy of the library and to establish relationships (or similarities) among isolates. Table II gives the similarity index for the first and second choice when screened against the newly created PC libraries. When tested against the four sub-group libraries, three of four isolates were correctly identified as *P. chrysosporium* with *P. sordida* as the second choice. These three identifications would be considered a "good" match. The fourth isolate selected *P. sordida* as its first choice and *P. chrysosporium* as its second choice. The distance between the first and second choice is small, indicating that this identification is "too close to call". Interestingly, the FAME profiles for the isolate with poorer identification is located in the larger A library. The better matches came from the smaller B and D libraries.

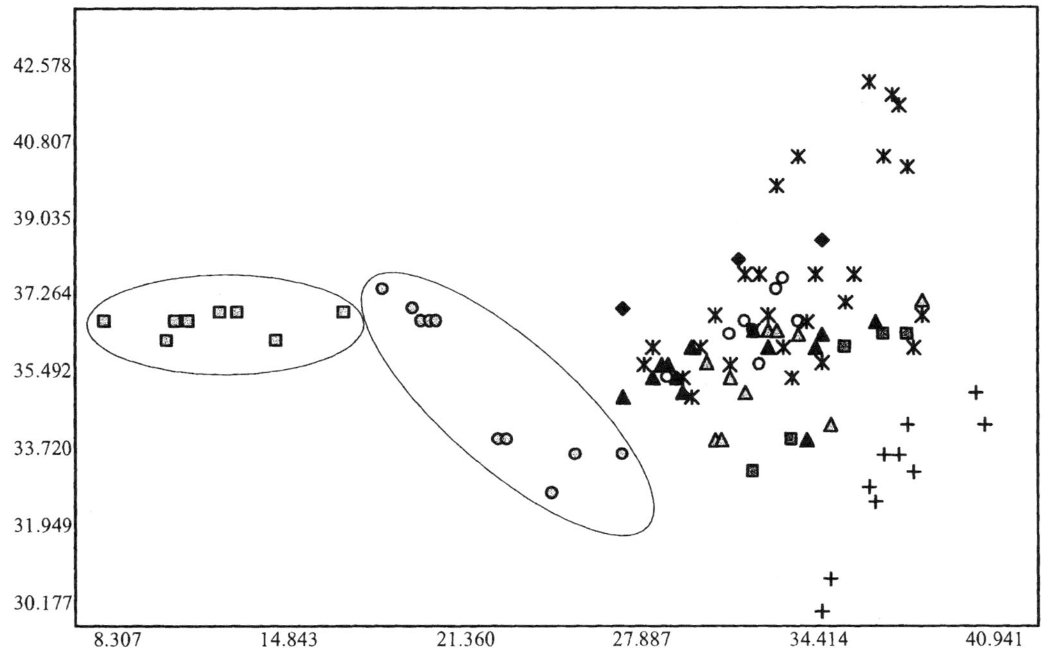

Figure 1. 2-D plot of all 96 data points generated for the Phanerochaete chrysosporium library. The two isolates which make up the P. chrysosporium library B are circled. X-axis is Principal Component 1 and y-axis is Principal Component 2. The ED^2 is 394.

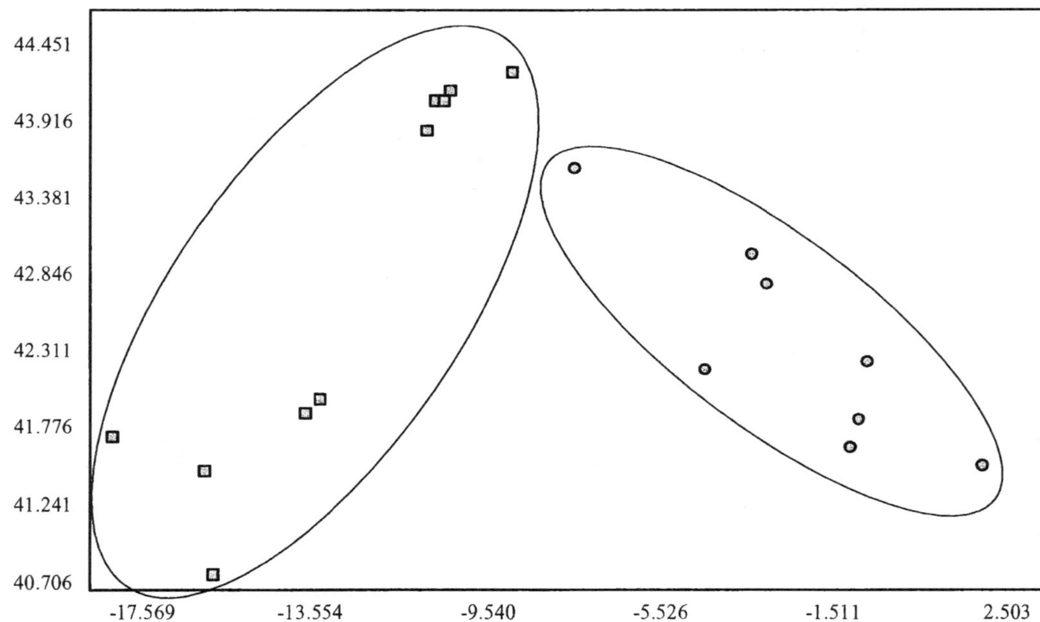

Figure 2. 2-D plot of Phanerochaete chrysosporium library B. The data for each of the two isolates that comprise this library are circled. X-axis is Principal Component 1 and y-axis is Principal Component 2. The ED^2 is 75.

Table I. Details of the identification libraries created for *Phanerochaete* and *Trametes*.

	Number of Profiles in Library	*ED^2*	*Number of Isolates Used*
P. chrysosporium	96	394.2	9
A	61	58.2	5
B	18	75.2	2
C	10	18.5	1
D	7	7.2	1
P. sordida	40	558.5	5
A	5	18.7	1
B	12	52.2	1
C	23	52.9	3
P. sanguinea	37	398.9	7
A	4	22.4	1
B	8	8.4	1
C	25	86.3	5
T. versicolor	49	106.0	7
T. hirsuta	71	736.5	10
A	7	19.5	1
B	9	24.6	1
C	48	62.7	7
D	7	24.5	1
T. pubescens	94	94.5	8

ED^2 is the Euclidean Distance Squared

Table II. Similarity Indices for the identification of blind *P. chrysosporium* samples screened against the newly created libraries.

	Similarity Index First Choice	*Similarity Index Second Choice*	*Difference between 1st and 2nd*
34541 (Located in Library B)	*P. chrysosporium* 0.893	*P. sordida* 0.804	0.089
62778 (Located in Library D)	*P. chrysosporium* 0.885	*P. sordida* 0.785	0.100
32629 (Located in Library B)	*P. chrysosporium* 0.900	*P. sordida* 0.822	0.078
48747 (Located in Library A)	*P. sordida* 0.899	*P. chrysosporium* 0.858	0.041

The Phanerochaete and Trametes Libraries

The total library ED^2 for *P. sordida* (Table I) was 558, indicating that some of these isolates may not be same species. Three sub-group libraries were created. Blind samples of *P. sordida* run against the libraries correctly identified *P. sordida* as the first choice, but the second choice of *P. chrysosporium* was too close (0.033) for a good species match. The genus *Phanerochaete* was correctly identified in all blind samples for both species tested. Three sub-group libraries were created for *P. sanguinea* (Table I). The total library ED^2 was 399. Two of the sub-group libraries contained a single isolate, with the remaining five isolates in the third sub-group library. The only sample of *Phanerochaete* analyzed by recent molecular systematic studies (*3, 4*) was a single isolate of *P. chrysosporium,* which was unfortunately not used in this study. These studies reclassified *P. chrysosporium* with two genera it had never been classified with before (*Bjerkandera* and *Pulcherricium*) (*3*) and was widely separated from the other corticoid fungus used in that study. The molecular relationships within the genus *Phanerochaete* and individual species is yet to be determined.

A single library was created for both *T. versicolor* and *T. pubescens* since their ED^2 were only 106 and 94, respectively (Table I). This indicates that all isolates tested for those species are the same species. Four sub-group libraries were created for *T. hirsuta*. The ED^2 for the total group of 736 was the largest seen thus far. Single isolates composed sub-group libraries A, B, and D. The remaining isolates were grouped in library C. One blind sample of *T. versicolor* was run against the library and identified correctly with a very good match (difference of 0.292) with *T. pubescens* a distant second choice. FAME analysis of a different isolate of *T. versicolor* choose *T. hirsuta* first. Analysis of this isolate and its location on a dendogram showed this isolate clustered separately from other *T. versicolor* isolates, which suggests it should be pooled into its own sub-group. Blind samples of *T. pubescens* and *T. hirsuta* could not distinguish between these two species. Molecular systematic studies found *T. versicolor* and *T. sauveolens* to be closely related, however, neither *T. pubescens* nor *T. hirsuta* were evaluated (*3, 4*). Based on the high ED^2 there appears to be a lot of variability within *T. hirsuta* samples. This may be the first series of isolates we attempt to compare by DNA sequencing.

There were some isolates that were very distinct outliers, and thus were not included in the library profiles. It will be interesting in the future to compare the DNA sequences of these outliers and the sub-group libraries to determine if the genetic information supports the separations seen in the FAME profiles. The authors are currently in the process of creating identification libraries for the brown-rot fungi *Gloeophyllum trabeum*, *G. sepiarium*, and *G. striatum*.

ACKNOWLEDGMENTS

The authors wish to acknowledge Dr. Dana Richter of Michigan Technological University and Rita Rentmeester of the USDA-FS Forest Products Laboratory for their generous contribution of some of the fungal isolates used in this study.

This work was supported by a Department of Forest Products WUR Grant 96-34158-2557 and the State of Mississippi. This manuscript is approved for publication as FWRC Publication No. FP233.

REFERENCES

1. Hibbett, D.S.; Thorn, R.G. In, *The Mycota VII Part B*; McLaughlin, D.J.; McLaughlin, E.G.; Lemke, P.A., Eds., Springer-Verlag: Berlin, **2001**; pp 121-168.
2. Gilbertson, R.L. *Mycologia*, **1980**, *72*,1.
3. Hibbett, D.S.; Donoghue, M.J. *Can. J. Bot.*. **1994**, *73(Suppl 1)*, S853.
4. Hibbett, D.S.; Donoghue, M.J. *Syst. Biol.*, **2001**, *50*, 215.
5. Tanesaka, E.; Masuda, H.; Kinugawa, K. *Mycologia*, **1993**, *85*, 347.
6. Davidson, R.W.; Campbell, W.A.; Vaughn, D.B. *Fungi causing decay of living oaks in Eastern United States and their cultural identification.* USDA Tech. Bull. No. 785, Washington, DC, **1942**.
7.. Nobles, M.K. *Can. J. Bot.*, **1948**, *26*, 281.
8. Jellison, J; Jasalavich, C.; Ostrofsky, A. Chapter XX, THIS BOOK **2002**
9. Böttger, E.C. *ASM News*, **1996**, *62*, 247.
10. Bentivenga S.P.; Morton, J.B. *Mycolog. Res.,* **1994,** *98*, 1419.
11. Bernard, K.A.; Bellefeuille, M.; Ewan, E.P. *J. Clin. Microbiol.*, **1991**, *29*, 83.
12. Moss, C.W. *J. Chromatography*, **1981**, *203*, 337.
13. Müller, M.M,; Kantola, R.; Kitunen, V. *Mycolog. Res.*, **1994**, *98*, 593.
14. Stahl, P.D.; Klug, M.J. *Appl. Environ. Microbiol,* **1996**, *62*, 4136.
15. Moore, F.W. Master's Thesis, Mississippi State University, Mississippi State, MS, **1997**.

Chapter 18

Detecting Decay Fungi with Antibody-Based Tests and Immunoassays

C. A. Clausen

Forest Service, Forest Products Laboratory, U.S. Department of Agriculture, Madison, WI 53705–2398

Early detection of wood decay can prolong the service life of wood. Antibodies are the ideal probe for detecting fungi that cause biodeterioration because they are highly specific and can quantitatively determine the fungal antigen concentration from highly complex structures, such as wood. Polyclonal antibodies recognize multiple chemical sites of the targeted molecule, in our case, a fungal glycoprotein, while monoclonal antibodies recognize one specific protein sequence on the targeted molecule. Both polyclonal and monoclonal antibodies have been utilized separately or in concert to design various assay formats to detect incipient wood decay; depending on the target organism and the test format, an assay can be designed to be specific or broad spectrum, quantitative or qualitative. Immunodiagnostic tests for wood decay fungi include, in order of least to most sensitive, the particle agglutination assay, dot-blot immunoassay, enzyme linked immunosorbent assays (ELISA), and a patented particle capture immunoassay.

Introduction

Wood decay caused by brown-rot decay fungi has typically been described as having four stages, though the stages overlap and cannot be clearly separated (*1*). Incipient decay occurs when decay fungi initiate colonization and release enzymes; there is no visible evidence of damage. During early decay, slight changes in color or texture occur, but decay is not obvious. Intermediate decay includes obvious changes to color and texture, although the wood structure still appears to be intact. Advanced decay is obvious because the wood structure is affected; the wood turns brown, and crumbly with a cubicle appearance. Sensitive methods for detecting decay in wood, particularly in the initial stages, have long been sought by those responsible for inspecting and maintaining wood in-service (*2*). Traditional field methods such as sounding of wood, visual inspection of borings, and mechanical probing are useful for detecting advanced decay (*1*). However, early stages of decay are difficult to detect and incipient decay cannot be detected by traditional field methods (*2,3*). Traditional laboratory methods such as culturing and microscopic examination are reliable. However, they are time consuming and require professionally trained personnel and laboratory facilities. If a field method can detect decay in a structure before strength reduction occurs, then remedial treatments can be utilized to arrest decay and prevent further damage to wood in-service (*2,4*).

Recognition of early decay is important for the inspection of wood in service. Early decay is difficult to recognize because it is visually subtle and often occurs on the interior of a timber or below ground. Visually, macroscopic changes in color or texture are presumptive evidence of early decay; only the physical presence of fungi is considered definitive. Microscopic examination revealing hyphae with clamp connections is definitive evidence of early decay (*1*). Nonvisual methods of identifying early decay include culturing the fungus from infected wood, and various physical tests, such as changes in strength, acoustic, and electrical properties (*3,4*).

The ideal detection system for diagnosis of fungal decay would incorporate several criteria (*5*):

- Simplicity-easy to use and understand.
- Rapid analysis and accurate results.
- Inexpensive-enables multiple sampling which more accurately diagnoses decay.
- Specific for decay fungi without cross-reaction to non-decay wood-inhabiting microorganisms.
- Small sample size and minimally invasive sampling procedure.

- Portable for use in the field.
- Automated-automation is often mutually exclusive with field testing.
- Quantitative versus qualitative

Selection of antigen and type of antibody are critical elements in the successful development of an immunoassay (*4*). Ideally, immunizing antigens should originate from the natural substrate, i.e., decayed wood, since there are demonstrated immunogenic differences between fungal enzymes derived from liquid culture versus decayed wood (*6*). Once a target molecule is identified and antibodies are prepared, determining the antigen extraction method for test samples and optimal antigen concentration are equally critical to the success of an immunoassay. Likewise, choice of assay method and test substrate will influence how to proceed in method design.

Immunological probes

Fungal proteins and polysaccharides elicit an immune response when introduced into a higher animal, typically a rabbit, resulting in antibody production (*7*). Because antibodies have the capacity to specifically recognize and bind to fungal antigen in a highly complex structure such as wood they are an ideal probe for the detection of fungi that cause biodeterioration. Two distinctly different types of antibodies, monoclonal (Mab) and polyclonal (Pab), have been produced (*8,9*). Polyclonal antibodies recognize multiple chemical sites, i.e. epitopes, on the antigenic molecule and are typically produced *in vivo.* Monoclonal antibodies recognize one specific epitope, often a protein sequence on an antigenic molecule (*10*) and are typically produced *in vitro* by cell culture. When used either separately or in concert, Mab's and Pab's have been utilized to design various immunological tests to detect fungal antigens extracted from wood during the incipient stage of decay (*4,8,10,11*).

For all methods described in this article, samples consist of wood shavings extracted in an aqueous solution of Triton X-100 (Sigma, St. Louis, MO), which solubilizes the hyphal membrane of the fungus and releases antigens recognized by the antibodies (i.e. xylanase). Fungal xylanases, which are glycoproteins, vary in the ratio of protein:carbohydrate from fungus to fungus. In laboratory tests, the antibodies used in the following methods recognized brown-rot and white-rot fungi. A broad-spectrum test is desirable for most field inspection applications.

Test Methods

Particle agglutination test (PAT)

Particle agglutination assays are rapid immunodiagnostic tests (*4,8,12*) that have been utilized in the medical field since 1956. Particulate antigens may be agglutinated directly by polyclonal antibodies to surface antigens. Similarly, submicron-sized polystyrene particles coated with polyclonal antibody (Pab) will visibly agglutinate in 30-60 seconds when an antigen is present. Experience is necessary to evaluate reactions to avoid misinterpretation of weak positive reactions. For example, an extracted sample of wood is mixed with antibody-labeled particles in the well of a glass slide and the reaction viewed through the slide with the aid of a light box. A negative result shows a smooth homologous appearance, while a positive test exhibits grainy aggregates of reagent (Figure 1). The strength of agglutination is ranked as negative or 1 to 4, with 4 being the strongest reaction. Particle agglutination has successfully detected infection by brown-rot fungi prior to weight loss, with minimal cross-reactivity to white-rot, mold or sapstain fungi (*8*). This method for decay detection is simple, rapid, specific, and fairly portable, but it is not currently quantitative nor can it be automated.

Quantitative results could, however, be obtained by reading the assay turbidimetrically. Variations are described by Bangs (*12*) , which both simplify the test portability and increase the reaction complexity. In one variation, antibody-coated particles are dried on a cardboard test strip, and rehydrated with the test sample. Having only one liquid reagent eases field portability. To increase test complexity, multiple antibodies to different fungi or fungal components, can be used to coat polystyrene particles of differing colors. The mixture can then be used either in solution or dried to test samples. Depending on which fungus or component is present in the aqueous test sample, only the corresponding color of antibody-coated bead will agglutinate. Another way to improve portability is to place dry antibody coated particles in the well of a closed-channeled cassette with a zigzag capillary channel. Liquid test sample is added to the particle-containing well and capillary action pulls the reagents through the channel, mixing the antibody-coated particles and test sample as it moves. Agglutination will cloud the cassette indicating a positive reaction and the cassette will remain clear in a negative reaction.

Figure 1. Particle agglutination test (PAT). Left: visible grainy aggregates indicate a positive reaction, right: smooth homogenous particles indicate a negative reaction.

Dot Blot Immunoassay

Blotting membranes offer high protein binding capacity and sensitivity for qualitative detection of decay fungi. Aqueous wood extracts bound to nitrocellulose paper are probed with enzyme-labeled antibody, and addition of substrate yields a visible reaction (Figure 2). A detailed method is described by Clausen and Green (*7*). Glancy *et al.* (*13*) utilized polyclonal antibodies to a whole cell antigen preparation of *Lepideus lentinus* and Clausen *et al.* (*8*) utilized a multivalent polyclonal antibody to six common brown-rot fungi to show that simple dot-blot immunoassays are useful for detecting incipient decay (*8*). This method does not lend itself to field portability, nor is it simple, rapid or automated. Dot blots can, however, be analyzed quantitatively by densitometry.

Enzyme-linked Immunosorbent Assay (ELISA)

ELISA's give quantitative measurements of antibody or antigen and are performed in 96-well polystyrene plates with a variety of configurations of antigen, antibody and detection systems. While this method cannot be adapted as a field test, it has the advantage of providing automated, quantitative measurements and multiple tests in a single plate. There are several formats of ELISA (*4,8*), including direct, indirect, and double antibody sandwich (DAS). These three methods differ as follows:

- Direct ELISA immobilizes a specific antibody on the polystyrene plate and incubates antigen and second specific antibody that is conjugated to an enzyme or other probe.
- Indirect ELISA immobilizes either antibody or antigen to the polystyrene plate and the enzyme-linked antibody is linked to the immobilized antigen or antibody.
- DAS-ELISA immobilizes a specific antibody to the polystyrene plate and incubates with antigen followed by a second specific antibody. The Ab-Ag-Ab sandwich is incubated with a secondary antibody conjugated with an enzyme.

In all types of ELISA, the last step is to add a chromogenic substrate which will form a visible color reaction when acted on by the enzyme. The reaction can be quantified spectrophotometrically. Alternate methods may use fluorescent or radioactive probes.

Of the above, indirect ELISA has been used most often to detect brown-rot decay (Figure 3). Breuil (*14*) and Breuil *et al.* (*15*) used ELISA to detect the sapstain fungus, *Ophiostoma* sp. C28 with a polyclonal antibody. Clausen *et al.*

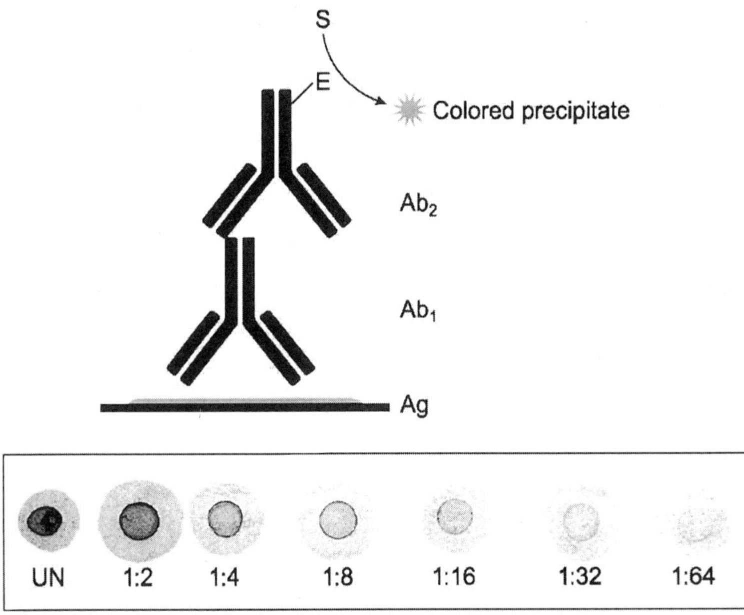

Figure 2. Dot blot immunoassay. Antigen, bound to nitrocellulose membrane, is probed sequentially with antibody(Ab_1) and enzyme-labeled antibody (Ab_2). Addition of substrate yields a colored precipitate. The amount of color is directly proportional to the amount of antigen in the test sample.

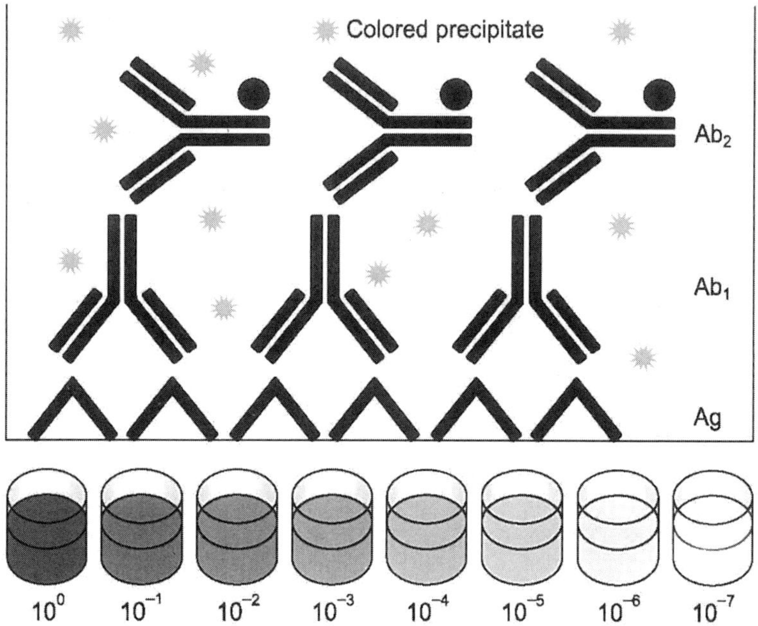

Figure 3. Indirect enzyme-linked immunosorbent assay (ELISA). Top: Schematic of reaction within each well-antigen, bound to polystyrene wells, is probed with antibody (Ab_1) and enyme-labeled (Ab_2). Addition of substrate yields a colored precipitate. Bottom: The amount of color in each well is directly proportional to the amount of antigen in the test sample.

(*5,8*) reported detection of six common brown-rot fungi from infected wood at 0-13% wood weight loss using either polyclonal or monoclonal antibodies prepared against xylanase from *P. placenta*. Goodell and Jellison (*16*) and Goodell *et al.* (*17*) detected *P. placenta* using polyclonal antibody in samples exposed to the fungus in a soil-block test after 10 days of incubation. A detailed ELISA method can be found in Clausen and Green (*7*).

Chromatographic Immunoassay

Numerous formats exist for chromatographic immunoassays (*12*). All formats rely on either an enzyme-substrate color reaction, such as that used for dot-blots and ELISA's or, more commonly, colored polystyrene particles coated with antibody, like those described in particle agglutination assays. The particle capture immunoassay combines the speed of a PAT with the sensitivity of a double antibody sandwich ELISA. Clausen (*10,11*) and Clausen and Green (*19*) developed and patented a dyed particle capture immunoassay that is superior in sensitivity and specificity to the indirect ELISA. Dyed particles eliminate the need for an enzyme-substrate reaction. In this test, antibody-coated particles move through a porous membrane when an aqueous wood extract is added to a sample well. If antigen is present in the test sample and binds to the antibody-coated particles, it becomes sandwiched by a second anti-xylanase antibody that is immobilized in another location on the porous membrane. The dyed particles also become trapped in the sandwich so that a positive reaction is visualized as a colored line (Figure 4). Unbound particles migrate to a negative control zone. This qualitative test is specific for decay fungi, sensitive to nanograms of xylanase and is able to detect decay in the field prior to weight or strength loss. Alternatively, utilizing an enzyme-substrate color reaction and portable spectrographic device could conceivably result in a quantitative version of the chromatographic immunoassay.

The particle capture immunoassay does not distinguish between active and inactive fungi. Decay fungi can remain inactive in wood for long periods of time when conditions are not favorable for growth; adding sufficient moisture can reactivate fungal growth and decay. By indicating the presence of both active and inactive decay fungi, the particle capture immunoassay can be particularly useful for the inspection of wood in-service, since it alerts the inspector to potential losses if steps are not taken to prevent exposure of the wood to moisture, enables the extent of colonization to be determined, and indicates and defines areas where remedial treatment or replacement are appropriate. Commercial lumber suppliers and mills could monitor the quality of their timber

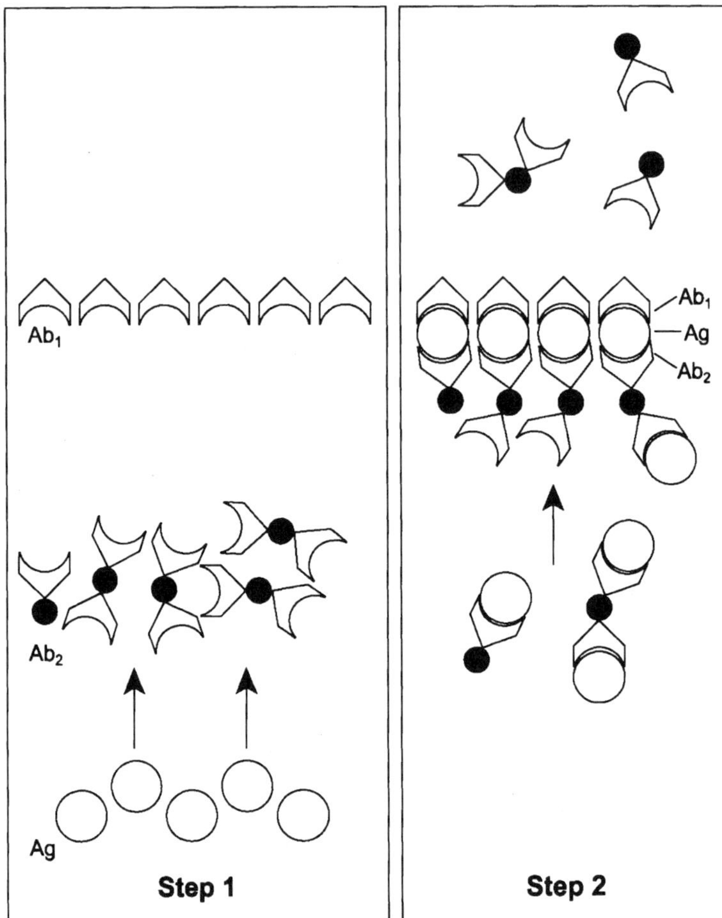

Figure 4. Diagram of the particle capture immunoassay. 1) antigen (liquid wood extract) is applied and migrates through the prepared test strip, 2) antigen captured by Ab_2 coated-particles, migrates to the immobilized antibody (Ab_1) and is sandwiched between Ab_1 and Ab_2, resulting in a colored line, i.e. positive reaction (4).

supply with this method. Since preservatives do not interfer with the antibody/antigen reaction, this assay is also suitable for monitoring treated wood in-service.

Summary

The particle agglutination test, dot blot, ELISA and chromatographic immunoassays are broad immunoassay categories having numerous variations within each category. Immunoassays have been developed within each category for the detection of incipient fungal decay from aqueous wood extracts. Each method meets most of the criteria for an ideal detection system, and the particle capture immunoassay is conducive to detecting incipient wood decay in the field. The field of immunodiagnostics has made advances through the development of sensitive immunoassays towards a practical field test to detect incipient wood decay.

References

1. Zabel, R.A.; Morrell, J.J. In *Wood Microbiology: Decay and Its Prevention;* Academic Press, Inc.: New York, NY, **1992**.
2. Highley, T.L.; Micales, J.A.; Illman, B.L.; Green III, F.; Croan, S.C.; Clausen, C.A. Research Paper FPL-RP-530, US Department of Agriculture, Forest Service, Forest Products Laboratory, Madison, WI, **1994**, 7 pp.
3. Clausen, C.A.; Ross, R.J.; Forsman, J.W.; Balachowski, J.D. Research Note FPL-RN-0281, US Department of Agriculture, Forest Service, Forest Products Laboratory, Madison, WI, **2001**, 4 pp.
4. Clausen, C.A. *Internat. Biodeter. Biodegrad.* **1997,** *39*(2-3), 133-143.
5. Clausen, C.A.; Green III, F.; Highley, T.L. In *Proceedings of the 8th Internat. Biodeter. Biodegrad. Symp.* Ed.; Rossmoore, H.W., Elsevier Applied Science, New York, NY, **1991**, pp.412-414.
6. Clausen, C.A.; Green III, F.; Highley, T.L. In *Biodeterioration Research 4: Mycotoxins, Wood Decay, Plant Stress, Biocorrosion, and General Biodeterioration:* Proceedings of the 4th Meeting of the Pan American

Biodeterioration Society, Eds.; Llewellyn, G.C.; Dashek, W.V.; O'Rear, C.E. Plenum Press, New York, NY, **1994,** pp.231-242.

7. Clausen, C.A.; Green III, F. In *Methods in Plant Biochemistry and Molecular Biology;* Dashek, W., Ed.; CRC Press, New York, NY, **1997**, pp. 69-85.
8. Clausen, C.A.; Green III, F.; Highley, T.L. *Wood Sci. Technol.* **1991,** *26*, 1-8.
9. Clausen, C.A.; Green III, F.; Highley, T.L. *Wood Sci. Technol.* **1993**, *27*, 219-228.
10. Clausen, C.A. *J. Immunoassay* **1994,** *15*, 305-316.
11. Clausen, C.A. *NASA Tech Briefs* **1994**, *18*, 23a.
12. Bangs, L.B. J. *Internat. Federation of Clinical Chem.* Sept. **1990**, 5 pp.
13. Glancy, H.; Bruce, A.; Button, D.; Palfreyman, J.W.; King,B. *Internat. Res. Group on Wood Preserv.* **1989,** Doc. No. IRG/WP/1422.
14. Breuil, C. Canadian Forestry Services Project Report **1987,** No. 39, pp. 43-46.
15. Breuil, C.; Rossignol, L.; Saddler, J.N. *Biotechnol Tech.* **1990**, *4*, 263-268.
16. Goodell, B.; Jellison, J. *Internat. Res. Group on Wood Preserv.* **1986,** Doc. No. IRG/WP/1305.
17. Goodell, B; Jellison, J.; Holsi, J.P. *For. Prod. J.* **1988,** *38*, 59-62.
18. Clausen, C.A. *Internat. Res. Group on Wood Preserv.* **1991,** Doc. No. IRG/WP/2378.
19. Clausen, C.A.; Green III, F. US Patent No. 5,563,040. **1996**.

Chapter 19

Synchrotron Applications in Wood Preservation and Deterioration

Barbara L. Illman

Forest Service, Forest Products Laboratory, U.S. Department of Agriculture, and University of Wisconsin, Madison, WI 53705

Several non-intrusive synchrotron techniques are being used to detect and study wood decay. The techniques use high intensity synchrotron-generated X-rays to determine the atomic structure of materials with imaging, diffraction, and absorption. Some of the techniques are X-ray absorption near edge structure (XANES), X-ray fluorescence spectroscopy (XFS), X-ray absorption fine structure (EXAFS), and X-ray computed microtomography (XCMT). Micro-fluorescence spectroscopy was used to map the accumulation and spatial distribution of elements around hyphae at the site of decay. MicroXANES determined the valence states of metals, such as manganese and iron, during fungal colonization of wood. Microtomography was used to characterize loss of wood structural integrity. The techniques are providing information about molecular structures and compositions in the heterogeneous matrix of wood.

Nondestructive methods are needed to analyze the chemistry and internal structures of wood without disturbing spatial integrity or producing structural artifacts (*1*). The methods are needed to study wood during attack by decay fungi and wood treatment with preservatives. To meet this need, we have successfully studied several systems using the X-ray facilities at the National Synchrotron Light Source (NSLS), Brookhaven National Laboratory (BNL), Upton, NY. Chemical mechanisms of fungal wood decay are dependent on transition metal redox reactions producing free radicals (*2-4*). Synchrotron methods are uniquely suited to detect and image metal oxidation states as probes of the decay process. The same methods have been applied to study metal-based preservatives that undergo redox reactions during fixation of wood. These methods include X-ray absorption near edge structure (XANES), X-ray fluorescence spectroscopy (XFS), extended X-ray absorption fine structure (EXAFS), and X-ray computed microtomography (XCMT). In addition to decay and preservation studies, the XCMT method has also proven to be invaluable as a tool to analyze insects and fungi that cause major diseases of forests worldwide (*5*). Microtomography was employed to study beetle structure and function, to locate fungal spores on or in beetles, to assist in identification of fungi, and to provide images depicting spatial relationships of tree–insect–fungi (*6,7*).

Synchrotrons

Synchrotrons are large scientific instruments that generate high intensity electromagnetic radiation that range from the infrared through the x-ray spectrum. They are particle accelerators consisting of an electron source; a circular storage ring and connecting beamlines that end in experimental work stations. Electrons are introduced into the vacuum storage ring, driven to velocities near the speed of light, guided by magnets to travel around the ring where the charged particles tangentially lose energy, emitting electromagnetic radiation known as synchrotron light. At NSLS, a third generation synchrotron, bending and insertion device magnets, wigglers and undulators control electrons in the storage ring, enhancing the intensity of the light (Figure 1). The emitted radiation is directed down a path known as a beam port, which is split into two to four beamlines (*8*). Light from the storage ring is a mixture of many wavelengths. Each beamline has a collection of lenses, special mirrors and filters to select specific wavelengths that direct and focus the synchrotron light to a specially designed experimental station. The station is outfitted for particular types of experiments with radiation detectors, recorders, computers, and other specialized equipment. Computer software, unique to each beamline, interprets data from the detector and images the results on a video screen in the form of graphs, element-specific spectra, or molecular models. Synchrotron radiation is used to determine the atomic structure of materials by diffraction, absorption, and imaging techniques at the beamline.

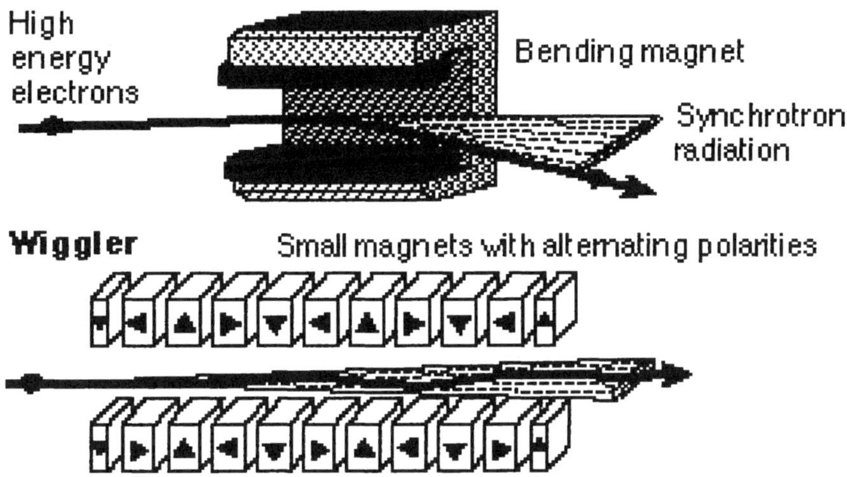

Figure 1. Schematic of synchrotron radiation created from high-energy electrons and radiation path in bending magnet and wiggler insertion devices.

The NSLS has an ultraviolet and an X-ray ring, which provide intense focused light at infrared and hard X-ray energies in excess of 100 keV and gamma rays at 200-400 MeV. The X-ray ring operates at 2.5 GeV to optimize radiation between 1 keV and 20 keV (*9*). The X-ray ring has over 30 beamlines extending from the ring to end at ports with multiple workstations. The beamlines are named from X1 to X30 with stations designated as letters. Wood deterioration and preservation experiments were conducted at beamlines X26A and X27A.

X-ray Fluorescence Spectroscopy

X-ray fluorescence spectroscopy (XFS) is a powerful and flexible technique that has long been available for the analysis and characterization of materials (*10*). The high intensity and brightness of synchrotron-generated X-rays make it possible to detect and quantities trace element distribution *in situ*. Beamline configuration of the X-ray microprobe at NSLS beamline X26A for white light has a spot size ranging from 10 square microns to 150-micron height x 350 microns vertical with an energy range from 3-30 keV. We have used XFS to scan for transition metal accumulation in decayed wood, tree rings of boreal forest spruce trees, brown-stained hemlock wood and hemlock tree rings for element distribution, especially for chromium, cadmium, and brown stain caused by manganese and iron oxides (Illman et al, unpublished data). We determined element location and distribution with XFS before obtaining spectroscopy spectra of trace elements in decayed wood or metal treated wood.

X-Ray Absorption Spectroscopy

Synchrotron-based XAS allows the study of elements at subatomic resolution. Synchrotron XAS techniques do not require a vacuum at the experimental station outside the storage ring, a distinct advantage over more established absorption techniques such as X-ray Photoelectron Spectroscopy (XPS). The XAS technique is based on element-specific absorption of electromagnetic radiation at distinct energies and, therefore it is highly dependent on an optical subassembly, a monochromator that accepts the polychromatic input radiation and outputs selectable monochromatic energy. An XAS experiment consists of irradiating a sample with a monochromatic beam of X-rays that is in the energy range above, and then below, the absorption edge of an element in question. The X-ray absorption coefficient is recorded as a function of energy. Two complimentary synchrotron XAS techniques are XANES and EXAFS. The XANES K-edge absorption spectrum for a representative transition metal, such as Mn, can be interpreted as follows. At low energies, the energy of the incoming radiation is not high enough to be absorbed by Mn in the sample, detected as the pre-absorption edge region of the spectrum. At sufficiently high energies, radiation is absorbed. The K-shell electron is ejected from the atom, released as photoelectron, giving rise to the absolute absorption edge of the spectrum. The absorption edge also contains a substructure of small peaks or shoulders. The EXAFS region of the spectrum is the result of outgoing photoelectron reflection by the atoms in the environment of the metal. Interference of outgoing and reflected parts results in oscillation modulation of the absorption coefficient at energies above the edge. Differences between the

XANES and EXAFS portions of a representative absorption spectrum are illustrated in Figure 2.

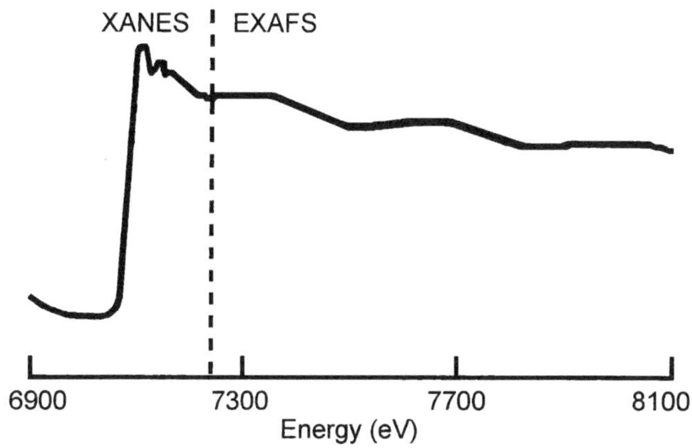

Figure 2. Representative XANES and EXAFS spectra

X-ray absorption near edge structure (XANES)

XANES spectroscopy is used to detect and determine oxidation states of metals that are provided by the absolute position of the absorption edge. The near edge absorption region, up to 40 keV above the edge, contains information about vacant orbitals, electronic configuration and site symmetry of the absorbing atom.

The microprobe beamline at NSLS X26A pioneered *in situ* detection of transition metal redox states in biological, environmental, and geological samples. The current X26A XANES sensitivity for trace element analyses is 10-100 ppm. The beamline is ideally suited for detection of metal oxidation states used to probe changes in chemical degradation of wood lignocellulose by decay fungi (*1,6,7*). Representative microXANES spectra of and Mn^{4+} in Southern yellow pine wood during decay by the white-rot fungus *Phanaerochaete chrysosporium* are illustrated in Figure 3. Peaks and shoulders in the rising edge can provide information about electronic configuration, ligand bonds, and atomic symmetry.

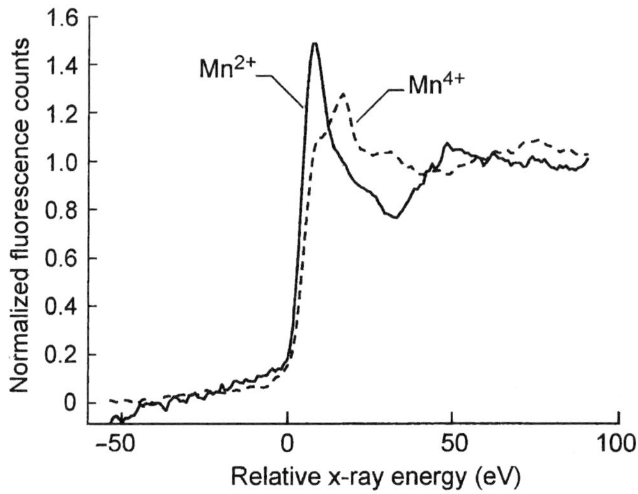

Figure 3. MicroXANES spectra of manganese oxidation in Southern yellow pine wood. Solid line is control wood: dashed line is wood inoculated with Phanaerochaete chrysosporium.

Metal-based wood preservatives have also been detected with the X26A microprobe. When lumber is pressure treated with chromated copper arsenate (CCA), the desired outcome is complete conversion of the toxic, more mobile Cr^{6+} species to the less toxic, less mobile Cr^{3+} species. Preliminary microXANES data of the ratio of Cr^{6+} to Cr^{3+} detected Cr^{3+} and not Cr^{6+} in wood stored for two years after CCA pressure treatment (*11*).

Extended X-Ray absorption fine structure (EXAFS)

EXAFS probes the local atomic and chemical environment of a selected element. By analyzing the modulations in the X-ray absorption coefficient at energies just above the X-ray absorption edge threshold, EXAFS measurements give quantitative information about coordination species, number, and distance. Extended *X*-ray absorption refers to the sinusoidal variation of the *X*-ray absorption coefficient as a function of *X*-ray photon energy, which occurs on a spectrum after each absorption edge of an element and extends for up to 1500 keV (Figure 2).

Bull et al. (*12*) measured copper and arsenic K-edge EXAFS of CCA-treated pine wood (Pinus radiata). They report that the data are consistent with arsenate anions bound to copper and chromium ions isolated from other heavy elements at all depths into the wood.

X-Ray Computed Microtomography

The XCMT instrumentation at NSLS beamline X27A has been described previously (*13,14*). Briefly, the facility is equipped with a single-crystal YAG:Ce scintillator with peak emission at 540 MM for converting X-ray an attenuation map to a visible image. The scintillator is coupled to a cooled charge-coupled device (CCD) by a mirror/lens combination. The CCD detector, with 1317 x 1035 pixels, can record data to reconstruct up to 1035 horizontal slices simultaneously. The beamline instrumentation can be illuminated by a filtered 'white' X-ray beam with energy of around 18 keV or by a monochromatized beam with a 1% bandpass and energy tunable between 4 and 14 keV. The monochromator is comprised of a highly efficient pair of W-B_4C multilayers deposited on Silicon substrates with properties well matched to the synchrotron source beam. Data for wood decay was collected with the monochromator tuned at 8.5 keV, which is in the range of the biologically important transition metals iron and manganese. A Si-III channel cut monochromator which replaced the multilayer monochromator on X27A, was used to verify results reported here. Microtomograms are obtained with resolutions down to 3-micrometer voxels and fields of view of over 5 millimeters.

A specimen is mounted in a vertical position on a motorized x-y stage for centering in the optical field of view. The specimen stage in turn is mounted to a rotational stage attached to a tilt and translational stage for prealignment of the rotational stage to the CCD. During data acquisition, the specimen is rotated over a full 180 degrees, producing angle-dependent views of the attenuation map digitally recorded by the CCD camera. The recorded views of the specimen are processed using a Fourier-based Fast Filtered Back Transform algorithm to generate transverse images or slices through the sample for every row of the CCD. The reconstructed slices are stacked to produce a volume representation of the specimen. The IBM Data Explorer software was used for volume rendering. Representative tomographic images of Southern pine wood are given in Figure 4. A reconstructed, smoothed and segmented slice of control wood is given in Figure 4a and a reconstructed volume in Figure 4b. Density differences were observed in the reconstructed images, reflecting differences in chemical composition of structures within the wood.

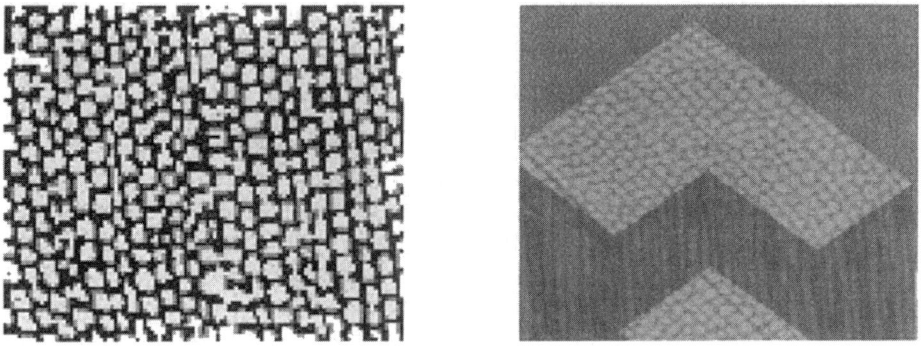

Slice Volume

Figure 4. Microtomographic images of control Southern yellow pine wood

Summary

Non-intrusive synchrotron methods have been successfully applied to detect and follow chemical mechanisms of fungal wood decay and metal-based wood treatments.

Acknowledgements

This research was supported by USDA Competitive Grant 94-36103-1016. Research was carried out (in part) at the National Synchrotron Light Source, Brookhaven National Laboratory, supported by the U.S. Department of Energy, Division of Materials Sciences and Division of Chemical Sciences. The author would like to thank research cooperators Steve Sutton, University of Chicago; Darrell Schultz, Purdue University; Sasa Bajt, Lawrence Livermore National Laboratory; and Betsy Dowd, National Synchrotron Light Source; and to thank Les Ferge, Rene Holiday, and Jason Jurd for their technical support.

References

1. Illman, B. L; Bajt, S. *Int. Biodeter. Biodeg.* **1997**, 39(2-3): 235-242.

2. Illman, B. L. In *Active Oxygen/Oxidative Stress and Plant Metabolism;* Pell, E. J.; Steffen, K. L.; Shannon, J. C., Eds.; Current Topics in Plant Physiology, American Society of Plant Physiologists, Rockville, MD, 1994; Vol. 6, pp 97-106.
3. Highley, T. L.; Clausen, C.A.; Croan, S.C.; Green III, F.; Illman, B. L.; Micales, J. A. *Research on Biodeterioration of Wood, 1987-1992.* USDA/FS Forest Products Laboratory Research Paper, FPL-RP-529; USDA/FS Forest Products Laboratory, Madison, WI, 1994.
4. Goodell et al., this book.
5. Werner, R. A.; Illman, B. L. *Physiol. Chem. Ecol.* **1994**. 23(2): 472-478.
6. Illman, B. L; Dowd, B.A. *Sync. Rad. News*, **1997**, 10(2): 18-23.
7. Illman, B. L; Dowd, B.A.; Holaday, R. In *Developments in X-Ray Tomography III*, Bonse, U., Ed.; Proc. SPIE, The International Society of Optical Engineering, Bellingham, WA, 2001; Vol. 4503, pp 172-177.
8. Robinson A. L. *Beam Line*; Spring, **1994**, 4(3): 17-28.
9. Duke, P. J. *Synchrotron Radiation: Production and Properties*; Oxford University Press, London, 2000.
10. Jenkins, R. *X-Ray Fluorescence Spectrometry* 2nd ed.; Wiley Publishing Co, New York, NY, 1999.
11. Illman, B. L.; Bajt, S. Sutton, S.R. In *National Synchrotron Light Source, Activity Report: Brookhaven National Laboratory and Associated Universities.* BNL-52455; National Technical Information Service, Springfield, VA, 1994, p B-183.
12. Bull, D. C.; Harland, P. W.; Vallance, C.; Foran, G. J. *J. Wood Sci.* **2000**, 46(3): 248-252.
13. Illman, B. L; Dowd, B.A. In *Developments in X-Ray Tomography II*, Bonse, U., Ed.; Proc. SPIE, The International Society of Optical Engineering, Bellingham, WA, 1999; Vol. 3772, pp 198-204.
14. B.A. Dowd, G. H. Campbell, R. B. Marr, V. Nagarkar, S. Tipnis, L. Axe, and D. P. Siddons. In *Developments in X-Ray Tomography II*, Bonse, U., Ed.; Proc. SPIE, The International Society of Optical Engineering, Bellingham, WA, 1999; Vol. 3772, pp 224-236.

Chapter 20

Detecting and Identifying Wood Decay Fungi Using DNA Analysis

Jody Jellison, Claudia Jasalavich, and Andrea Ostrofsky

Biological Sciences Department, University of Maine, Orono, ME 04469

Amplification of genomic DNA by the polymerase chain reaction (PCR) is a sensitive and specific tool that can be used to detect degradative fungi in wood. PCR coupled with other molecular methods has been used to detect specific fungal products and to identify and survey fungi in the environment. Researchers have used PCR methods to detect and monitor specific fungal degradative genes, to detect and identify basidiomycetes in culture, and to sequence ribosomal DNA for taxonomic studies of selected wood decay fungi. In our work basidiomycete specific primers derived from fungal ribosomal DNA sequences were shown to detect the presence of fungal DNA in spruce wood, even at very early stages of decay. When used in conjunction with restriction fragment length polymorphism (RFLP) analysis, this procedure was able to provide reliable fungal species identification. The assay is currently being developed for use in wood of other species and in wood composite materials. The ultimate goal of this work is to develop a reliable and sensitive DNA-based assay to detect incipient decay in wood and identify the fungi involved.

Introduction

The primary decomposers of wood in non-aquatic environments are the white and brown rot fungi. These fungi are capable of causing significant modification of the wood lignocellulose and also cause dramatic reductions in wood strength properties. With brown rot fungi significant strength loss can occur during early stages of decay, before visual indications of infection and degradation become apparent. Thus, it is necessary to identify the presence of incipient decay in structures for control and/or remediation or replacement purposes. However, the detection and identification of wood degrading fungi can be both difficult and time consuming. Earlier methods include isolation and culturing of the fungus, sonics, electrical resistance, dyes, ergosterol and chitin assays, immunoassays and other techniques (*1*). However, many methods are either unreliable or necessitate extensive destructive sampling. More recently our laboratory and others have explored the use of the polymerase chain reaction (PCR) (*2*) and other molecular techniques to aid in the detection and identification of degradative fungi in the early stages of fungal attack.

PCR amplification of genomic DNA can be both specific and sensitive. Useful strategies include RAPD analysis (3), targeting specific degradative genes, *e.g.* genes coding for enzymes involved in wood decay or bioremediation, or targeting the ribosomal DNA (rDNA). Ribosomal DNA is found universally and exists in high copy number as tandem repeats of several hundred copies; thus it is a good target for the detection of dilute or partially degraded DNA. Ribosomal DNA contains both variable and conserved regions; this makes it useful for comparing organisms at different taxonomic levels. Each tandemly repeated unit of nuclear rDNA contains the coding regions for the 18S, 5.8S and 28S rRNAs separated by the internal transcribed spacers (ITS), with ITS I lying between the 18S and 5.8S rDNAs, and ITS II between the 5.8S and 28S rDNAs. These tandemly repeated units are separated by the intergenic spacers (IGS). The 18S, 5.8S and 28S rDNAs are generally highly conserved, while the ITS and IGS show greater variability and, therefore, are potentially useful for interspecific and intraspecific studies, respectively. In particular, a portion of the tandem repeat, the ITS region (ITS I-5.8S rDNA-ITS II) has been extensively used for detection and identification of fungi. White *et al.* (*4*) established the utility of rDNA amplification and sequencing. Additional researchers have developed rDNA PCR protocols for use in combination with restriction fragment length polymorphism (RFLP) analysis (*5*, *6*, *7*, *8*), temperature gradient gel electrophoresis (*9*), ELISA amplicon detection (*10*), automated fluorescent capillary electrophoresis (*11*), and single-strand-conformation polymorphism (SSCP) analysis (*12*).

Detection of Specific Degradative Genes

Detection of the genes or messenger RNAs (mRNA) of specific fungal degradative enzymes in culture and in soil has been reviewed by Reddy and D'Souza (*13*). Degradative enzymes of interest include lignin peroxidase, manganese peroxidase, laccase and selected cellulases. Lamar *et al.* (*14*) demonstrated the use of reverse transcription PCR (RT-PCR) to quantify specific fungal mRNAs for lignin peroxidase (*lip*) genes and cellobiohydrolase (*cbh*) genes in soil colonized by *Phanerochaete chrysosporium*. Subsequently, the expression of all ten *lip* genes in soil (both mRNA transcripts and translated proteins) and concurrent oxidation of anthracene by *P. chrysosporium* were studied (*15*). Bogan *et al.* (*16*) monitored manganese peroxidase mRNA and enzyme levels in soil during bioremediation studies of polycyclic aromatic hydrocarbons. D'Souza *et al.* (*17*) used primers based on conserved sequences around two copper binding regions to isolate laccase genes from white rot fungi. Our laboratory has used degenerate laccase primers to amplify DNA from white rot, brown rot and assorted wood-inhabiting ascomycetes (Jasalavich, unpublished). Catechol 2,3-dioxygenase specific primers have also been used to monitor bioremediation processes in petroleum-amended soil by *Pseudomonas* sp. using competitive quantitative PCR (*18*).

Environmental Sampling

PCR technology has been used to sample the environment for the presence of specific fungi and fungal products. Johnston and Aust (*6*) amplified ligninase H8 DNA and the nuclear ITS region to monitor *P. chrysosporium* in the soil. Lamar *et al.* (*14*) quantified fungal mRNAs for lignin peroxidases, cellobiohydrolases, and B-tubulin using RT-PCR from soil colonized by *P. chrysosporium*. They also used this method to monitor the differential expression of complex gene families in soil versus liquid culture in organopollutant degradation studies (*15*). Bacterial and fungal soil communities have been characterized using ribosomal DNA. Ranjard *et al.* (*18*) exploited length polymorphisms in the nuclear ITS region to profile soil fungal communities and IGS fingerprints to profile soil bacterial communities. Van Tuinen *et al.* (*20*) used taxon-specific primers based on two variable regions in the 28S rDNA to monitor patterns of colonization of onion and leek roots by four different species of arbuscular mycorrhizae from soil. Smit *et al.* (*9*) coupled PCR amplification using fungal specific primers based on the 18S rDNA with temperature gradient gel electrophoresis (TGGE) to profile fungal communities in the wheat rhizosphere. Schmalenberger *et al.* (*12*) studied how primer selection and the targeting of different variable regions of the 16S rDNA (analogous to the 18S rDNA in fungi) affect the detected composition of bacterial profiles in soil in PCR based microbial community analysis. Amplification of the nuclear ITS region has also been used to identify ancient fungi trapped in the ice of the Greenland glacier (*21*).

Detection and Identification of Basidiomycetes in Culture

The use of PCR amplification specifically for the detection and identification of basidiomycete fungi was aided by the early work of Gardes and Bruns (*22*) who reported the use of a fungal specific primer ITS1-F in combination with a basidiomycete specific primer ITS4-B for the detection of mycorrhizal fungi. The tree parasites and wood-attacking basidiomycetes *Heterobasidium annosum* and *Armillaria ostoyae* have been identified by taxon specific PCR (*8, 23, 24*). Zaremski (*25*) characterized tropical wood-decaying fungi by RFLP analysis of the ITS region amplified with the universal primers ITS 1 and ITS 4 from DNA isolated from pure fungal cultures. Since differences in the original amplicon lengths were not sufficient to distinguish among the five brown rot and three white rot species examined, amplicons were digested with restriction enzymes and electrophoresed in gels to generate RFLP profiles of the fungi. The eight species were separated by their RFLP profiles. Subsequent work allowed the design of species-specific primers.

Moreth and Schmidt (*26, 27*) have used species-specific PCR primers as a diagnostic tool for the European dry rot fungus *Serpula lacrymans*. They amplified and sequenced the ITS region from several isolates of each of seven species of indoor rot basidiomycetes. Species-specific primers based on the sequence of internal transcribed spacer II (ITS II) were designed for differential diagnosis of *Serpula lacrymans*, *Serpula himantioides*, *Donkioporia expansa*, *Coniophora puteanna*, *Antrodia vailantii*, *Tyromyces placenta* and *Gloeophyllum sepiarium*.

Taxonomic Relationships

Taxonomic relationships among fungi have also been examined using similar PCR techniques. Ribosomal DNA has been used to estimate phylogenetic relationships in plant pathogenic fungi (*28, 29*) and wood decay fungi (30, 31, 32). Lieckfeldt *et al.* (*33*) used phylogenetic analyses of DNA sequences of the nuclear ITS region and the 5'-end of the 28S rDNA, along with RFLP analysis of the endochitinase gene and PCR fingerprinting, to separate two morphological types of the wood-inhabiting ascomycete *Trichoderma viride* into two species. Harrington and Wingfield (*34*) developed a PCR-based method to identify isolates of the tree pathogen *Armillaria* to species. They separated 74 isolates into 11 *Armillaria* species by RFLP analysis of the amplified intergenic spacer (IGS). Chillali *et al.* (*35*) detailed the delineation of European *Armillaria* species based on ITSI and ITSII polymorphisms. Phylogenetic analysis of nuclear 28S rDNA sequence data supports division of the genus *Paxillus* into three genera, *Paxillus, Tapinella* and a new genus *Austropaxillus* (30). The genus *Leucogyrophana* is polyphyletic based on analysis of the nuclear 28S rDNA (31). Hibbett and Donoghue (32) have generated a phylogenetic framework based on sequences of the mitochondrial small subunit rDNA and the nuclear 18S rDNA for wood decay fungi in several families of Homobasidiomycetes.

Significant work has also been done on the taxonomic characterization of the European dry rot fungus *Serpula*. Theodore *et al.* (36) reported on RAPD polymorphisms among isolates of *S. lacrymans*. Schmidt and Moreth (37) found RAPD homogeneity within *S. lacrymans* and polymorphism within *S. himantioides*. More recently White *et al.* (*38*) reported on the molecular analysis of intraspecific variation between building and wild isolates of *S. lacrymans* and their relationship to *S. himantioides*. Random amplified polymorphic DNA (RAPD) analysis did not distinguish between wild Himalayan and building isolates of *S. lacrymans*, as there was very little intraspecific variability in this species. However, the two species *S. lacrymans* and *S. himantioides* were clearly resolved by both RAPD and ITS rDNA sequence analysis. ITS sequence data suggest that an American isolate of *S. lacrymans* from Mount Shasta in California is intermediate between European *S. lacrymans* and *S. himantioides* (39).

Detection of Degradative Fungi in Wood

Detection and Identification in Spruce Wood

Much of the work detailed above focuses on the detection of gene transcripts of specific fungal products, or the identification and analysis of fungal species and/or isolates in culture. The use of universal primers makes detection and identification in the field problematic due to potential amplification of DNA from microbial contaminants (*40*) or constitutive members of the microbial community in the wood or soil environment being sampled. Our recent work has focused on the use of basidiomycete specific primers to detect degradative fungi in the wood substrate. Basidiomycete specific primers were used because of their greater utility in diagnosing decay organisms in general versus targeting a specific species with the risk of missing infection by numerous other degradative organisms. The ITS region of the rDNA was selected because its high copy number makes it a more sensitive and appropriate target for detection in environments such as wood where total fungal DNA levels could be low during incipient decay as well as late term decay when DNA degradation can occur.

Protocols for DNA extraction, PCR amplification and restriction digestion to generate RFLPs have been previously outlined (*5, 41*). Initial work indicated that primer pairs ITS1-F (specific for higher fungi) and ITS4 (universal primer) and ITS1-F and ITS4-B (specific for basidiomycetes) could be used effectively to detect and differentiate decay fungi and non-degradative wood-inhabiting ascomycetes both in pure culture and in spruce wood. Fifty seven isolates representing 14 species of decay fungi and 25 species of wood-inhabiting ascomycetes were used to demonstrate assay specificity with DNA isolated from pure culture. Seven brown rot fungi, 13 white rot fungi and 10 wood-inhabiting ascomycetes were used to test the utility of this approach for the detection and identification of the fungi in spruce wood. Primer pair ITS1-F and ITS-4 successfully amplified DNA from all samples, as expected. Primer pair ITS1-F and ITS4-B amplified DNA from only the white and brown rot fungi and not from the wood-inhabiting ascomycetes (Figure 1). Although amplicon length, as estimated by gel electrophoresis, ranged from 850 to 1,460 bp, this characteristic was not diagnostic for individual fungal species, since several fungal species had amplicons of the same or similar length. The identities of the white and brown rot fungi were confirmed by comparing RFLP profiles (Figure 2). Time studies demonstrated that the fungi could be detected very early in the decay process, before significant weight and strength loss to the wood had occurred (*5*).

Detection in Oak Wood and in Wood Composites

Subsequent work has suggested that caution may be warranted when attempting to amplify DNA directly out of some wood species, as the DNA extraction method developed for red spruce (*5*) does not always yield amplifiable DNA from other wood species. For example, attempts to detect both brown rot and white rot fungi in red oak using our original DNA extraction method were poor and inconsistent. Recovery of amplifiable DNA from red oak can be improved by modification of this DNA extraction protocol to reduce phenol polymerization (*42*). However, additional work is needed to make detection of fungi in red oak reliable.

Detection of decay in oriented strand board (OSB) was also examined because of the increased utilization of composite wood products, and concerns that glues and other additives might inhibit the detection of fungal DNA. The OSB contained a phenolic resin. Poplar wood comprised 80-90% of the wood component of the OSB with the remaining 10-20% a mix of northern conifer

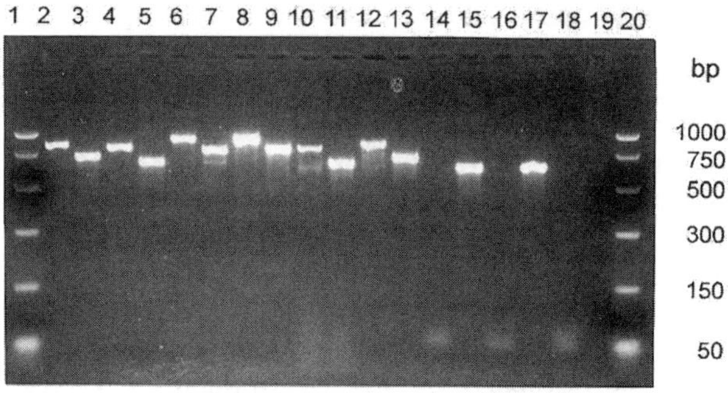

Figure 1. PCR amplification of nuclear rDNA from total DNA isolated from red spruce wood blocks colonized by wood decay fungi or endophytes. Electrophoresis in 2% (w/v) agarose in 1 X TBE. The two outer lanes contain molecular weight markers. Inner even-numbered lanes contain samples amplified by the primer pair ITS1-F and ITS4-B, and the odd-numbered lanes contain samples amplified by primers ITS1-F and ITS4. Lanes 2 to 7, brown rot basidiomycetes; lanes 8 to 13, white rot basidiomycetes; lanes 14 to 17, endophytic ascomycetes. Lanes 1 and 20, PCR markers (Promega); lanes 2 and 3, *Postia placenta* Mad-698-R; lanes 4 and 5, *Gloeophyllum trabeum* Mad-617-R; lanes 6 and 7, *Leucogyrophana pinastri*; lanes 8 and 9, *Lentinula edodes* 117=1t(d); lanes 10 and 11, *Trametes versicolor*; lanes 12 and 13, *Scytinostroma galactinum* ATCC 64896; lanes 14 and 15, *Hormonema dematiodes*; lanes 16 and 17, *Pestalotiopsis* sp.; lanes 18 and 19, no template DNA (*i.e.*, negative controls).

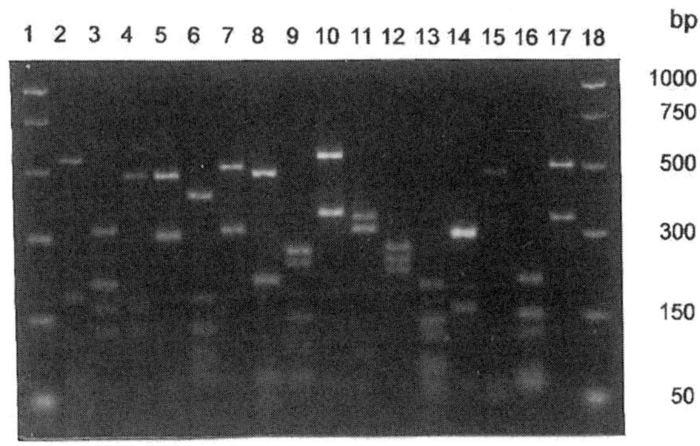

Figure 2. *Taq*I restriction digests of the PCR product amplified by the primer pair ITS1-F and ITS4-B from DNA isolated from pure cultures of basidiomycetes. Electrophoresis in 2% (w/v) Sepharide Gel Matrix (Gibco-BRL) in 1 X TAE. The two outer lanes contain molecular weight markers. Each inner lane contains a different fungal species; lanes 2 to 8 contain brown rot fungi, and lanes 9 to 15 contain white rot fungi. Lanes 1 and 18, PCR markers (Promega); lane 2, *Coniphora puteana* Fp-90099-Sp; lane 3, *Fomitopsis pinicola* K8Sp; lane 4, *Gloeophyllum sepiarum* 10-BS2-2; lane 5, *Gloeophyllum trabeum* Mad-617-R; lane 6, *Leucogyrophana pinastri*; lane 7, *Postia placenta* Mad-698-R; lane 8, *Serpula lacrymans* Harm-888-R; lane9, *Irpex lacteus* KTS 003; lane 10, *Lentinula edodes* 117=1t(d); lane 11, *Phanerochaete chrysosporium* ATCC 24725; lane 12, *Resinicium bicolor* ATCC 64897; lane 13, *Scytinostroma galactinum* ATCC 64896; lane 14, *Trametes versicolor* Fp-101664-Sp; lane 15, *Trichaptum abietinum* 1247 MJL; lane 16, *Pisolithus tinctorium* ATCC 38054, an ectomycorrhiza; lane 17, *Rhizoctonia solani* 1AP, a pathogen of herbaceous plants.

wood. Initially, we used the DNA extraction method developed for red spruce (*5*). The brown rot fungi *G. sepiarium*, *Gloeophyllum trabeum* and *S. lacrymans*, and the white rot fungi *Irpex lacteus*, *P. chrysosporium* and *Trametes versicolor* were detected from colonized OSB blocks with both primer pairs. We were unable to detect *Postia placenta* and *Leucogyrophana pinastri* using either primer pair. These problematic samples were heavily degraded and darkly colored. Modification of the original DNA extraction procedure to reduce phenol polymerization (*42*) was required to achieve reliable detection of all wood decay fungi tested in OSB (Table I). A time course study of OSB blocks infected with the white rot fungus *T. versicolor* or the brown rot fungus *G. trabeum* showed reliable detection at all time points for the representative white rot and reliable detection of the brown rot at weeks 1, 2, 3 and 4, but variable detection at week 8 with difficulty amplifying fungal DNA from the OSB samples which had heavy degradation. Diagnostically, this is not a practical problem as wood and manufactured wood products at these later stages of fungal degradation show extensive structural modification, weight loss and dark coloration which are easily detected.

Table I. Detection of wood decay fungi in oriented strandboard (OSB)[a]

Species	*% Weight Loss*	*PCR Amplification*	
	Mean ± SD[b]	*ITS1-F/ITS4*	*ITS1-F/ITS4-B*
Brown rot basidiomycetes			
Gloeophyllum sepiarum	32.4 ± 1.8	-+++[c]	-+++[c]
Gloeophyllum trabeum	36.1 ± 4.0	++++	++++
Leucogyrophana pinastri	44.3 ± 6.9	++++	++++
Postia placenta	48.8 ± 7.5	++++	++++
Serpula lacrymans	19.9 ± 4.5	++++	++++
White rot basidiomycetes			
Irpex lacteus	27.9 ± 8.1	++++	++++
Phanerochaete chrysosporium	20.3 ± 14.3	++++	++++
Trametes versicolor	25.6 ± 4.5	++++	++++

[a]OSB blocks were harvested after 12 weeks of colonization, and the DNA was isolated using an extraction procedure modified to reduce phenol polymerization (*35*).
[b]Mean of four replicate blocks.
[c]Each plus or minus sign represents the PCR amplification results for one OSB block. Negatives confirmed by reamplification.

Sequence Data and Specific Primers

We have sequenced the nuclear ITS region for a number of wood decay basidiomycetes for use in specific primer design and taxonomic analysis. Amplicon lengths obtained using primers ITS1-F and ITS4-B ranged from 800 to 911 base pairs. All insertions and deletions were observed in the internal transcribed spacers ITS I and ITS II, with a few base substitutions occurring in the 5.8S rDNA. Sequences from our isolates of *S. lacrymans* are similar to those of isolates sequenced by Palfreyman's group (*43*) and Schmidt's group (27). Both of these laboratories have recently submitted DNA sequences of the ITS region for a number of wood decay basidiomycetes to international nucleotide databases (26, 27, 38, 43, 44). We have also sequenced a portion of the mitochondrial small subunit rDNA for several wood decay basidiomycetes.

Applications and Limitations

Primers of varying specificity can be designed to suit differing applications. Our work using basidiomycete specific primers clearly indicates the specificity and sensitivity of this approach. Particularly encouraging was our ability to detect degradative organisms in wood at the incipient decay stage, suggesting that the sensitivity of molecular assays could be utilized to allow decay detection in in-service wood products before significant strength loss or other damage had occurred. Where necessary, individual degradative species could be identified based upon amplification and subsequent restriction digest patterns.

Sequence information or amplification by species-specific primers was done by directly sampling the wood products without the need for culturing and traditional identification procedures. Work by us and many others has demonstrated the utility of molecular approaches in the detection and identification of wood decay fungi and fungal metabolites in their natural environments of soil (*14*) and wood (*5*). DNA-based molecular techniques provide a powerful tool to examine wood degradative processes (*6, 13*) and taxonomic relationships (*38*) among the wood degrading fungi.

Acknowledgements

The authors would like to acknowledge the support of the Maine Agricultural and Forestry Experiment Station and USDA-WUR grants #: 95-34158-1347 and #: 99-34158-8489. This is publication # 2527 of the Maine Agricultural and Forestry Experiment Station.

References

1. Jellison, J.; Jasalavich, C. *Int. Biodet. & Biodeg.* **2000**, *46*, 241-244.
2. Mullis, K. B.; Faloona, F. A. *Methods Enzymol.* **1987**, *155*, 335-350.
3. Schmidt, O. *Holzforschung* **2000**, *54*, 221-228.
4. White, T. J.; Bruns, T.; Lee, S.; Taylor, J. In: *PCR Protocols. A Guide to Methods and Applications*, Innis, M. A.; Gelfand, D. H.; Sninsky, J. J.; White, T. J.; Eds.; Academic Press, San Diego, CA, USA, **1990**; pp. 315-322.
5. Jasalavich, C.; Ostrofsky, A.; Jellison J. *Appl. Environ. Microbiol.* **2000**, *66*, 4725-4734.
6. Johnston, C. G.; Aust, S. D. *Appl. Environ. Microbiol.* **1994**, *60*, 2350-2354.
7. Schmidt, O.; Moreth, U. **1998**, *International Research Group on Wood Preservation* Series Document 98-10245.
8. Schulze, S.; Bahnweg, G.; Möller, E. M.; Snadermann, H. *Eur. J. For. Path.* **1997**, *27*, 225-239.
9. Smit, E.; Leeflang, P.; Glandorf, B.; van Elsas, J. D.; Wernars, K. *Appl. Environ. Microbiol.* **1999**, *65*, 2614-2621.
10. Loeffler, J.; Hebart, H.; Sepe, S.; Schumcher, U.; Klingebiel, T.; Einsele, H. *Med. Mycology* **1998**, *36*, 275-279.
11. Turenne, C. Y.; Sanche, S. E.; Hoban, D. J.; Karlowsky, J. A.; Kabani, A. M. *J. Clin. Microbiol.* **1999**, *37*, 1846-1851.
12. Schmalenberger, A.; Schwieger, F.; Tebbe, C. C. *Appl. Environ. Microbiol.* **2001**, *67*, 3557-3563.
13. Reddy, C. A.; D'Souza, T. M. In: *Applications of PCR In Mycology*. Bridge, P. D.; Arora, D. K.; Reddy, C. A.; Elander, R. P.; Eds.; CAB International: New York, NY, USA, **1998**; Ch. 10, pp.205-242.
14. Lamar, R. T.; Schoenike, B.; Vanden Wymelenberg, A.; Stewart, P.; Dietrich, D. M.; Cullen, D. *Appl. Environ. Microbiol.* **1995**, *61*, 2122-2126.
15. Bogan, B. W.; Schoenike, B.; Lamar, R. T.; Cullen, D. *Appl. Environ. Microbiol.* **1996**, *62*, 3697-3703.
16. Bogan, B. W.; Schoenike, B.; Lamar, R. T.; Cullen, D. *Appl. Environ. Microbiol.* **1996**, *62*, 2381-2386.
17. D'Souza, T. M.; Boominathan, K.; Reddy, C. A. *Appl. Environ. Microbiol.* **1996**, *62*, 3739-3744.
18. Mesarch, M. B.; Nakatsu, C. H.; Nies, L. *Appl. Environ. Microbiol.* **2000**, *66*, 678-683.
19. Ranjard, L.; Poly, F.; Lata, J. C.; Mougel, C.; Thioulouse, J.; Nazaret, S. *Appl. Environ. Microbiol.* **2001**, *67*, 4479-4487.
20. Van Tuinen, D.; Jacquot, E.; Zhao, B.; Gollotte, A.; Gianinazzi-Pearson, V. *Mol. Ecology* **1998**, *7*, 879-887.
21. Ma, L.; Rogers, S. O.; Catranis, C. M.; Starmer, W. T. *Mycologia* **2000**, *92*, 286-295.

22. Gardes, M. ; Bruns, T.D. *Mol. Ecology* **1993**, *2,* 113-118.
23. Garbelotto, M.; Ratcliff, A.; Bruns, T. D.; Cobb, R. W.; Ostrosina, W. J. *Phytopathology* **1996**, *86,* 543-551.
24. Schulze, S.; Bahnweg, G. *Forstw. Cbl.* **1998**, *117,* 98-114.
25. Zaremski, A.; Ducousso, M.; Prin, Y.; Fouquet, D. **1998,** *International Research Group on Wood Preservation* Series Document 98-10251.
26. Moreth, U.; Schmidt, O. *Holzforschung* **2000**, *54,* 1-8.
27. Schmidt, O.; Moreth, U. *Mycol.Res.* **2000**, *14,* 69-72.
28. Jasalavich, C. A.; Morales, V. M.; Pelcher, L. E.; Séguin-Swartz, G. *Mycol. Res.* **1995**, *99,* 603-614.
29. Morales, V. M.; Jasalavich, C. A.; Pelcher, L. E.; Petri, G. A.; Taylor, J. L. *Mycol. Res.* **1995**, *99,* 593-603.
30. Bresinsky, A.; Jarosch, M.; Fischer, M.; Schönberger, I.; Wittman-Bresinsky, B. *Plant Biol.* **1999**, *1,* 327-333.
31. Jarosch, M.; Besl, H. Plant Biol. **2001**, *3,* 443-448.
32. Hibbett, D.; Donoghue, M. *Syst. Biol.* **2001**, *50,* 215-241.
33. Lieckfeldt, E.; Samuels, G. J.; Nirenberg, H. I.; Petrini, O. *Appl. Environ. Microbiol.* **1999**, *65,* 2418-2428.
34. Harrington, T. G.; Wingfield, B. D. *Mycologia* **1995**, *87,* 280-288.
35. Chillali, M.; Wipf, D.; Guillaumin, J-J.; Mohammed, C.; Botton, B. *New Phytol.* **1998**, *138,* 553-561.
36. Theodore, M. L.; Stevenson, T. W.; Johnson, G. C.; Thornton, J. D.; and Lawrie, A.C. *Mycol. Res.* **1995**, *99,* 447-450.
37. Schmidt, O.; Moreth, U. *Holzforschung* **1998**, *52,* 229-233.
38. White, N. A.; Dehal, P. K.; Duncan, J. M.; Williams, N. A.; Gartland, J. S.; Palfreyman, J. W.; Cooke, D. E. L. *Mycol. Res.* **2001**, *105,* 447-452.
39. Palfreyman, J. W.; Gartland, J. S.; Sturrock, C. J.; Lester, D.; White, N.A.; Low, G. A.; Bech-Andersen, J.; Cooke, D.E.L. The relationship between 'wild' and 'building' isolates of the dry rot fungus *Serpula lacrymans.* FEMS Letters, **2002**, in press.
40. Loeffler, J.; Hebart, H.; Bialek, R.; Hagmeyer, L.; Schmidt, D.; Serey, F. P.; Hartmann, M.; Eucker, J.; Einsele, H. *J. Clin. Microbiol.* **1999**, *37,* 1200-1202.
41. Jasalavich, C., Ostrofsky, A.; Jellison, J. **1998**, *International Research Group on Wood Preservation Series Document* 98/10279.
42. Jasalavich, C., Ostrofsky, A.; Jellison J. Detection of wood decay fungi in red oak and oriented strand board by PCR amplification with specific primers. **2002**. In preparation.
43. Palfreyman, J. Personal communication. 2001.
44. Schmidt, O.; Grimm, K; Moreth, U. On the molecular identitiy of species and isolates of the *Coniophora* cellar fungi. *Holzforschung,* **2002**, *56,* in press.

Chapter 21

Molecular Genetic Methods: New Approaches to Termite Biology

C. Husseneder[1], E. L. Vargo[2], and J. K. Grace[1]

[1]Plant and Environmental Protection Sciences, University of Hawaii at Manoa, 3050 Maile Way, Gilmore 310, Honolulu, HI 96822
[2]Department of Entomology, North Carolina State University, Campus Box 7613, Raleigh, NC 27695

Termites (Isoptera) are of global economic and ecological importance as decomposers of lignocellulose matter. Yet little is known about their biology due to their cryptic feeding and nesting habits. The advent of a variety of molecular genetic techniques provides a powerful method to elucidate many aspects of termite ecology and social organization. We present an overview of currently applied molecular genetic methods, including analyses of proteins, mitochondrial and genomic DNA, in a variety of termite species. These methods complement each other in the description of termite population structure, the identification of colonies and their breeding systems. The increasing application of these and other methods will yield a much improved understanding of termite ecology and social evolution as well as more effective means of controlling pest species.

A Short Introduction to Termite Biology

Termites are of immense ecological and economic importance. Ecologically, termites are the primary insect group adapted to consume wood and dry plant matter (lignocellulose), and to turn this nutritive-poor matter into protein biomass. With similar efficiency termites consume urban wooden structures and agricultural crops. Although only about 83 among the 2700 termite species are pests, they severely impact tropical and subtropical regions. In the USA subterranean termites of the genus *Reticulitermes* spp. and *Coptotermes* spp. are the most destructive. Subterranean termites are established in every region of the contiguous United States and Hawaii, costing an estimated $2 billion per year for damage repair and management efforts (*1*). There is growing concern about the economic impact of termites on urban structures and agricultural crops, calling for a combination of preventative and remedial measures to control colonies, as well as regulatory actions to reduce the chance of new introductions and limit the further spread of termite populations. Improvement on both levels requires a thorough understanding of termite biology.

The main features of termite biology, such as their colonial life-style, cooperative tasksharing and the monopolization of reproduction by a limited number of breeders (kings and queens), are linked to their evolution as social insects. Compared to research on other social insects (ants, bees and wasps), termite sociality has been somewhat neglected. Recent studies, however, have proven that termites provide valuable alternative and complementary models to the understanding of social evolution (*2*). Description of termite sociality requires detailed knowledge of the genetic variation within and between colonies, i. e. the population's genetic structure. These genetic patterns depend on, as well as reflect, the dispersal mode, distinctness of colonies and the breeding system (*3*).

A frequent problem in termite research is the cryptic life-style of termites. Because most termites live entirely hidden underground or inside wooden timbers or trees, direct observation under natural conditions is at best limited. Sometimes the only indication of their presence are swarming events, when a considerable number of winged reproductives are released from their mother colony to found new colonies. Besides swarming, termite colonies can proliferate by "budding", i.e., mature colonies containing multiple kings and queens form satellite nests that may eventually become independent colonies. This creates complicated colony structures consisting of widespread interconnected foraging areas and multiple nests containing a variable number of reproductives. Moreover, different levels of inbreeding in colonies, depending on the degree of genetic relatedness of reproductives, in turn affects the genetic make-up of the population (*3*). Due to the termites' hidden nesting and foraging habits these fundamental patterns and processes in termite biology are still literally hidden in the dark (*2*, *4*). However, important features of insect biology, such as population structure, dispersal patterns, colony identity and social organization, can be inferred from the genetic structure of colonies and populations (*3*, *5*).

Molecular Genetic Research in Termites

Recently, molecular genetic methods have been employed to gain insight into termite phylogeny (e.g., *6*, *7*) and population genetic patterns (*4*, *8-20*). In this chapter we review the molecular genetic approaches to population and colony genetic structure of termites. We first explain the basic features of the molecular methods used in termite research. We then address what information these methods provide to answer questions concerning termite population genetics and colony organization, drawing on examples from a variety of termite species. We will emphasize subterranean termites, especially *Reticulitermes* spp. and the Formosan subterranean termite, *Coptotermes formosanus* Shiraki (Isoptera: Rhinotermitidae), because these are the most severe and broadly distributed pest species in the US and worldwide and are therefore primary targets for improved research and management.

The Methods

In the last decade, a rapidly increasing palette of molecular techniques has been developed to address a wide range of questions (*21*, *22*). Which technique is most appropriate depends on the amount of genetic variability and "resolution power" needed to answer a particular question. Molecular markers currently used in termite research are (1) alloenzymes, (2) mitochondrial DNA and (3) genomic DNA, all of which have different characteristics and inheritance patterns and therefore target different but overlapping research areas.

(1) Diploid organisms, such as termites, have two copies of each genetic region (locus) on homologous pairs of chromosomes, called alleles. Mutations in genes coding for enzymes create a variety of different alleles in a population. The resulting gene products (proteins) have the same function but different structure and are called alloenzymes. Due to their structural differences and enzymatic activity, alloenzymes can be separated through electrophoresis and detected by enzyme specific stains (*23*). Alleles are inherited according to Mendel's rules and are presumed to be selectively neutral. Therefore the

particular pattern of allelic distribution within and among colonies reflects relatedness and inbreeding of populations and colonies as well as the colonies' social structure, i.e., numbers and relatedness of reproductives (e.g., *9*, *15, 16, 24*).

(2) Mitochondrial DNA is a double-stranded ring of extrachromosomal DNA located in the mitochondria, the organelles of cell respiration. Mitochondrial DNA reaches about 16-20kb in size (*25*). It lacks efficient DNA repair mechanisms, so populations contain a variety of different types of mitochondrial DNA, called haplotypes. Differences between haplotypes can be detected through length heterogeneity, restriction fragment analysis, double strand conformation polymorphism or sequencing of mitochondrial DNA regions (*26*). Mitochondrial DNA is inherited in a non-Mendelian, cytoplasmic fashion, generally through maternal transmission only. This marker type is therefore used for identifying the distribution of matrilines within populations and colonies (e.g., *14*, *16*, *19*, *24*).

(3) Genomic DNA provides powerful markers for fine-grained analysis of genetic variation. Termite genomic DNA is currently analyzed by two methods, multilocus DNA fingerprinting and microsatellite genotyping (*13*, *18*, *20, 27-31*). Multilocus DNA fingerprinting and microsatellite genotyping have excellent resolution power because they target simple repetitive sequences with high mutation rates in non-coding regions of the genome. DNA fingerprinting detects barcodelike banding patterns by restriction of genomic DNA and hybridization with labeled probes visualized by autoradiography or chemiluminescence. Microsatellite genotyping uses polymerase chain reaction to amplify alleles containing repetitive sequences. Alleles vary in length according to the variable number of repeats and are co-dominant. Thereby the proportion of heterozygotes (individuals carrying different alleles at gene loci on corresponding chromosomes) in colonies and populations can be detected and analyzed using the standard statistical procedures of population genetics (e.g., *F*-Statistics). DNA fingerprints, as well as microsatellite genotypes, are inherited according to Mendelian rules and therefore contain information about ancestries and relationships of individual termites, colonies and populations (methods reviewed by 22, *26*).

Population Differentiation and Dispersal of Termites

With increasing globalization due to trade, commerce, and tourism, alien species rapidly spread worldwide. The introduction of exotic species to an area can pose threats to the local ecology, e.g., through displacement of native species, as well as cause significant economic problems. Exotic termites, such as the Formosan subterranean termite, have invaded over a dozen states of the

USA, inflicting costs of at least $1 billion/year for damage repair and control. Regulatory efforts, such as inspections and quarantines, need to be targeted at the most common routes of introduction. Therefore, it is essential to understand the sources and routes of termite introduction, as well as the means by which subsequent spread through a habitat takes place. To date, there is much speculation but little known about the numbers and sources of introduction and ways of spread of termites. The key to understanding how termite colonies establish and proliferate lies in molecular genetic research on termite populations, including native as well as introduced species. Yet, in contrast to other insect pests, only a few such studies have been conducted on termites (*3, 5*).

The first genetic marker used to investigate genetic differentiation of termite populations were alloenzymes. Depending on the species and geographic range, the percentage of polymorphic alloenzyme loci and, therefore, their resolution power and applicability to particular questions differ [*Nasutitermes nigriceps* 8% (*15*); *C. formosanus* 12%-17% (*10, 11*); *Incisitermes schwarzi* 17% (*32*); *Reticulitermes flavipes* 29% (*9*); *Reticulitermes* spp. 52% (*33*); *Hodotermopsis* spp. 49% (*34*); *Zootermopsis* spp. 63% (*35*)].

In most cases the degree of polymorphism has been sufficient to describe the geographic differentiation between species or subspecies, e.g., of *Reticulitermes* spp. in Europe (*33*), *Hodotermopsis* spp. of Japan and China (*34*) or populations of *Coptotermes acinaciformis* with different nesting behaviors (mound-building versus subterranean) in Australia (*36*).

Beyond geographic distribution, description of genetic differentiation within and among populations is important for investigating the patterns and processes of how termites disperse through an area. Alloenzyme analysis in a native subterranean termite, *R. flavipes,* detected no significant genetic differentiation among sites in central Tennessee. However, alloenzyme frequencies differ between colonies (*9, 37*). Therefore the lack of increasing genetic differentiation over geographical distance was not attributed to a overall lack of genetic diversity but to considerable dispersal potential of winged reproductives resulting in homogenization of the population genetic structure. High dispersal capability might also explain the low mitochondrial DNA divergence among sites spanning across Georgia (*19*).

Alloenzyme studies of *C. formosanus*, an introduced invasive termite pest in the USA, revealed moderate genetic differentiation between geographically separated populations, such as Hawaii, Florida and Louisiana (*10*). In a subsequent study, Wang and Grace (*38, 39*) included southern China, the presumed origin of *C. formosanus,* in their analysis. Based on the genetic distances between the populations, multiple introductions on the US mainland were suggested with Hawaii and Lake Charles (Louisiana) being genetically similar to the native Chinese populations while New Orleans (Louisiana) and

Florida seem to have remained genetically isolated from the other group. However, the small sample size of colonies per population and populations per region in this study makes it difficult to pinpoint the actual source of introduction or ways of spread.

On a smaller geographic scale, such as the Hawaiian islands, no genetic differentiation was detected between colonies and populations of *C. formosanus*. All 8 investigated alloenzyme loci were monomorphic (*11*). A similar lack of genetic differentiation between Hawaiian populations was reported by Broughton and Grace (*12*) based on restriction analysis of mitochondrial DNA. Because *C. formosanus* is an introduced species, this lack of genetic variation in Hawaii might be due to a single introduction, multiple introductions from related sources, or bottlenecks (e.g., through successful termite control efforts). However, because of the limited polymorphism of alloenzymes and of mitochondrial DNA restriction sites, conclusions based on these methods are tentative. The application of methods with higher resolution power, such as genomic DNA markers is needed for a more detailed understanding of the genetic relationships within and among populations.

In contrast to alloenzymes and mitochondrial DNA, genomic DNA markers (DNA fingerprints and microsatellite genotypes) have revealed considerable genetic polymorphism even in introduced termite species on small spatial scales, such as *C. formosanus* on the island of Oahu, Hawaii. We have recently found that colonies were clearly genetically differentiated, yet the genetic distance between them showed no correlation with geographic distance, as in *Reticulitermes* (see above). Given the presumed weak flight capacity of *C. formosanus,* hindered further by mountain ranges dividing the island of Oahu, this lack of isolation by distance is surprising. It suggests that the predominant method of spread even on a fairly small scale is not by natural means but is human mediated (Husseneder, Vargo, Grace unpubl.). In contrast to the findings in *Reticulitermes* spp. and *C. formosanus*, natural populations of a termite species from the same family, the African subterranean termite *Schedorhinotermes lamanianus*, do show isolation by distance in spatial scales below their swarming distance of about 1 km apparently due to limited alate dispersal (*18*).

To trace ways and speed of dispersal and to track possible source populations of *C. formosanus*, we are currently assembling a large data base, measuring genetic interrelationships within and among introduced and native populations from the USA, Japan and China. First data show a lower genetic diversity in introduced populations, probably due to genetic bottlenecks reflecting the introduction event and ongoing control efforts. Nevertheless, genetic differentiation between populations is sufficient to classify individuals to source populations (Husseneder, Vargo, Grace unpubl.).

Population genetics in termites is still in its early days and only a few aspects in a few species have been studied so far. These studies indicate a wide range in intraspecific genetic variability, degrees of population differentiation and dispersal modes. A more complete picture will require detailed comparative studies involving natural and introduced populations of a variety of termite species.

Genetic Identification of Termite Colonies

To interpret genetic differentiation at the level of the population it is important to understand how a termite colony is organized, because colony identity and breeding systems shape the local population structure. Moreover, from a termite management point of view, the colony is the target unit. Therefore, it must be clearly identified and its structure described. Yet, because mark-release-recapture of dyed termites to connect termites from different collection sites to foraging areas of colonies is time consuming and requires large numbers of termites at each site, the affiliation of collection (or infestation) sites to particular colonies prior to control efforts is rarely known. In addition, if termites re-appear after control in the same location it is not known from which colony they originate. Molecular methods make it possible to assign termite collection sites to colonies and to determine the origin of re-appearing termites.

In termite species that exhibit sufficient genetic variation within populations colonies can be segregated by alloenzyme and mitochondrial DNA analysis. For example, alloenzyme studies found that colonies of *R. flavipes* are genetically isolated from adjacent colonies (*4, 9*). Using genetic differentiation between colonies, Bulmer et al. (*24*) were able to determine colony affiliation of sampled termites using alloenzymes and mitochondrial DNA haplotypes.

However, in other termite species alloenzymes and mitochondrial DNA markers failed to reveal genetic variation between colonies. For example, restriction fragment analysis of mitochondrial DNA, showed no genetic differentiation between subspecies of *Zootermopsis nevadensis* nor between colonies within a proposed subspecies (*40*). Especially in introduced species, such as *C. formosanus,* alloenzyme and mitochondrial DNA markers are not polymorphic enough to reveal colony differentiation on a small scale. In Hawaiian *C. formosanus* populations no restriction fragment variation of mitochondrial DNA and only one polymorphic alloenzyme locus was discovered (*11, 12, 34*). Therefore more variable DNA regions, such as repetitive sequences of genomic DNA, had to be employed to detect genetic differentiation between colonies.

Husseneder et al. (*18*) used multilocus DNA fingerprinting to differentiate colonies of the African subterranean termite *S. lamanianus* even on a small geographic scale. We employed this same method to describe colonies of *C. formosanus*. In *C. formosanus* populations from Hawaii and Louisiana genetic similarities within colonies were significantly higher than genetic similarities between termites from different colonies. Colonies can thus be genetically differentiated. Due to this differentiation, termites can be assigned to colonies by looking for the closest match with the highest genetic similarity reflected by the proportion of shared bands. In terms of colony assignment molecular methods proved to be superior to behavioral tests and morphometric studies in *C. formosanus* (*41-43*). In addition, the genetic profile accumulated from pools of colony members is specific for a colony. Provided there is sufficient genetic differentiation between colonies of a local population, diagnostic bands distinguish the genetic profile of the target colony from the profiles of other colonies. Termites that possess these bands can be assigned to a particular colony. In addition to assigning termites to colonies, genetic profiles can be used to "tag" colonies prior to elimination for re-infestation studies. In cases where termites appear in the same location after elimination through baiting, it can be decided - by comparing genetic profiles - if they are remnants from the same colony, invaders from neighboring colonies or – if the profiles do not match known colonies - new infestations from outside the area. Genetic profiles combining multiple microsatellite loci can be used for the same purpose (Vargo and Husseneder unpubl.).

In sum, depending on the genetic variation in a termite population, alloenzymes, mitochondrial DNA and/or genomic DNA markers can be used to group termites from different collection sites to colonies and thus give an estimation of the foraging area of a colony. Moreover, it is possible to monitor the success of colony elimination efforts by showing if activity at all collection sites belonging to a colony ceases and verifying the origin of re-appearing termites. Lastly, defining a colony is a necessary first step for investigating breeding systems and intracolonial structures.

Social Organization and Intracolonial Structures

Surprisingly little is known about social and spatial organization of termite colonies. Termite colonies are not always simple families consisting of one breeding pair. Polygamy can arise either from multiple adults co-founding a colony (pleometrosis) or from multiple reproductives reared within a colony. The origin and relatedness of breeders determines the degree of inbreeding within a colony. Molecular methods make it possible to assess whether a single pair or multiple kings and queens are reproducing within a colony, the degrees of inbreeding between them and whether the kin composition leads to genetic structure within the colony.

Mitochondrial DNA is passed on through matrilines only. Therefore, it can be used to detect the number of different matrilines in colonies and thus the number of maternally unrelated females. Mitochondrial DNA haplotypes were investigated to determine the origin of co-occurring kings and queens in colonies of several termite species. Distinct haplotypes can occur due to several reasons. Heteroplasmy (different mitochondrial haplotypes within the same individual) and paternal contribution of mitochondrial DNA are considered rare (*40*). Queen adoption or mixing of different colonies resulting in multiple haplotypes at the same collection site are discussed (*44*), however not yet unambiguously proven for termites. Thus, colony foundation by multiple maternally unrelated queens remains the most likely hypothesis. Different mitochondrial DNA haplotypes, for example, were found among workers at the same collection site in *Z. nevadensis* (*40*) and *Reticulitermes* spp. (*44*) and *R. flavipes* (*24*). In colonies of *Nasutitermes corniger* with a small numbers of queens, the queens had different haplotypes and were thus maternally unrelated. In colonies with large numbers of reproductives, however, all reproductives shared the same haplotype, suggesting replacement of a single founder pair by several offspring recruited by their natal colony (*14*). Similarly, in *Nasutitermes nigriceps* and *Nasutitermes costalis* all individuals of a colony shared the same mitochondrial DNA haplotype (*15, 16*). In these cases it cannot be determined whether a single reproductive queen or multiple kings and queens from the same matriline (secondary replacement reproductives) reproduce in a colony.

To decide if a colony contains a single pair or has multiple reproductives, even if they stem from the same matriline, an allelic approach has to be employed. Using alloenzymes and microsatellite genotypes, one can test whether the distribution of the homo- and heterozygote genotypes in a colony deviate from the Mendelian ratios expected for a single pair (e.g., *15*).

Alloenzymes are appropriate for analyzing the colony social organization of termite species with sufficient genetic variability. In some termite species Mendelian distribution of genotypes in the majority of colonies show that monogamy is the predominant mode of reproduction. For example, alloenzymes have shown that colonies of *N. nigriceps* are headed predominantly by single unrelated pairs, which mate randomly. Colonies therefore show little inbreeding. Only in 7 in 136 colonies did alloenzymes show offspring genotype distribution indicating multiple reproductives (*15*). As mentioned above, individuals shared the same mitochondrial DNA haplotype, which confirms that multiple reproductives have developed within their natal colony. Similarly, colonies of *Incisitermes schwarzi* are generally simple genetic families headed by one reproductive pair (*45*). Yet, a quarter of the colonies in the field showed multiple related reproductives and sometimes the alloenzyme patterns of the offspring were a mixture of the founder pair and the replacement breeders. This reveals the natural turnover in the colony cycle of many termite species, which prolongs a colony's life far beyond the lifespan of its founders. After the death of the founder pair in mature colonies, reproduction is continued by multiple replacement reproductives, which are offspring of the colony founders and therefore related (*Nasutitermes* spp.: *14*, *15*; *Hodotermopsis spp.*: *34*, *Reticulitermes* spp.: *9*, *33*).

For species with low alloenzyme variation genomic DNA has been employed to determine number and relatedness of breeders. Multilocus DNA fingerprinting has shown that about half of the colonies of *S. lamanianus* were headed by an unrelated pair, probably the colony founders, while the other colonies contained multiple related reproductives. Comparison of the degrees of relatedness among multiple breeders and their offspring to the expections assuming various numbers of generation turnovers suggests that only one inbreeding cycle took place (*46*).

Microsatellite genotypes show up to 50 alleles per locus and thus have a much higher resolution power than alloenzymes (*21*). To date, microsatellite primers have been developed for *Macrotermes michaelensi* (*29*), *R. flavipes* (*27*), *Coptotermes lacteus* (*30*) and *C. formosanus* (*28*), and studies on the social organization in most of these species are currently under way. First studies on the social organization of *C. formosanus* show an intriguing difference in the reproductive structure of colonies in introduced versus native populations. In introduced populations of Hawaii and Louisiana the proportion of colonies headed by multiple reproductives ranges from 36-65%. However, in a native population from Guangdong, China, all 14 investigated colonies were headed by multiple reproductives (Husseneder, Vargo, Grace, unpubl.).

The prime example of the possible variety of alternative breeding systems depending on species, location, or ecological factors are the *Reticulitermes* species in the USA and Europe. A number of studies employing alloenzymes and mitochondrial DNA have shown that colonies can be headed by single outbred pairs, multiple inbred reproductives, as well as multiple maternally unrelated breeders (*4*). Clément's enzymatic studies showed geographical variation in breeding systems for *Reticulitermes santonensis* and the *Reticulitermes lucifugus* species complex in Europe (*8*, *33*). Social organization in these species correlated with season and location. Living in humid climates with no resource limitation, colonies of *R. santonensis* and subspecies of *R. lucifugus* were generally highly polygynous. In contrast, in geographical regions with dry climate, monogyny combined with high aggression between colonies was prevalent. Alloenzymes and mitochondrial DNA studies on *R. flavipes* in the USA demonstrated considerable variation in colony and breeding structure with colonies headed by single unrelated pairs, colonies containing multiple inbred replacement reproductives and large multiple nest colonies containing unrelated queens (*9, 24, 44*).

Molecular methods have so far revealed a wide range of possible breeding systems in termites. This reproductive plasticity facilitates adaptation to different ecological challenges and makes termites successful in establishing themselves in a variety of habitats (*2, 4*). On the one hand, colonies headed by single unrelated pairs are common. The fact that the majority of colony founders are unrelated suggests that mating with closely related nestmates during the process of swarming and colony foundation is limited. Shellman-Reeve (*20*) has provided the first evidence based on multilocus DNA fingerprinting that adults of *Z. nevadensis* actively avoid pairing with nestmates when given a choice.

On the other hand, colonies can be genetically heterogenous societies containing multiple reproductives with varying degree of relatedness. This can create structures within colonies. For example, in the African termite *S. lamanianus,* genetic differentiation was found between individuals from different collection sites of one single large colony. This suggests multiple nests with reproductives within a colony that are spatially segregated, yet still connected by worker exchange (*18*). In colonies containing multiple reproductives, termites from the same tunnel were genetically more similar than termites from different tunnels and termites within the nest. This suggests kin-biased foraging, i.e., termites segregate according to kin lines during foraging (*13*). We are currently investigating the possibility of such intracolonial structuring in several subterranean termite pest species. Separate breeding units and kin-biased foraging could influence penetration of a colony by bait toxin, and has implications for the effect of kin structure on cooperative behavior within spatially complex termite colonies.

Conclusion

The application of molecular markers to the study of termites is a fairly young science, and, until recently, has received little attention compared to research on ants, bees and wasps. Nevertheless, the use of molecular methods has already provided important new insights into many areas of termite biology, including population structure, colony identification and breeding systems of various termite species. Comparative research on a variety of termite species will bring us closer to understanding the genetic and ecological determinants of life in the colony, including conflict and cooperation, optimal levels of inbreeding and reproductive strategies. Understanding the constraints and possibilities in the life of termites will also help to control pest species.

References

1. Jones, S. C. In *Handbook of structural insect pests*; Gold, R. E.; Jones, S. C. Eds.; Entomol. Soc. America: Lanham, MD, **2000**, pp 119-124.
2. Shellman-Reeve, J. S. In *Social behaviour in insects and arachnids*; Choe, J. C.; Crespi, B. J., Eds.; Cambridge University Press: Cambridge, **1997**, pp 52-93.
3. Ross, K. G. *Mol. Ecol.* **2001**, *10*: 265-284.
4. Thorne, B. L.; Traniello, J. F. A.; Adams, E. S.; Bulmer, M. *Ethol. Ecol Evol.* **1999**, *11*, 149-169.
5. Pamilo, P.; Gertsch, P.; Thorén, P.; Seppä, P. *Annu. Rev. Ecol. Syst.* **1997**, *28*, 1-25.
6. Kamphampati, S.; Kjer, K. M.; Thorne, B. L. *Ins. Mol. Biol.* **1996**, *5*, 229-238.
7. Miura, T.; Maekawa, K.; Kitade, O.; Abe, T.; Matsumoto, T. *Ann. Entomol. Soc. Am.* **1998**, *91*, 515-523.
8. Clément, J.-L. *Sociobiology* **1986**, *11*, 311-323.
9. Reilly, L. M. *Am. Nat.* **1987**, *130*, 339-349.
10. Korman, A. K.; Pashley, D. P. *Sociobiology* **1991**, *19*, 41-50.
11. Strong, K. L.; Grace, J. K. *Pan-Pac. Entomol.* **1993**, *69*, 51-56.
12. Broughton, R. E.; Grace, J. K. *Sociobiology* **1994**, *24*, 121-126.
13. Kaib, M.; Husseneder, C.; Epplen, C.; Epplen, J. T.; Brandl, R. *Proc. Royal Soc. Lond. B.* **1996**, *263*, 1527-1532.
14. Atkinson, L.; Adams, E. S. *Proc. R. Soc. Lond. B.* **1997**, *264*, 1131-1136.
15. Thompson, G. J.; Hebert, P. D. N. *Heredity* **1998a**, *80*, 48-55.
16. Thompson, G. J.; Hebert, P. D. N. *Ins. Sociaux.* **1998b**, *45*, 289-299.
17. Husseneder, C. Ph.D. Diss., Univ. of Bayreuth, Bayreuth, Germany, **1998**.
18. Husseneder, C.; Kaib, M.; Epplen, C.; Epplen, J. T.; Brandl, R. *Mol. Ecol.* **1998**, *7*, 983-990.
19. Jenkins, T. M.; Basten, C. J.; Dean, R.; Mitchell, S. E.; Kresovich, S.; Forschler, B. T. *Sociobiology* **1999a**, *33*, 239-263.
20. Shellman-Reeve, J. S. *Anim. Behav.* **2001**, *61*, 869-876.
21. Parker, P. G.; Snow, A. A.; Schug, M. D.; Booton, G. C.; Fuerst, P. A. *Ecology* **1998**, *79*, 361-382.
22. Hoy, M. *Insect molecular genetics*. Academic Press: San Diego, CA, **1994**.
23. *Molecular Systematics*; Hillis, D. M.; Moritz, C., Eds.; Sinauer Associates, Inc.: Sunderland, MA, **1990**.
24. Bulmer, M. S.; Adams, E. S.; Traniello, J. F. A. *Behav. Ecol. Sociobiol.* **2001**, *49*, 236-243.

25. Simon, C.; Frati, F.; Beckenbach, A.; Crespi, B.; Liu, H.; Flook, P. *Ann. Entomol. Soc. Am.* **1994**, *87*, 651-701.
26. *Molecular tools for screening biodiversity;* Karp, A.; Isaac, P. G.; Ingram, D. S. Eds.; Chapman and Hall: London. **1998**.
27. Vargo, E. L. *Mol. Ecol.* **2000**, *9*, 817-820.
28. Vargo, E. L.; Henderson, G. *Mol. Ecol.* **2000**, *9*, 1935-1938.
29. Kaib, M.; Hacker, M.; Over, I.; Hardt, C.; Epplen, J. T.; Bagine, R. K. N.; Brandl, R. *Mol. Ecol.* **2000**, *9*, 502-504.
30. Thompson, G. J.; Lenz, M; Crozier, R. H. *Mol. Ecol.* **2000**, *9*, 1932-1934.
31. Husseneder, C.; Vargo, E. L.; Grace, J. K. Proc. XXI. Int. Congress Entomol. *Sociobiology* **2001**, in press.
32. Santos, O.; Luykx, P. *Biochem. Genet.* **1985**, *23*, 729-740.
33. Clément, J. L. In *Biosystematics of social insects;* Howse, P. E.; Clément, J. L., Eds.; Academic Press: London, **1981**; pp 49-61.
34. Wang, J.; Grace, J. K. *Hawaii Agriculture: Positioning for Growth. Conference Proceedings. CTAHR University of Hawaii, Honolulu.* **1995**, pp 168-169.
35. Korman, A. K.; Pashley, D. P.; Haverty, M. I.; La Fage, J. P. *Ann. Entomol. Soc. Am.* **1991**, *84*, 1-9.
36. Wang, J.; Grace, J. K. *Sociobiology* **2000a**, *36*, 21-31.
37. Reilly, L. M. Ph. D. Diss., Vanderbilt University, TN, **1986**.
38. Wang, J.; Grace, J. K. *Sociobiology* **2000b**, *36,* 7-19.
39. Wang, J., and J. K. Grace. *Sociobiology* **2000c,** *36*, 1-6.
40. Broughton, R. E. *Ann. Entomol. Soc. Am.* **1995,** *88*: 120-128.
41. Husseneder, C.; Grace, J. K. *IRG.* **2000,** WP 00-10371.
42. Husseneder, C.; Grace, J. K. *Env. Entomol.* **2001a**, *30*, 262-266.
43. Husseneder, C.; Grace, J. K. *J. Ins. Behav.* **2001b,** *14*, 173-186.
44. Jenkins, T. M.; Basten, C. J., Kresovich, S.; Forschler, B. T. *Sociobiology* **1999b**, *34*, 161-172.
45. Luykx, P. *Insectes Soc.* **1993**, *40,* 191-205.
46. Husseneder, C.; Kaib, M.; Epplen, C.; Epplen, J. T.; Brandl, R. *Behaviour* **1999**, *136*, 1045-1063.

New Preservation–Protection Systems

Chapter 22

Can Understanding the Mechanisms of Biodegradation Help Preservative Development?

Alan F. Preston

Chemical Specialties, Inc., 200 East Woodlawn Road, Suite 250, Charlotte, NC 28217

The development of preservative treatments for wood products has historically been a process of trial and error. Formulations considered to have potential, usually through previous use in related biocide applications, are tested against wood destroying organisms, either in laboratory culture challenge tests, or in treated wood specimens in laboratory or field tests. This is usually a very long and often unsuccessful process, where experimental science plays the predominant role.

Understanding biodeterioration mechanisms can assist in the preservative development process. The scope of such understanding can be quite broad, including fungal mechanisms, insect attack modes, weathering effects, chemical degradation related to biological mechanisms, potential for preservative loss in service, preservative fixation mechanisms, etc. This paper addresses the various mechanisms that affect biodegradation of wood products, and illustrates these with examples of their use in furthering preservative development.

Biodeteriogens of Wood

The primary biodeteriogens of wood are fungi and insects. As a general rule, with wood products exposed in a natural environment for an extended service life fungi present the primary hazard to the wood and are generally more difficult to control than insects. However, in specific environments, for example inside the framing of houses, insects can present the predominant hazard.

Insects

The use of fundamental knowledge of biodeterioration mechanisms for vector control has lead to the development of toxicant baits for termites. This is a clear example of a biodeteriogen suppression strategy developed from knowledge of causal agent's mode of action. While this application is not directly within the realm of wood preservation in the sense of applying a treatment to wood, it is nevertheless an example of wood protection through control of the organism in the environment prior to its potential attack on wood.

Fungi

Much of the research in wood preservation over the last half century has focused on developing advanced knowledge of mechanisms of action of wood-destroying fungi. In this regard, significant advances have been made in the understanding of specific brown rot fungi, for example *Postia placenta*. In some fields, for instance control of short life cycle crop diseases in agriculture, knowledge of specific fungal mechanisms can provide useful insights. In contrast, the multiplicity of organisms that wood can be subject to during its long service life makes knowledge of specific fungal mechanisms less useful in wood preservation than in other fields.

One area where fundamental understanding has played an important role is in the understanding of the mode of action of copper tolerant fungi. With the change in the wood preservative market in some regions from copper arsenate preservatives to copper-based systems, the understanding of when, where and how such fungi work has led to strategies for the development of secondary biocides to be used in copper-based preservative systems. Copper tolerant fungi are considered to be of little importance in above-ground exposures, especially when low moisture contents are maintained in the wood. However, they can present a hazard to wood in ground contact, although their presence is highly sporadic and largely unpredictable.

Degradative Processes

While it can be argued that knowledge of specific mechanistic processes of particular fungi is of limited value in wood protection when wood is subject to degradation by a plethora of organisms, depending of geographic, construction, treatment, meteorological and other factors, the key fungal degradation processes of wood products in service are oxidative. In a similar manner, wood weathering is also an oxidative process. Beyond the need for an oxidative atmosphere, most wood destroying fungi also require the presence of appropriate moisture conditions for optimal growth. Increasing moisture content of wood generally increases the propensity for fungal attack, although this becomes limiting at very high moisture contents.

Wood destroying fungi utilize oxidative processes, and it has been found that anti-oxidants may synergize fungicide performance in wood preservative formulations. While anti-oxidants are unlikely to provide control as a sole wood protection treatment at economic levels, this ability to provide secondary protection may allow enhance the performance of protection concepts or chemicals that currently provide marginal performance, or allow the development of systems having acceptable cost effectiveness and protection. At this time anti-oxidants do not appear to be significant as additives in metal-based preservatives, although scant data exists, as with copper-based systems the phenolic group of the anti-oxidant and the copper(II) will likely form a complex. This complex would negate the anti-oxidant properties of the phenolic compound. Thus it appears that anti-oxidants are more likely to be important with pending moves toward water-based organic preservative and non-biocide systems.

As mentioned above, optimal moisture contents provide optimal conditions for fungal decay of wood products. The converse is also true, and moisture control strategies can play a significant role in the future of wood protection. Wood moisture content reduction strategies have received increased attention in recent years, although the concept dates back to antiquity. It could be argued that to some extent the oil-borne preservative pentachlorophenol utilizes a water controlling strategy in the synergism seen between oil carrier type and preservative performance in pole treatments. Similarly, millwork treatments have long been known to be enhanced by the presence of water repellents in the solvent treatment. For the water-based preservatives, water repellents were initially developed to enhance the long-term physical weathering aspects of chromated copper arsenate (CCA) treated wood and provide little, if any, benefit to CCA in enhancing preservative performance. However, there is evidence that water repellent additives can enhance the performance of copper-based preservatives, allowing lowering of preservative retentions in some situations. Similarly, water repellents appear to enhance the performance of organic preservatives in above ground applications, although further development is required to meet expectations for ground contact applications.

Recent research in Europe has seen the commercialization of heat treatments for wood products. The concept of protecting wood by heating in a non-oxidative atmosphere has been known for many years, with the work of Stamm in the 1940's being particularly important in this regard. The heating process largely impacts the hemi-celluloses in wood and interferes with the wood's susceptibility to decay. Questions remain as to the broad applicability of the technology, but in today's changing environment this technology certainly merits further development. Issues that need to be addressed include strength losses, emissions during processing, weathering performance, insect protection, ground contact applications, etc, but heat treatments in many respects are not inconsistent in properties with the wood-plastic composites that are currently in vogue in some quarters.

In a similar vein, interest in wood modification processes has seen something of a revival as the wood preservative market changes and diversifies. For solid wood products, the costs of an approximately 15-20% weight add-on of a modifying agent remain a deterrent to commercialization, but clearly wood modification is a technically sound approach to wood protection. The fundamental strategy behind wood modification is one of making the wood substrate inert to fungal attack through reaction of the active hydroxyl sites in wood to provide ester or ether functionalities that are essentially impervious to enzymatic or chemical attack.

The strategies described above all have their weaknesses in providing cost effective protection, but combinations of such treatments may improve the commercial likelihood of success. At the same time enhanced properties are commanding increased value in the marketplace and will likely allow for treatments that would not be commercially viable in today's market for wood preservatives.

Realities of Developments in Wood Preservation

In many respects wood preservation appears to have highly attractive commercial potential for biocide applications, but there are a number of development constraints that impact this potential. The first of these is the long product service expectations for the products used. The second is the relatively small value ascribed to the wood protection agent relative to the value of the raw material (wood) used. Overall, the value of treated wood sold is large, but the economic value of the chemical treatment is a minor component of this treated wood cost. Furthermore, the implied liability of the failure of the treatment involves the entire value of the product. That is, the risk to reward profile is not as attractive as in other biocides businesses, and the time to failure, and hence total product risk, is long.

In the world of biocides, wood preservation is a relatively low value sector. For new chemical biocide developments, specifically fungicides and insecticides, the wood protection industry is dependent on the agricultural and pharmaceutical businesses to provide new products. On top of that, newly developed fungicides in the agricultural and pharmaceutical areas are highly organism-specific, while wood protection requires broad-spectrum protection. This dependence is due to the fact that registration costs for active ingredients is nowadays very high, and the rapid returns necessary to sustain such costs are a large constraint to the development of specific biocides for use solely in the slow-to-change field of wood protection. The chances of success in wood protection are low, due to the multiplicity of degrading organisms. Furthermore, broad-spectrum biocides may display less than optimal properties in their toxicological and/or ecotoxicological profile. In other words, there is an unbalanced up-front risk for a hard-to-quantify long-term potential gain.

In developing new protection agents for wood products, it is interesting to consider the amount, or retention, of various agents in protecting wood. Current wood preservatives are used at levels around 1% mass/mass of wood. At these levels distribution in wood is generally acceptable although distribution gradients can be steep. To be competitive, newer fungicides will have to be used in the 0.01 - 0.1% mass/mass add-on range in wood. At these retentions, macro distribution can become problematic, especially with biocides that react with the wood substrate. Conversely, of course, biocides that don't react with the wood substrate may be more leachable than desired.

At the other end of the scale, investigation of wood extractives responsible for natural durability effects has had a rebirth in some quarters. Difficulties here include the fact that utilization of natural biocides as biocides will still require highly expensive toxicological testing with, at best, dubious patent protection, and that the effective use levels of such extractives in wood are often well in excess of 1% mass/mass. Thus a combinatorial strategy from those described is probably the most likely to provide a cost effective protection product at acceptable levels of financial risk.

In some jurisdictions, and Europe in particular, performance criteria for product approvals are based on laboratory test protocols. Such laboratory tests often give results which significantly under-estimate the level of preservative required for sound protection of treated wood products in service. Over the last decade this has led to a rapid, and probably irreversible, decline in product performance and expectations. Most of the rest of the world believe that laboratory results provide valuable insight into performance against specific known organisms but that field evaluations are far more likely to provide assurance of performance in various environments where treated wood products are subject to attack by multiple organisms. This is especially important where the performance expectations are subject to litigation against the producers rather than being seen as being established by some governmental authority. In the former environment, it is prudent that one take into account expected worst case usages in designing test protocols. Also, of increasing importance is the need to take into account fit-for-purpose aspects of the products, rather than just protection from biodegradation.

Conclusions

Fundamental scientific research into understanding termite behavior has led to the development of commercial baiting systems for protection of structures in a low environmental impact treatment. However, fungal mechanisms are more challenging as a broad-spectrum protection strategy is necessary for wood products with long-term service expectations. It is clear that stopping, modifying or slowing oxidative processes can provide a positive impact in enhancing the performance of some wood preservative chemicals and that reducing moisture content in wood also can provide positive benefits.

Further developments may provide the key to organic or non-biocidal treatments for long term service life of wood products.

Chapter 23

New Environmentally-Benign Concepts in Wood Protection: The Combination of Organic Biocides and Non-Biocidal Additives

Frederick Green III[1] and Tor P. Schultz[2]

[1]Forest Service, Forest Products Laboratory, U.S. Department of Agriculture, Madison, WI 53705
[2]Forest Products Laboratory, Mississippi State University, Mississippi State, MS 39762

The principal wood preservative for residential use is currently a broad-spectrum biocide which contains As, Cr and Cu. However, the use of totally organic wood preservative systems in certain applications or localities may be required in the future. This will greatly increase the price of treated wood. The combination of organic biocides with non-biocidal additives to give enhanced efficacy is one possible means to lower the cost of future wood preservative systems. Possible non-biocidal additives which could be mixed with organic biocides include water repellents, antioxidants, and chelators for specific elements, including metals required by wood-decaying fungi such as iron or manganese. Selected laboratory and field exposure results obtained by combining organic biocides with one or more non-biocidal additives are presented in this chapter. Future wood preservative systems, either based on this chapter's concept or other possible techniques, will likely be used for relatively specific applications rather than today's one broad-spectrum preservative. This will require users of treated wood to be more educated in the future.

INTRODUCTION

Wood is a natural and renewable resource used extensively in home construction, decks, fences, utility poles, etc. Unfortunately, if sufficient moisture is present lumber, poles, and wood composites can be attacked by many organisms, principally brown-rot fungi and termites. In the U.S. alone the wood damaged yearly in residential structures by fungi and insects amounts to about $500,000,000 per year, and the labor involved further increases this to about $5 billion per year (Amburgey, Miss. State Univ., personal communication). Fortunately, wood can be treated with biocides to prevent damage caused by fungi or insects. Use of treated wood not only helps homeowners save money but also conserves our forests.

The major wood preservative used today in the U.S. is water-borne chromated copper arsenate (CCA). Based on wood volume, about 80% of all treated wood in the U.S. is preserved with CCA (*1*), and CCA is - by far - the major preservative used to treat lumber for residential construction, the major market for treated wood. However, there are some concerns with CCA including a public perception of possible arsenic exposure, leaching of the metal oxides and the question of the ultimate disposal of CCA-treated wood. Thus, use of CCA-treated lumber will be restricted starting by 2004 to only non-residential applications. An alternative is borate-treated lumber (*2-4*) which is commercially available in Hawaii and provides cost-effective and benign protection in certain applications. However, borate use is limited to non-leaching applications.

Alternative, second generation wood preservatives for residential applications are copper-organic mixtures (*2*), such as ammoniacal copper quat (ACQ) and copper azole mixtures, with or without boron (CBA, CA), and both ACQ and CA are commercially available. Although copper is not as toxic as arsenic or chromium accumulation of copper does have some negative environmental effects, especially in aquatic systems, and disposal of any metal-treated wood product may be expensive and difficult in the future. Thus, copper-organic mixtures may also face future restrictions in the U.S. Indeed, several countries in Europe are already moving towards totally organic wood preservative systems. Consequently, third-generation wood preservative(s) for residential use, based solely on non-metallic biocides (*2*) or a fixed borate system, need to be developed. A number of organic biocides are already commercially available as agrochemicals and have been examined as potential wood preservatives (*5*). The relatively high cost of these organic biocides (mostly $15 - 25/lb versus $1.50/lb for CCA), environmental regulations, and public concerns with bioactive compounds will undoubtably result in efforts to minimize the amount of biocide used. In addition to effectively and economically protecting wood against a wide variety of wood decaying organisms, alternative preservative systems must have good weathering and UV protection properties, minimal leaching of the active compound, and be noncorrosive to metal fasteners. Finally, water-based emulsion formulations for the organic biocides (*2*) will likely need to be developed.

Two potential methods for reducing the level of a biocide, and thus the cost and environmental impact of treated wood, are to combine two or more biocides (fungicides and/or insecticides) to give a synergistic mixture [covered in Leightley's chp.], or to combine biocides and non-biocidal additives to give increased efficacy. This second option is the objective of this chapter. Specifically, we are examining non-biocidal additives which by themselves offer little or no protection but when combined with a biocide give wood greater and/or broader protection against fungi and/or termites than achieved with only the biocide. While many of the compounds discussed in this chapter are ineffective or require unreasonably high levels when used alone, when combined with a biocide the mixture may provide increased and/or broader efficacy and improved economics. A second advantage of this approach is that a synergistic mixture may be patentable.

NON-BIOCIDAL ADDITIVES EXAMINED

For fungi to survive and grow in wood they need to: 1) Be able to live in and colonize the woody substrate; 2) generate the enzymes and reagents necessary to initially disrupt and break down lignocellulose; 3) have free water in the lumen so that the enzymes and small reagents can diffuse from the fungal mycelium to the cell wall and then penetrate and partially degrade the lignocellulose; and, 4) digest the partially degraded wood components as a food source. For example, adding a fungicidal compound can prevent fungi from living in treated wood; this is the approach currently used to preserve wood.

Alternatively, non biocidal compounds might affect one or more of the above steps. For example, it may be possible to tie up essential minerals, elements, or compounds such as amino acids needed by fungi to colonize and grow in wood. Thus, one control method would be to prevent fungi from obtaining essential elements such as N, P, or Ca. Positive results have already been reported using only the calcium chelating compound Na-*N',N*-napthaloylhydroxamine (NHA, sodium salt) as a wood preserving agent (*6,7*). [Alternate mechanisms, beside calcium chelation, might also explain NHA's effectiveness against termites and fungi]. In another example, researchers have long examined methods to make wood more hydrophobic so that lumber used in an above-ground application such as decking would absorb less water. This approach includes the commercially available and relatively inexpensive wax emulsion formulations (*2,8*). An alternate but expensive approach is to add monomeric compounds to wood which are then polymerized *in situ* to make wood hydrophobic (*8*). These compounds might also covalently bond with, and thus alter, the chemical structure of the wood components such that the fungal enzymes would no longer be capable of degrading the woody material, thus inhibiting both steps 3 and 4 above. While this approach has been extensively studied in the laboratory, in discussing this and other novel wood preservation approaches Suttie - in a British understatement - noted that "The scaling-up of laboratory techniques to the pilot plant is not without problems" (*8*). Alternatively, to degrade wood fungi need to first disrupt and open up the cell wall. Fungi accomplish this by generating compounds which diffuse into the cell wall and generate reactive radicals which then disrupt the wood lignocellulose (*9-12*) [also see chps. by Aust, Enoki, Messner, and Goodell]. Consequently, the presence of free radical scavengers (antioxidants) in the cell wall would protect the cell wall from becoming more porous (*9,13*) [and may also help protect an organic biocide from being biodegraded]. An alternative and potentially elegant method would be to identify an additive which would disrupt the initial generation of the pre-radical oxidant in the acidic region of the fungal mycelium, and thus prevent the wood cell wall from being perturbed while possibly also causing the oxidant to form a radical near the fungus. Finally, metals such as Fe or Mn are well known to be involved in fungal degradation mechanisms, either as part of an enzymatic system or as a free metal. Thus, addition of appropriate metal chelators might prevent the metals from being available to the fungi (*8,9,14*). Other mechanisms, besides chelating essential metals, might also explain the enhanced efficacy obtained when a biocide and metal chelator are combined, *e.g.* (*15*).

In reviewing the above potential additives, it is worthwhile to examine the properties of the heartwood extractives in naturally durable woods. The fungicidal properties of the vast majority of extractives have been found to be mediocre - at best - when compared to commercial biocides (*16*). However, the various phenolic extractives are well known to be excellent antioxidants (*17-19*), and many of these phenolics also have metal chelating properties (*19*). Thus, the combination of a biocide, antioxidant and/or metal chelator might simply mimic nature's approach to make wood durable. For example, the combination of various antioxidants and/or metal chelators increases the efficacy of a wide variety of organic biocides (*14,20*). Furthermore, gallic acid derivatives, derived from the tannic acids found in heartwood, enhances the efficacy of the relatively expensive biocide propiconazole (*14*). Finally, many extractives, such as the terpenoids, are hydrophobic. The high level of water-repelling rosin in SYP lightered wood makes this wood extremely durable, even in ground contact (*21*). Adding extremely high levels of wax alone (about 26 pcf) to ground contact stakes mimics this and increases their average life to about 19 years as compared to less than 3 years for untreated stakes (*22*). While adding such high wax levels is unrealistic, we believe that the co-addition of smaller amounts of a wax along with a biocide and other possible additives might improve the biocide's efficacy so that relatively low biocide levels could protect wood. Indeed, the combination of a biocide and water repellent is already used in window joinery.

While less is known about the mechanisms by which termites and their symbiotic microorganisms degrade the holocellulose in wood, many of the above additives may also control termites. For example, the heartwood of naturally decay-resistant woods are also usually resistant to termites and, thus, the combination of biocides, antioxidants and/or metal chelators may be successful in protecting wood against both termites and fungi. SYP lightered wood, such as found in old pine stumps, has both decay and termite resistance, and the stakes with very high levels of wax described above were also resistant to termites even though paraffin wax has no termiticidal properties.

LABORATORY AND FIELD TEST RESULTS

Initial Considerations

The development of wood preservative systems requires considerable time and expense. This is especially true with systems comprising two or more components, where the number of tests to be run is much greater than for single component systems. While the most realistic test of a system's efficacy is an outdoor exposure study, the extensive time required [years] and the multitude of potential combinations makes initial screening using outdoor testing impractical. Thus, initial evaluation requires selection of an appropriate laboratory decay method to rapidly test the many different blends **and greatly reduce the number of the potential systems - hopefully without "tossing the baby out with the bath water".** That is, initial testing should be conducted quickly and provide data which explicitly separates promising from inferior systems.

The selection of an appropriate laboratory test(s) is more difficult than generally realized. The most common laboratory decay test used in wood preservation studies in the U.S. is the AWPA E10 12-week incubation soil block test, but this test may not necessarily be the optimal screening assay and also takes 16 or more weeks in total to perform. For example, when developing a new system for above-ground use the presence of minerals in the soil media may overwhelm certain additives and give a negative result which would not necessarily be observed during actual outdoor tests, as shown below. Another example is water repellents; minimizing the amount of water adsorbed by wood will obviously reduce the decay potential. However, wood blocks are typically first steam sterilized and, therefore, already have sufficient moisture for decay to occur at the start of the laboratory soil-block test. Finally, the soil-block test is unrealistically harsh and is initiated by actively growing fungal mycelium. In contrast, outdoor above-ground tests are initiated by delicate spores. Other bioactivity laboratory tests include the fungus cellar (*23,24*), agar-block and agar-plate tests (*25*); each test has particular advantages and limitations. In this chapter we describe results obtained using soil- and agar-block tests using compression strength loss to measure extent of fungal degradation after 5-6 weeks of incubation.

Biocide/Antioxidant/Metal Chelator Combinations

As discussed above, heartwood extractives have biocidal, antioxidant and metal chelating properties, all of which may influence natural durability. Consequently, we have combined various commercial organic biocides with different antioxidants and/or metal chelators (*13,14,20*). In laboratory tests, an antioxidant or metal chelator alone often had little or no protective effect (*12,26*) but when combined with a biocide enhanced or synergistic efficacy can be observed. For example, the antioxidant propyl gallate at 2% levels provided no protection to southern yellow pine (SYP) exposed to the brown-rot fungus *Gloeophyllum trabeum* for 5 weeks in the soil-block test and, similarly, no protection is observed with 3% of the metal chelator EDTA (Figure 1). [The concentrations given in this and other experiments are the active ingredients (% a.i.) used to treat the wood by a full-cell process for laboratory or outdoor exposure tests]. The biocide propiconazole, at treating solution concentrations up to 0.12% a.i., provided essentially no protection, and no benefit was observed when the same treating concentrations of propiconazole were combined with either 2% propyl gallate or 3% EDTA. However, when all three components are present increased efficacy at all four biocide levels is readily apparent. The results obtained here and previously with relatively low levels of propyl gallate (*14*), compared to the higher levels of other antioxidants used such as 4 to 5% butylated hydroxytoluene (BHT), is intriguing. It is possible that the good results in Table 1 are due to the dual antioxidant and metal chelating properties of gallate derivatives (*14*); other mechanisms might also account for the promising results.

We mentioned above that the particular laboratory test used can affect the results and thus the conclusions. For example, when aspen sapwood is exposed to the white-rot fungus *Trametes versicolor* for 6 weeks in the agar-block test, 3% EDTA alone was surprisingly effective (100% strength loss for untreated samples vs. only 5.4% loss for aspen samples treated with 3% EDTA). However, in the soil-block test EDTA alone offers little protection (Fig. 1). Other researchers have also reported that a metal chelator alone offers little protection to wood when a soil media test is used but is very effective in an agar media test (*14,26*). Presumably, the large amounts of minerals in a soil-containing test can quickly overwhelm and inactivate a metal chelator, while in agar media (which has relatively low mineral levels) a metal chelator alone may be quite effective at protecting wood. Similarly, when testing antioxidants better results are often obtained with an agar-block than soil-block test, perhaps because of the different levels of radicals produced (*13*) with the different substrates (*27*). Finally, the substrate used also has an effect. For example, a fungicide's effect on protecting cotton cellulose against a fungus is not comparable with the results obtained using SYP wood (*28*).

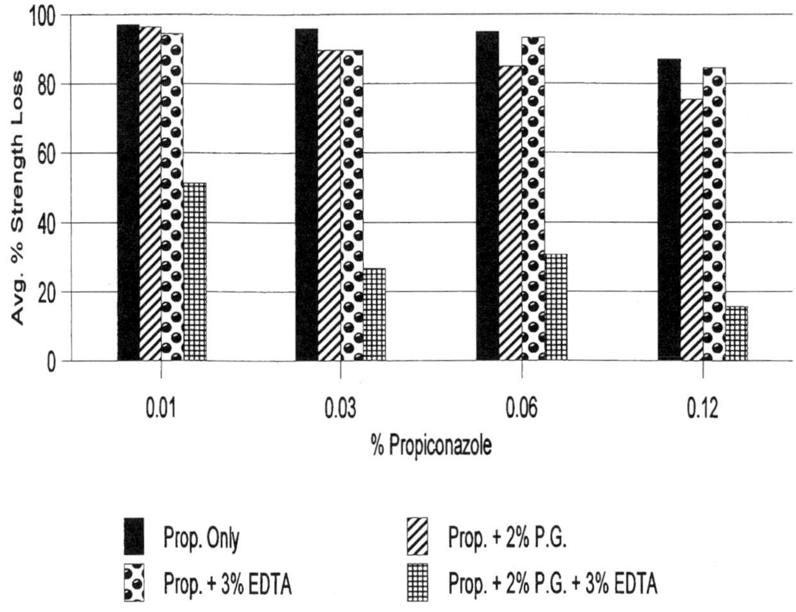

Figure 1. Efficacy of a biocide (propiconazole), antioxidant (propyl gallate, P.G.), and metal chelator (EDTA) systems. The soil block test was run using southern yellow pine sapwood with the brown-rot fungus G. trabeum and 5 weeks of incubation. The average strength loss with only 2% propyl gallate was 93.4%, 3% EDTA alone had 98.5% strength loss and untreated controls had an average of 98 and 96% strength loss.

The combination of the antioxidant BHT with the commercial biocide chlorothalonil also gave enhanced efficacy in ground-contact field tests conducted at two sites. [The Dorman Lake plot is located in northeast MS near Mississippi State Univ., has a heavy clay, poorly-drained soil and is in a high (Zone 4) deterioration zone. The Saucier test plot is located in the Harrison National Forest near the town of Saucier, MS, and has a sandy loam, well-drained soil and is in a severe (Zone 5) deterioration zone. Since this site is near the Gulf Coast it has a relatively mild winter and wet summer as compared to the Dorman Lake site]. For example, better protection was observed against both fungi and termites when BHT was combined with the chlorothalonil as compared to chlorothalonil alone (Table 1). It is interesting that 4% BHT alone provided some protection to wood in outdoor exposure as compared to untreated controls, even after 33 months of exposure. By contrast, in the soil block test 5% BHT alone gave no protection after four weeks of incubation (*13*). Thus, laboratory decay tests are unrealistically harsh but data is obtained within weeks as compared to the years needed for outdoor exposure tests. We use laboratory decay tests to quickly determine if a particular combination is synergistic, and outdoor exposure to judge the efficacy of a particular system under "real life" conditions.

Table 1. Average decay and termite ratings for SYP field stakes treated with chlorothalonil (CTN) alone, or a mixture of CTN and BHT, after 33 months of exposure at two field test sites.

Treatment	Average Retention, kg m^{-3}	Dorman Lake		Saucier	
		Decay	Termite	Decay	Termite
0.15% CTN	0.74	7.4	7.2	7.4	7.9
0.15% CTN/2%BHT	0.72/9.5	9.8	9.8	9.7	7.3
0.15% CTN/4%BHT	0.70/18.8	9.8	9.9	9.5	8.4
0.30% CTN	1.47	8.8	8.9	8.4	7.8
0.30% CTN/2%BHT	1.54/10.1	9.9	9.9	9.9	9.6
0.30% CTN/4%BHT	1.41/18.8	10	9.9	9.4	8.7
0.50% CTN	2.42	10	10	8.0	8.0
0.50% CTN/2%BHT	2.41/9.6	10	10	10	9.9
0.50% CTN/4%BHT	2.43/19.5	10	10	10	9.9
4% BHT	19.4	9.4	9.4	4.9	5.5
Controls	—	0.7	0.4	0	0

Average of 10 stakes per treatment per site. A "10" rating is no attack, "9" trace of attack, etc., as per AWPA Standard E7-93.

Water Repellents

The addition of a water repellent with a biocide not only improves the decay resistance of wood in above-ground applications (Table 2), including samples treated with only a water repellent, but also greatly improves the weathering and dimensional stability of exposed lumber (*2*). Furthermore, paraffinic waxes may be the most cost-effective additive for improving durability (A. Preston, CSI, personal communication) and are environmentally benign. Consequently, various companies already have commercialized water-repellent systems for applications such as CCA-treated decking, and formulations for ground-contact applications may be available in the future. Other researchers (*29*) have also examined the combination of a water repellent and metal chelator to protect wood against decay and mold fungi.

Table 2. Average decay ratings of above-ground L-joint samples, with and without a co-added water repellant, after four years of exposure at CSI's Hilo, HI test site.

Treatment, % a.i. (DDAC:Na Omadine)	Average Decay Rating[a]	
	without water repellent	with water repellent[b]
0.1 : 0.02	3.4	7.7
0.2 : 0.05	4.7	6.6
0.4 : 0.1	4.9	9.1
0.6 : 0.15	8.6	9.9
0.8 : 0.2	9.5	10.0
Controls	5.3	7.7[c]

[a] Average of 10 replicates, with a "10" being no attack, etc.
[b] Water repellent was 5% palmitic acid, 3% butyl amine, and 3% butyl carbitol dissolved in water, with the biocides co-added.
[c] These control samples were treated with only the water repellent.

NHA Systems

The polycyclic organic compound NHA as its sodium salt chelates calcium, and has been examined as a possible stand-alone wood preservative (*6,7,26*). [As mentioned earlier, however, NHA might protect wood by other mechanisms]. When used at a level below the threshold [~0.5% a.i.], the protonated form of NHA alone provides no protection against the brown-rot fungus *G. trabeum* in the soil-block test (Fig. 2), or the white-rot fungus *Irpex lacteus* in the agar-block test (data not shown). When low levels of NHA are combined with low levels of the biocides DDAC, propiconazole, or 3-iodo-2-propynyl butyl carbamate (IPBC), synergism is clearly evident with IPBC against *G. trabeum* (Fig. 2) and synergism is likely against *I. lacteus*. However, no synergism was observed when NHA was combined with DDAC or propiconazole in these wood decay tests.

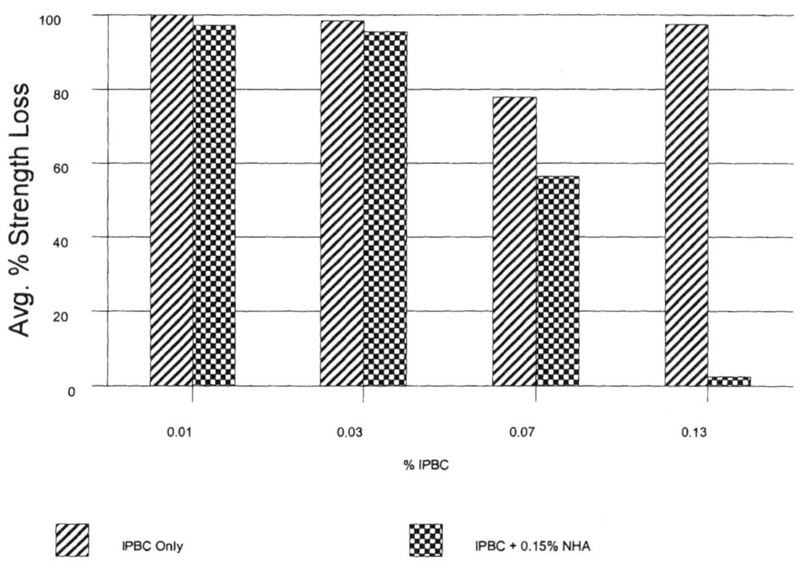

Figure 2. Synergism between IPBC and NHA, in the soil-block test using the brown-rot fungus G. trabeum for 5 weeks of incubation and SYP sapwood, with 5 replicates per set. The untreated controls lost an average of 97.9 and 97.6% strength, 0.15% NHA alone had a 97.5% loss, and the sodium salt form of NHA had 94.4 and 97.8% average strength loss at 0.15 and 0.30% concentration, respectively.

A preliminary synergism test against termites was also examined using NHA combined with DDAC or disodium octaborate tetrahydrate. However, no synergism was observed with either biocide.

CONCLUDING REMARKS

Future wood preservatives for certain applications or localities may be required to use non-metallic (organic) biocides, which will greatly increase the cost of treated wood. The combination of one or more organic biocide(s) with various non-biocidal additives might be one method to reduce the cost. Possible additives might include water repellants, free-radical scavengers (antioxidants), and chelators for specific metals required by the wood-destroying fungus and/or termites. Development of future totally organic wood preservative systems, based on ideas from this chapter or other techniques, will require extensive time and effort. It is extremely unlikely that under harsh conditions these systems will prove to be as effective as CCA in protecting wood. Furthermore, these future systems may be limited to specific applications, such as above-ground-use only. Thus, consumers will need to be aware of any limitations inherent in the forthcoming non-metallic targeted wood preservatives.

REFERENCES

1. Micklewright, J.T. *Wood Preservation Statistics 1997.* Am. Wood Pres. Assoc.: Granbury, TX, 1999.
2. Preston, A.F. *Forest Prod. J.* **2000,** *50(9)*, 12-19.
3. Grace, J.K. *Proceedings: The Second International Conference on Wood Protection with Diffusable Preservatives and Pesticides*; Forest Products Soc.: Madison, WI, 1996; 85-92.
4. Preston, A.F.; Jin, L.; Archer, K.J. *Am. Wood Pres. Assoc. Proc.* **1996**, *92*, 205-220.
5. Nicholas, D.D.; Schultz, T.P. *Wood Preservation in the '90s and Beyond*; Forest Products Soc.: Madison, WI, 1995; 169-173.
6. Green III, F.R.; Kuster, T.A.; Ferge, L.; Highley, T.L. *Int. Biodeterior. Biodegrad.* **1997,** *39*, 103-111.
7. Green III, F.G.; Crawford, D.; Lebow, S.; Yoshimura, T. *Proceedings: Second Annual Conference on Durability and Disaster Mitigation in Wood-Frame Housing*; Forest Products Soc.: Madison, WI, 2001; 235-236.
8. Suttie, E. *Chem. Ind. (London)*, **1997,** *18*, 720-724.

9. Goodell, B.; Jellison, J.; Liu, J.; Daniel, G.; Paszcynski, A.; Fekete, F.; Krishnamurthy, S.; Jun, L.; Xu, G. *J. Biotechnol.* **1997,** 53, 133-162, and references therein.
10. Tanaka, H.; Itakura, S.; Enoki, A. *Holzforschung* **1999,** *53*, 21-28.
11. Jensen Jr., K.A.; Houtman, C.J.; Ryan, Z.C.; Hammel, K.E. *Appl. Environ. Microbiol.* **2001,** *67*, 2705-2711.
12. Kerem, Z.; Jensen, K.A., Hammel, K.E. *FEBS Lett.* **1999,** *446*, 49-54.
13. Schultz, T.P.; Nicholas, D.D.; *Phytochemistry* **2000,** *54*, 47-52, and references therein.
14. Schultz, T.P.; Nicholas, D.D. *Phytochemistry*, in press.
15. Masayuki, K. *Bokin Bobai* **1987**, *15*, 281-286, Chem. Abst. 107:112507.
16. Schultz, T.P.; Harms, W.B.; Fisher, T.H.; McMurtrey, K.D.; Minn, J.; Nicholas, D.D. *Holzforschung* **1995,** *49*, 29-34, and references therein.
17. Larson, R.A. *Phytochemistry* **1988,** *27*, 969-978, and references therein.
18. Hadi, S.M.; Asad, S.F.; Singh, S.; Ahmad, A. *IUBMB Life* **2000,** *50*, 167-171.
19. Cooper-Driver, G.A.; Bhattacharya, M. *Phytochemistry* **1998,** *49*, 1165-1174, and references therein.
20. Schultz, T.P.; Nicholas, D.D. **2001,** U.S. Patent 6,231,651.
21. Koch, P. Utilization of the Southern Pines. U.S. Dept. of Agric. Forest Service, Washington D.C., 1972, pp. 644.
22. Gutzmer, D.I.; Crawford, D.M. *Comparison of Wood Preservatives in Stake Tests. 1995 Progress Report.* U.S. Dept. of Agric. Forest Service, FPL Research Note FPL-RN-02, Madison, WI, 1995, pp. 58.
23. Nilsson, T.; Edlund, M.-L. *Int. Research Group/Wood Preservation Document* **1995,** IRG/WP 95-20076.
24. Nicholas, D.D.; Archer, K.J. *Wood Preservation in the '90s and Beyond*; Forest Prod. Soc. Madison, WI, 1995; pp. 110-115.
25. Archer, K.; Nicholas, D.D.; Schultz, T.P. *For. Prod. J.* **1995,** *45(1)*, 86-89.
26. Green III, F.; Kuster, T.A., Highley, T.L. *Int. Research Group/Wood Preservation Document* **1997,** IRG/WP 97-10203.
27. Backa, S.; Gierer, J.; Reitberger, T.; Nilsson, T. *Holzforschung* **1992,** *46*, 61-67.
28. Green III, F. *Int. Biodeterior. Biodegrad.* **2000** *46*, 77-82.
29. Ritschkoff, A-c.; Mahlberg, R.; Viikari, L. **2001**, Int. Patent Appl. Chem. Abst. 134:267953.

Chapter 24

Protection of Wood Using Combinations of Biocides

Liam E. Leightley

Rohm and Haas Company, 100 Independence Mall West, Philadelphia, PA 19106–2399

Traditional wood preservatives provide a broad spectrum of activity but are being increasing regulated due to perceived environmental concerns. New organic biocides, developed for applications other than protecting wood, possess low toxicity and high selectivity against some organisms. However, the high selectivity also means that a particular biocide may be relatively inactive against a few of the many wood-destroying or staining/molding organisms. One possible solution is to develop new formulations based upon synergistic blends of biocides which are already registered for other non-wood applications. These formulated products can possess a broad spectrum of activity and low environmental impact. Examples of several synergistic combinations which are effective against stains and molds, or wood-destroying organisms, are given. This approach offers a good opportunity to develop new, relatively low cost and environmentally-benign wood protection systems.

Wood enjoys universal use as a versatile construction material. However, wood can be attacked and deteriorated through the action of a wide variety of fungi, insects and marine boring animals. This deterioration can be prevented if construction practices are followed which create unsuitable conditions for the activities of these destructive agents, such as keeping wood dry. However, if the wood product cannot be kept dry, or is in ground contact or an aquatic environment, then wood must be protected by treatment with biocides. Treated wood is resistant to fungi and insect attack and helps homeowners protect their property while conserving our nation's forests.

Mold or stain fungi can also inhabit wood. While these microorganisms do not degrade the wood, a moldy wood surface is unsightly, can stain the surface of lumber, and will cause any stain or paint applied later to quickly fail. Furthermore, the surface of painted wood surface can sometimes quickly become moldy, especially in humid environments.

The global market for wood protection chemicals is about US$120 million. Typical end uses for these types of chemicals are for treating wood/lumber intended for decking, fencing and utility poles. In addition to these end uses, biocides are applied to wood surfaces to inhibit stain and mold fungi. Biocides can also be added to wood-coating products such as stains, clears, varnishes and primers to prevent molds and stains on the painted surface. The global value for these second type of biocide products is approximately US$80 million. Thus, the total value of biocides used to treat wood products, or added to stains and paints, is relatively modest compared to the market for agrochemicals or drug development. Consequently, the market is simply unable to support the extensive research and development costs associated with developing entirely new biocides specifically for wood products.

Biocides used to protect wood are coming under increasingly stringent regulatory requirements. These requirements have increased the cost of supporting these chemicals. Furthermore, as discussed above it is unlikely that new biocides specifically for wood protection will be developed. Thus, an opportunity exists to develop new wood protection systems using blends of two or more biocides which are already developed and labeled for uses other than wood protection, such as agrochemicals used to protect plants against fungal or insect attack.

Requirements for the Ideal Wood Protection Biocide

Wood protection biocides protect wood from attack and subsequent degradation by fungi or insects. Typically, to be considered satisfactory as a wood protection chemical the product should possess the following characteristics

- Good efficacy at low cost against a wide variety of wood-destroying organisms
- benign to humans and other non-target organisms

- permanent/stable for the guaranteed life of the treated wood
- good penetration
- easy and safe to use, and acceptable to regulatory agencies
- not deleterious to wood
- readily available
- capable of being formulated into a commercial system (see below)
- allow the treated wood to be easily disposed, or recycled, at the end of the treated wood's life

Few of the newer biocides possess all of the above characteristics, especially the broad activity against the wide variety of organisms that degrade wood. However, mixtures of two or more biocides might offer broader protection. A classic product in this regard would be copper-chromium-arsenic (CCA), the major wood protection system currently used in the U.S. The presence of two biocides (Cu and As), and the Cr which fixes the metal biocides in wood and minimizes corrosion of metal fasteners, offers an economical system with permanency and broad activity against fungi and insects.

To protect wood against mold and stain fungi, a biocide needs many of the same characteristics as listed above but only has to be effective for a relatively short time. However, when used as a mold/stain inhibitor in coatings, a biocide must be safe and effective for a long time, and stable against photodegradation.

Not only must a biocide have the above characteristics, it also must be capable of being formulated into an effective and economical wood protection system. Formulations used to protect wood, or prevent staining/molding, need the following properties:

- have low volatility and flammability
- penetrate the wood to sufficient depth
- non-corrosive to metal fasteners
- capable of being concentrated (liquid or solid) for shipment
- capable of formulating into a solution, preferably water-based
- be chemically stable during shipment and at a treating facility
- not deleterious to wood
- not hazardous at a plant or in the treated product
- economical
- allow wood to be painted after treatment

Furthermore, formulations for stains/coatings must be resistant to water leaching or degradation by UV light, and have abrasion (for decks), checking and cracking resistance.

New wood protection formulations need to satisfy the demands of both regulators and the market. Local, national or global regulations must be complied with, and the use of registered ingredients is mandatory. The marketplace also demands manufacturers provide novel and economical solutions to the many requests of homeowners. For example, a wood protection system is expected to effectively protect wood exposed to a wide variety of different environments, such as the relatively benign conditions in the northeast U.S. **and** the high humidity and temperatures favorable to decay fungi and the introduced aggressive Formosan termites along the Gulf Coast. Furthermore, the homeowner not only expects the price of treated wood to remain competitive with other non-wood construction materials, but also demands the performance of any new environmentally-benign system to be equal or superior to the traditional wood preservatives. Finally, the durability of current exterior finishes, clears and stains on wood surfaces is viewed as inadequate by homeowners. However, at the same time VOC issues and regulations are requiring the use of water-borne instead of solvent-based formulations which generally has a negative impact on performance.

Biocides

The classic wood protection biocides consist of three basic types:

1. Tar oil preservatives, e.g. creosote
2. Light organic solvent preservatives, e.g. pentachlorophenol
3. Water-borne arsenicals, e.g. chromated copper arsenate

These historical biocides already have some restrictions and will undoubtedly face more restrictions in the future. As use of these very economical and efficacious biocides is further limited, biocide manufactures will need to offer the wood protection industry a variety of formulated products for many different wood-protection applications.

Today there is an extensive list of agrochemicals that are already available and registered for non-wood use, and which could possibly be used to create new wood protection products. Table I lists a few organic chemicals that already are, or could be, used to protect wood against decay and insect attack. However, many of these compounds are only active against insects but not decay fungi, such as the synthetic pyrethoids, or active against decay fungi but not molds and stain fungi such as the triazoles. Also, while the cost of didecyldimethyammonium chloride (DDAC) is an order of magnitude cheaper than the triazoles, for example, it is considerably less effective at controlling decay fungi than the triazoles.

Furthermore, these organic biocides are non-water soluble and, thus, cannot be formulated using a water-borne system like today's CCA system [except for DDAC, which is water soluble but then fixes in wood]. Thus, use of these biocides will likely require the development of oil-in-water emulsion formulations.

Many of the biocides listed in Table 1 are also capable of controlling stain and mold fungi. For example, a combination of DDAC and 3-iodo-2-propyl butyl carbamate (IPBC) is used commercially, as are 2,4,5,6-tetrachloroisophthalonitrile, a mixture of isothiazolinones, or 2-(thiocyanomethylthio) benzothiazole (Busan 30).

Table I. Potential Organic Biocides for Wood Protection

Chemical	*Target*
3-Iodo-2-propnylbutyl carbamate	Fungi
Didecyldimethylammonium chloride	Fungi and Insects
Triazoles	Fungi
Isothiazolones	Fungi
Synthetic Growth Inhibitors	Insects
2-(Thiocyanomethio)benzothiazole	Fungi
Synthetic Pyrethroids	Insects
2,4,5,6-Tetrachloroisophthalonitrile	Fungi and Insects

Synergism Studies

The benefits of combining two or more biocides has long been recognized and utilized. These benefits include activity against a broader range of organisms than is possible with only one biocide, and the possibility of synergistic action. Synergism is where a mixture of two or more bioactive compounds give greater activity than would be predicted from each components' individual activity (*1,2*), and is detected for a mixture of two biocides using the equation below:

$$\frac{Ca}{CA} + \frac{Cb}{CB} = \text{Synergy Index (SI)}$$

where Ca is the concentration of biocide A in a mixture of biocides' (A and B) necessary to give a specific response, Cb is the concentration of biocide B in a mixture of A and B necessary to obtain a specified response, and CA and CB are the concentrations of biocides A and B, respectively, which when used alone give the same specified biological response.

A SI of less than 1 indicates a synergistic effect from combining biocides. It is also possible to obtain a value greater than 1, which indicates antagonism (an inhibitory effect) from combining biocides. Most commonly, however, an additive effect is observed (SI = 1). While the above equation is for two biocides, a synergistic mixture of three or more biocides is also possible.

A desired combination biocide formulation is apparent when synergism is observed. A synergistic combination allows less total biocidal material to be used than when a single biocide used alone. The commercial value here is that it is easier to comply with regulations while also offering a more cost-effective formulation to customers. In addition, the mixture may provide a product with broader spectrum activity. Furthermore, a synergistic combination can be patented to protect a company's development costs.

Identification of synergistic mixtures is thus a key objective in designing new formulations. This means understanding the target organisms and their modes of action. Once this difficult task is accomplished, it might be thought that synergy could be easily predicted from understanding the mechanism(s) of action. However, such predictions have often not been demonstrated in practice. Basically, the antibiotic literature generally suggests only that synergy is most likely to occur between chemicals that have dissimilar modes of action and that an additive action is the most probable results of combining biocides having the same action. Thus, finding combinations that are synergistic usually requires extensive and tedious trial-and-error laboratory studies.

As a cautionary note, observation of enhanced efficacy by combining two components does not always mean that the combination is synergistic. For example, combining copper(II) with a carboxylic acid or phenolic will form a copper(II)-organic ligand complex. This new compound might be more active than either of the two starting "reagents" but, since a new compound is formed by the addition of the copper and organic ligand, the observed increase in bioactivity is not strictly synergistic. One example is the large bioactivity increase observed against a wide variety of wood-destroying organisms using various laboratory and outdoor exposure tests with the combination of copper(II) with oxine copper [Cu-8] (*3*). However, the authors suggested that the reason for the greatly enhanced activity was not synergism but rather the formation of mono-Cu-8 by the combination of copper(II) with the bis-Cu-8, and prior literature indeed suggests that the mono form of Cu-8 has greater fungicidal activity than the bis form. In addition, it is possible to observe synergism by combining a biocide with a non-biocidal additive; for example, see Chp. by Green and Schultz in this book.

Synergistic Examples

Combinations to Control Molds/Stains

Using laboratory test methods, the combination of 4,5–dichloro-2-n-octyl-3-isothiazole (DCOIT) and 3-iodo-2-propynylbutyl carbamate (IPBC) was evaluated for synergy (*4*). The following values were used to determine synergy, with MIC (minimal inhibitory concentration, ppm, or the minimal biocide level necessary to completely inhibit growth of a particular fungus grown in an agar plate) used as the specified biological response:

CA - Concentration needed to achieve MIC when only DCOIT is present
CB - Concentration needed to achieve MIC when only IPBC is present
Ca - Concentration of DCOIT, in a mixture of IPBC and DCOIT, to obtain MIC
Cb - Concentration of IPBC, in a mixture of IPBC and DCOIT, to obtain MIC
A:B - Ratio of DCOIT:IPBC in the mixture

The results obtained from these evaluations are presented in Table II, and show that the combination is synergistic against a wide variety of stains and molds.

Table II. Results between DCOIT and IPBC Using Various Stain and Mold Fungi

MICROBE	CA	CB	Ca	Cb	SI	A:B
A. niger	16	4	4	2	0.75	2:1
A. niger	16	4	8	0.5	0.62	16:1
A. pullulans	16	2	8	0.25	0.62	32:1
C. albicans	32	16	4	8	0.62	1:2
C. albicans	32	16	16	2	0.62	8:1
E. coli	8	250	4	62	0.75	1:16

The application of the synergistic combination of DCOIT and IPBC has been described by Tsunoda et al. (*5*). An emulsified mixture of 2% IPBC and 1.5% DCOIT controlled stain fungi on wood better than the biocide trichlorophenol. The mixture is currently used in commercial anti-sapstain formulations, and could be used for other wood applications.

Other examples of synergistic combinations to control stain/molds are the combination of DDAC and IPBC (*6*), a commercial product, and the combination of an oxathiazine and benzothiophene-carboxamide-S,S,-dioxide (*7*), a non-commercial system.

Synergistic Wood Preservative Mixtures

The combination of copper(II) and the quat DDAC was found to be synergistic against various wood-destroying fungi, including a copper-tolerant fungus (*8*). This combination is now sold commercially as ammoniacal copper quat (ACQ-B) or amine copper quat (ACQ-D), depending on the particular formulation (*9*). The combination of copper(II) with the triazole tebuconazole is also synergistic against wood-destroying fungi, especially a copper-tolerant fungus (*10*). This latter system has been examined further (*11-12*), and the above systems are now commercial products, principally in Europe and Asia. Both systems are also listed in the American Wood Preservers' Association Standards (*9*) and will likely replace CCA for treating residential construction lumber as the use of arsenically-treated lumber is phased out by 2004. However, the presence of copper in treated wood products has several potential disadvantages (environmental concerns with copper, especially in aquatic systems, and questions on the ultimate dispose of treated wood products which contain a metal). Thus, several European countries are already moving towards totally organic wood preservative systems, a move that the U.S. may follow in the future. Other non-commercial examples of synergistic mixtures for protecting wood include 2,4,5,6-tetrachloroisophthalonitrile and chlorophyrifos (*13*), and a metal/carboxylate acid with an isothiazolone (*14*).

Conclusions

Novel wood protection formulations, based upon mixtures of already-developed biocides, offer some real advantages to the wood protection industry. Through careful design, formulations can be developed relatively quickly and at a much lower cost than for developing entirely new biocides. The demonstration of synergy in mixtures of biocides can provide broader activity, lower overall biocide concentrations, make it easier to register the final formulation, and provide patent protection to the developing company. As the current wood protection biocides are removed from the marketplace, the use of mixtures of commercially-available and labeled environmentally-benign chemicals presents excellent opportunities for manufacturers of wood protection formulations.

References

1. Hodges, N.A.; G.W. Hanlon. In: *Mechanisms of Action of Chemical Biocides*; Blackwell Science, **1990**; 297-310.
2. Schultz, T.P., D.D. Nicholas. Utilizing Synergism to Develop New Wood Preservatives; In: *Wood Preservativation in the 90's and Beyond*, FPS Symposium Proceedings, Forest Products Soc.: Madison, WI, **1994**; 187-191.

3. Schultz, T.P.; Nicholas, D.D. Enhanced Efficacy from the Combination of Copper(II) and Cu-8: Laboratory and Field Studies; In: *Enhancing the Durability of Lumber and Engineered Wood Products*, FPS Symposium Proceedings, in press. Forest Products Soc.: Madison, WI.
4. Hsu, J.C. U.S. Patent 5,292,763 **1994**.
5. Tsunoda, K.; Kumagai, H.; Sakurai, M. Internat. Res. Group Paper IRG/WP/94-30035 **1994**.
6. Ward, H.A. U.S. Patent 4,950,685 **1990**.
7. DeWitte, L.A.; Valcke, A.R.A.; Van der Flass, M.A.J.; Willems, W.M.L. U.S. Patent 6,242,440 **2001**.
8. Findlay, D.M.; Richardson, N.G. U.S. Patent 4,929,454 **1990**.
9. Anomyous. American Wood-Preserver's Association Standards **2001**. Granbury, TX.
10. Williams, G.; Cornfield, J.A.; Brown, J.; Ryan, N.P. U.S. Patent 5,527,384 **1996**.
11. Williams, G.; Cornfield, J.A.; Borwn, J.; Ryan, N.P. U.S. Patent 5,634,967 **1997**.
12. Williams, G.; Cornfield, J.A.; Borwn, J.; Ryan, N.P. U.S. Patent 5,916,356 **1999**.
13. Woods, T.L. *Wood Preservation in the '90s and Beyond*; Forest Products Soc.: Madison, WI, **1995**; 192-197.
14. Grove, S.L. U.S. Patent 4,783,221 **1998**.

Chapter 25

Wood Composite Protection

Douglas J. Gardner, Cihat Tascioglu, and Magnus E. P. Wålinder

Advanced Engineered Wood Composites Center, University of Maine, Orono, ME 04469

Wood composites should be protected against microbial and insect attack when used outdoors, especially in construction applications with prolonged exposure to moisture. Preservative systems and treatment processes affect composite properties, especially adhesive/wood bonding and mechanical properties. Several common systems for preservation of composites include: 1) the use of pretreated wood, applicable particularly to some solid lumber laminates; 2) in-process preservative treatments favored for composites made from flakes, particles, and fibers where the preservative treatment is incorporated during the manufacturing process; 3) post-process preservative treatments which are generally favored for wood composites made from lumber and veneer; and 4) the use of recycled treated wood elements in manufacturing or the use of wood species with a high natural resistance against biodegradation. This chapter discusses these four preservative methods and presents a general overview of current research concerning preservation practices and techniques in North America including the effect of preservatives on composite properties, durability issues, and degradation modes.

Introduction

The production of wood-based composites has increased dramatically over the past three decades due to a number of factors including the changing wood supply, development of new composite technologies, and the widespread acceptance by architects and builders of wood composites for use in construction. Depending on the particular application, wood composites require protection from the effects of moisture, weather, biological infestation (decay, insects, marine borers), and fire. It must be emphasized, however, that decay problems are in most cases a result of faulty design or construction, incorrect installation, lack of reasonable care in the handling of the wood, or improper maintenance of the structure (*1*). In fact, with a good design, including applications of surface coatings and the addition of wax, some wood composites, such as fiberboard, have been utilized for several decades as exterior siding with good results.

The preservation or protection of wood composites is a topic that continues to receive considerable attention from researchers and manufacturers (*2, 3*). The public perception of wood composites is both good and bad. Unfortunately, much of the recognition of wood composites in the popular media has centered on class action lawsuits resulting from product failures such as the thermal degradation of fire-retardant treated plywood (*4*) and the oriented strandboard siding failures (*5*). The topic of wood composite protection remains important as the impact of energy efficient construction practices (*6*), new preservative chemistries (*7*), and environmental issues concerning the use and disposal of treated wood (*8*) provide focus for research and development efforts. This chapter provides an overview of wood composite preservation in North America including current preservation practices, effect of preservatives on properties, durability issues, and a synopsis of current research topics.

Wood Preservatives Used for Wood Composite Protection

Table I lists the chemical names and acronyms of preservatives used in wood composite protection. Included in Table I are columns indicating whether a particular chemical is used as a composite additive or an adhesive additive. Wood composites are classified into product types depending on the broad application, e.g. either structural or non-structural. In addition, wood composites can be classified according to the particular wood element(s) used to manufacture the composite. For instance, glulam beams are manufactured from lumber elements, plywood from wood veneer, and particleboard from particles. Secondary manufacturing processes use oriented strandboard and

Table I. Partial list of chemical treatments used for Wood Composite protection.

Acronym or common name	*Chemical Name*	*Composite Additive*	*Adhesive Additive*	*Reference*
ACA	Ammoniacal copper arsenate	x		*10, 11*
ACZA	Ammoniacal copper zinc arsenate	x		*11*
CCA	Chromium copper arsenate	x		*11*
ACQ	Alkaline copper quaternary ammonium compounds	x		*11–13*
ACC	Ammoniacal copper citrate	x		*11*
	Arsenic oxide		x	*14*
Borates	Disodium octaborate tetrahydrate, zinc borate, boric acid	x	x	*11, 15, 16 17*
Penta	Pentachlorophenol	x	x	*10, 11, 18*
	Chlordane		x	*19*
Creosote	Coal tar creosote	x		*10, 11*
Copper	Copper metal	x		*20, 21*
Azole	Azaconazole, tebuconazole		x	*14*
Copper-8	Copper-8-quinolinolate	x	x	*10, 11*
CuNap	Copper naphthenate	x		*11*
	Fenvalerate	x	x	*14*
	Fenitrothion	x	x	*14*
IPBC-CPF	3-iodo-2-propynyl butyl carbamate, chlorpyrifos	x		*7, 22*
FR	Fire-retardant	x		*23*
NaF	Sodium floride	x	x	*18*
	Pirimiphos methyl		x	*14*
Azole	Azaconazole, tebuconazole		x	*14, 24*
CTL	Chlorothalonil		x	*24*
	Cypermethrin (EC)		x	*25*
DNBP	2-sec-butyl-4,6-dinitrophenol (EC)		x	*25*

EC = emulsifiable concentrate

laminated veneer lumber to produce wooden I-joists. There are many possible wood element combinations available for creating wood composites (*9*). Table II lists some of the most common wood composites and preservatives used. A variety of adhesives are used in the manufacture of wood-based composites, including urea-formaldehyde (UF), melamine-formaldehyde (MF), melamine-urea-formaldehyde (MUF), phenol-formaldehyde (PF), phenol-resorcinol-formaldehyde (PRF), and polymeric diphenylmethane diisocyanate (pMDI).

The application of preservatives to wood composites is a standard industrial practice for many of the composites listed in Table II depending on the particular application. In most instances, wood composites used in interior or protected applications require the addition of wax or water repellents to provide dimensional stabilization. Glulam beams are treated with CCA, penta in light or heavy solvents, creosote, IPBC-CPF, and ACQ in both pre- and post-manufacture (*11*). In most cases, in large laminated timbers, waterborne treatments are applied pretreatment to eliminate moisture related distortion in the final product, whereas solvent and oil borne treatments may be preferentially applied post-fabrication. Plywood is commercially treated for a wide variety of service applications using both organic and inorganic preservatives (*10*). Parallam is treated with creosote for use in railroad applications, penta for highway bridges, and CCA for residential applications.

Oriented strandboard and laminated strand lumber can be treated with zinc borate for use in areas with the potential for high Formosan subterranean termite activity. A solvent-based preservative system containing 3-iodo-2-propynyl butyl carbamate (IPBC) (fungicide) and chlorpyrifos (CPF) (insecticide) is used to treat engineered wood products such as I-joists, which are typically specified for structural applications where the material is protected from the weather. I-joists used in Hawaii are being treated with this preservative system (*26*). Currently, there is not much commercial activity in the treatment of hardboard with preservatives, but one product (Miratec) containing zinc borate is being produced (*27*). Both particleboard (PB) and medium density fiberboard (MDF) products can be produced with moisture resistance or fire retardant properties. For moisture resistance, the UF resin used in PB and MDF board manufacture might be replaced by MUF co-polymer resins, PF resin, or pMDI. Fire retardant PB and MDF products may incorporate inorganic salts such as borates or coatings to reduce flame spread. Extruded wood plastic composites may include zinc borate to provide protection against decay fungi (*28*).

Table II. Common wood composites, and typical preservatives used.

Composite Type	*Wood Element*	*Treatment Chemical*
Glu-lam Timbers	Lumber	ACQ, CCA, Creosote, Penta, IPBC-CPF
Plywood	Veneer	ACA, ACZA, ACQ, CCA, Creosote, Penta, copper-8, Borates, FR
Laminated Veneer Lumber (LVL)	Veneer	CCA
Parallel Strand Lumber (Parallam)	Veneer	CCA, Creosote, Penta
Oriented Strandboard (OSB)	Flakes or strands	Zinc Borate, Wax
Laminated Strand Lumber (Timberstrand)	Flakes or strands	Zinc Borate, Wax
Particleboard	Particles	FR, Wax, Permethrin
Fiberboard (Medium density fiberboard)	Thermomechanical fibers	FR, Wax
Hardboard	Thermomechanical fibers	FR, Wax
I-joists	Flanges (Lumber, LVL) Webs (Plywood, OSB)	IPBC-CPF in light organic solvent
Wood plastic composites	Wood (Particles, fibers) Plastics (polyethylene, polypropylene, polyvinyl chloride)	Zinc Borate

Preservative Treatment Application Processes for Wood Composites

Preservative type, wood species, treatment method, adhesive type, and active ingredient retention are important parameters for the treatment of wood composites and the resulting quality of the final product. Because of the many types of wood composite products and manufacturing processes, there are a number of ways to apply preservative treatments to these materials (Table III). In most instances, wood composites manufactured from lumber or veneer elements are usually preservative treated post-manufacture while, composites manufactured from flakes, particles, and fibers have the preservative treatment applied during the manufacturing process. Exceptions to this rule are found in Table III. A comparison of the advantages and disadvantages of the in-process and post-process treatment methods is shown in Table IV.

Conventional treatment processes used for solid wood are also used for wood composites including vacuum/pressure treatment, dipping, spraying, and brushing. A commonly asked question in wood composite preservation is: when should the preservative be applied to the wood composite in the manufacturing process? To simply answer this question, it depends on whether the preservative treatment will have a negative impact within the manufacturing process or on the final product. For example, glulam timbers are usually pressure-treated with oil-borne preservatives such as creosote after manufacture; while conversely CCA-treated glulam is manufactured from CCA-treated lumber that is glued together to produce the timbers. The reason for this difference in preservative treatment application is that oil-borne preservatives can interfere with the adhesive bonding process while pressure treating a glulam beam with a water-borne preservative like CCA may result in warpage and/or unacceptable dimensional changes in the resulting product.

In-process Treatment

Perhaps the most cost effective and efficient way to produce preservative treated composites is to incorporate the preservative during the manufacturing process (*3*). Laks and Palardy (*24, 29*) summarized the different methods of preservative incorporation in flake- and wafer-type wood composites during the manufacturing process. These methods are also applicable to fiber and particle composites.

Table III. Examples of preservative treatment applications in wood composite manufacture.

Composite Type	Treatment method: In-process	Treatment method: Post-process
Glulam Timbers	Waterborne preservatives (CCA), penta in light solvent	Oil-borne preservatives
Plywood		Oil- and waterborne preservatives
Laminated Veneer Lumber (LVL)	Water and organic solvent systems	Waterborne (ACQ, CCA)
Parallel Strand Lumber (Parallam)		Waterborne (CCA), Creosote, Penta
Oriented Strandboard (OSB)	Zinc Borate	
Laminated Strand Lumber (Timberstrand)	Zinc Borate	
I-Joist		IPBC-CPF
Particleboard	Wax, Permethrin	
Fiberboard (MDF)	Wax, ZB	
Hardboard	Wax, ZB	
Wood plastic composites	Borates	

Table IV. Comparison of advantages and disadvantages of wood composite treatment methods.

Treatment Method	Advantages	Disadvantages
In-process	-Protection throughout the cross section of composite -Can be cut, drilled and /or notched anytime	-Chemical interaction and inhibition with adhesive and adhesion -Effects on other properties (physical, mechanical) of composite -Emissions and high treatment chemical losses during hot pressing of treated furnish -Treated wood must be machined and associated concerns with treated waste shavings, sawdust must be addressed
Post-process	-Easy to apply after manufacturing done. -Does not require any modification of manufacturing process. -Machining of treated wood not an issue	-Envelope protection only -No cutting, drilling, or notching should be done without additional field treatment - Irreversible thickness swelling and dimensional change -Other potential effects on mechanical properties

These methods are:

- Treatment of the wood furnish (flake, fiber or particle) before or after drying using an additional sprayhead or blender system in the production process.
- Mixing the preservative chemical with the adhesive or wax which is subsequently applied to the wood furnish in the blender.
- Spraying the wood furnish with a preservative solution or emulsion, or mixing the powdered chemical with the furnish in the blender.
- Metering a powdered preservative onto the dried wood furnish immediately before the blender. The powder and furnish is then mixed together in the blender.

In-process preservative treatment has economic advantages since post-treatment requires additional process steps that add cost to the product. An additional benefit of in-process treatment is the homogeneous distribution of preservative in the resulting wood composite product. However, problems can result if the preservative chemical negatively impacts the adhesive curing process or contributes to hazardous volatile organic compound emissions. As pointed out by Laks (*3*), a preservative system for in-process treatment of wood composites should have the following characteristics:

- High stability and limited volatility during the processing conditions.
- No interaction with the adhesive which could have an unfavorable effect on the bond formation.
- No adverse effect on the mechanical properties of the resulting composite.
- Low leaching during in-service.

In this case, it is obvious that influences on the adhesive properties, such as gel time and adhesive wettability, may also impact the bondability of the constituents and the physical and mechanical properties of the resulting composite. Depending on the preservative and resin-adhesive used, it is also expected that different preservative-resin interactions will occur.

In recent years, a great deal of research and development work has been performed with regard to the incorporation of borate as a preservative into flakeboard and oriented strandboard (OSB) furnish during the blending process (*30–32*). Moreover, zinc borate (ZB), applied in powder form, is currently the most commonly used preservative for in-process treatment of wood composites in North America (*3*). One critical problem in the use of borates as a preservative for wood composite panels bonded with PF resins is the interaction between the borate and the resin during the manufacturing process of the panel, which could significantly decrease the resin gel time. A reduced ability of the resin to flow and cure properly could severely impact the resulting properties of the panel. Laks *et al.* (*30*) reported that the addition of biologically effective levels of the water-soluble sodium borates or boric acid in PF-bonded waferboard had a detrimental effect on the bond strengths. However, no such effects were observed using a pMDI resin. Lee *et al.* (*32*) showed that the gel time of phenolic resin used for oriented strandboard decreased as the amount of zinc borate increased. It was, however, also shown that the reduced gel time could be recovered by using polyethylene glycol (PEG) in combination with the borate. Correspondingly, Sean *et al.* (*31*) investigated the effect of zinc borate treatment on the properties of phenolic bonded aspen OSB, and found that the adverse effect of borate on the adhesive fluidity and the resulting mechanical or rheological properties of the composite can be reduced by the addition of PEG into the panel composition. This effect was attributed to the PEG hydroxyl groups, which effectively can act as a flowing agent for the resin. It should, however, be noted that borate treatments are only suitable for protection against decay and insects attacks in nonleaching environments and not for ground contact or similar environments (*33*).

Laks *et al.* (*7*) tested the compatibility of newer organic preservative chemicals in aspen waferboard bonded with pMDI. The following preservative combinations were used: IPBC/chlorpyrifos; tebuconazole/chlorpyrifos; and tebuconazole/propiconazole/chlorpyrifos (TEB/PROP/CPF). It was observed that some loss of tebuconazole, propiconazole, and chlorpyrifos occurred during the manufacturing process with 83, 82, and 57% average recoveries of the target loading. Incorporation of these organic based preservatives did not introduce any negative effect on the strength or thickness swell stability properties of the waferboards tested. Overall, the TEB/PROP/CPF combination showed the best performance in soil block tests with an active threshold ratio of 150/150 ppm TEB/PROP (*7*).

Bridaux *et al.* (*34*) investigated the effect of boric acid addition in MUF-bonded poplar LVL on the mechanical properties of the resulting composite and also boron leaching effects. Interestingly, they showed that the incorporation of a boric-acid-water solution in the LVL glueline increases bonding strength when compared to LVL without the addition of boric acid. This effect was possibly due to water decreasing the viscosity of the glue, which improves its penetration ability. It was also found that the MUF glueline appears to be a good barrier against water and boron migration. Additionally, Padmanabhan *et al.* (*19*) incorporated 1% sodium octaborate tetrahydrate and 1% chlordane as a glue additive during manufacture of bamboo mat board bonded with PF adhesive. This treatment technique gave good resistance to biological deterioration and had no adverse effect on the bond strength of the mat boards.

Wood-thermoplastic composites have traditionally been viewed as building materials that are not prone to biodegradation. However, it has recently been reported that both brown- and white-rot fungi attack occurred after four-year-exposure in Florida on a recycled wood-plastic composite (*35*). Also, Laks (*3*) noted that their research group has observed (in unpublished work) rapid fungal degradation of such composites at their test site in Hawaii. Accordingly, it may therefore be desirable to incorporate inorganic and thermally stable preservatives, such as zinc borate, during the extrusion of wood-plastic composites. Work on adding zinc borate to polyethylene-wood extrusion formulations has been done at Washington State University (*28*). Field testing of these materials in Hawaii is ongoing with early inspections indicating promising results.

Post-process Treatment

Wood composites such as oriented strandboard (OSB), medium density fiberboard (MDF), and particleboard are not suitable for post manufacturing treatment with waterborne preservatives because of problems with thickness swelling and/or surface roughness of the panels. For housing construction, the OSB panels are also often exposed to rain at the job site which could cause leaching of the preservative and therefore decrease the decay resistance. A good solution in this case may be pressure treatment with mineral spirits-based preservatives. Plywood, on the other hand, can be successfully treated with waterborne preservatives such as CCA, ACA and ACZA. Treatments applied under AWPA standards should not greatly affect plywood strength and stiffness (*10*).

Kimmel *et al.* (*36*) reported several characteristics of creosote treated hardwood and softwood LVL. Creosote treated LVL showed very uniform preservative distribution through the treated section when compared to solid lumber. This reduction in preservative gradient will ensure a more consistent penetration pattern throughout the cross-section of larger structural members (decks, bridge members etc.). They also reported that the thickness dimension was most sensitive to the preservative-related swelling and that the thickness swell for yellow-poplar LVL was 85% greater than its solid lumber counterpart, which still appears minor when compared to the volumetric swelling with water. Moreover, the flatwise and edgewise flexural modulus of elasticity and the shear strength in either parallel or perpendicular to adhesive line test orientation were not affected by the creosote treatment.

Merrick *et al.* (*37*) presented data that suggested parallel strand lumber did not require incising to obtain AWPA acceptable retention or penetration of creosote when utilizing the Bethel (full-cell) process. They also found that no significant dimensional changes resulted from the commercial treatment schedule. Additionally, Manbeck *et al.* (*38*) found that post-treatment with creosote of glulam beams, made from red oak, red maple and yellow-poplar and face bonded with resorcinol-formaldehyde resin, had no adverse effect on bond quality and resulting mechanical properties.

Non-conventional Treatments

Non-conventional preservative application processes for wood composites that have been examined, but are not widely used commercially, include vapor-phase and supercritical fluid treatment. Preservative and flame retardant treatment of wood composites using a vapor boron (VB) method has been thoroughly evaluated (*33, 39, 40*). The VB technique is carried out in a pressure vessel and employs a partial vacuum and elevated temperature (50°C) to vaporize methyl borate liquid to impregnate the dry wood panel (about 4% MC). The gaseous boron ether then reacts with the residual moisture within the wood to form unfixed boric acid and methanol. Compared with liquid diffusion treatment using solutions of boron salts, which normally requires several weeks of storage time, the boron gas phase technique could significantly reduce the treatment time to less than 30 hours. Complete borate penetration of various wood composites, including OSB, MDF, particleboard, Parallam® and Timberstrand® was achieved using this method. Also, the vapor treatment did not negatively impact their mechanical and physical properties, such as modulus of rupture (MOR), internal bond strength (IB), bond durability, dimension stability and water absorption (*40–42*).

Supercritical fluids (SCF), solvents generated under very high pressures (2000-4500 psi) and relatively low temperatures, act like gases in terms of easy diffusivity but have high solvating properties like liquids. They have been proposed as potential carriers for biocides (*43*). SCF treatment might prevent post-wetting of wood composites with wood preservatives and volatilization losses during the pressing if *in-situ* treatments are used. It has also been shown that the SCF treatment does not negatively impact the mechanical properties of plywood, flakeboard, MDF, and particleboard (*43*). However, the high capital cost for the sophisticated high-pressure equipment required for the SCF treatment presents a major obstacle in commercializing this process.

Variations in Treatability

Mitchoff and Morrell (*44*) highlighted the variations in preservative treatability of plywood panels. Although the presence of lathe checks, core gaps and incomplete gluelines should promote better preservative penetrability, none of the panels tested were completely penetrated even at high retention levels. The envelope protection provided by preservative treatment might fail in severe exterior environments, leaving an untreated core of plywood that is prone to fungal and insect attack. Ruddick and Louwers (*45*) reported that 95% of Douglas-fir veneer strands were well penetrated by CCA in lightly incised Parallam®. In general, when sampling parallel to the wide face of veneer strands, the CCA retention was significantly influenced by both void volume content and assay depth. They also pointed out that a variation up to 6 kg/m^3 could occur in adjacent sapwood and heartwood veneer strands.

Manufacture of Wood Composites from Durable Wood Species

An environmentally desirable manufacturing practice being examined in wood composite research is the production of composites from durable wood species. Research of this nature has been reported in the literature, and commercial manufacture of wood composites from durable species is being practiced to a limited extent commercially. Heartwood of species like cedar would have a high potential to be used as furnish for flake- and particleboards for specialized exterior applications where enhanced resistance properties are desirable, e.g., siding. Haataja and Laks (*46*) evaluated the strength and biodegradation-resistance of flakeboards made with northern white cedar furnish and bonded with pMDI. Compared with controls of conventional flakeboards made with aspen furnish, the cedar boards proved to have excellent termite and decay resistance after long-term exposure in and above ground in Hawaii. These boards also retained their bending properties better after aging. However, the MOE for the cedar boards was significantly lower than for aspen, reflecting the higher density of the latter. Hiziroglu and Kamdem (*47*) reported on improved decay resistance of black locust hardboard subjected to laboratory soil block tests. The mechanical properties of the boards were acceptable and, additionally, low thickness swelling was observed which was attributed to the high extractive content of the species. However, a high extractive content of the wood furnish could cause some processing difficulties, especially related to the bonding properties of the panels.

Important issues in the manufacture of wood composites from durable wood species include a constant wood supply and markets for the product.

There is a manufacturer of wood plastic composite lumber that uses western red cedar in the formulation (*48*).

Recycling Preservative Treated Wood into Composites

Over the past 10 years, the issue of preservative treated wood disposal has prompted research in the recycling or re-use of preservative treated wood (*8*). Material processing issues in the recycling of preservative treated wood in composites include the impact of preservatives and wood weathering on adhesive bonding. The adhesive bonding of CCA treated wood was studied by Vick and Kuster (*49*), and Vick and Christiansen (*50*). Bonding problems may be experienced when producing composite materials from CCA-treated wood because of the presence of insoluble metal deposits that physically block the force of attraction between the wood and adhesive. On the positive side, the recycling potential of CCA-treated wood in flakeboard and particleboard is high (*51–54*). Vick and Geimer (*51*) found that adhesive bonding of southern pine treated with CCA preservative resulted in less durable bonds than those of untreated wood. Similarly, Mengeloglu and Gardner (*53*) found that flakeboards manufactured from CCA-treated lumber exhibited lower overall physical and mechanical properties than untreated controls. An interesting approach in this case is the application of the coupling agent hydroxymethylated resorcinol (HMR), which has been found to radically improve the bond durability between adhesives, such as resorcinolic, epoxy, or polymer isocyanate, and CCA-treated wood (*55*). Munson and Kamdem (*52*) investigated the mechanical and physical properties of PF-bonded particleboards manufactured from mixtures of untreated wood and CCA-treated red-pine utility poles furnishes and found that addition of 50 percent treated furnish did not significantly affect the board properties. Additionally, Zhang *et al.* (*54*) concluded that artificial weathering had little influence on bondability of CCA-treated southern pine. The manufacture of cement-bonded particleboard using CCA-treated wood removed from service has also been examined (*56*).

The material properties of wood composites manufactured from CCA-treated wood are somewhat lower than from untreated wood, but the technical feasibility of manufacturing has been demonstrated. It should be pointed out that according to the results of a survey by Smith and Shiau (*57*), most respondents (including wood composite manufacturers and end users) were not in favor of using recycled treated wood particles in conventional wood composite manufacturing. Their primary concerns were the safety of workers and environmental problems that may arise with composite products that were made from treated wood. The respondents suggested uses for these products as energy fuel (incineration), specialty treated wood composites, and wood-non-wood composites

The recycling of oil-borne preservative treated wood in composites can be considered more problematic compared to water-borne preservatives because of the potential for hazardous volatile organic compound (VOC) emissions during manufacture. The recycling of creosote-treated wood in particleboard has been examined (*58, 59*). Creosote-treated wood particles can be incorporated up to 15 weight percent in particleboard furnish, with a positive effect on dimensional stability of resulting panels but a negative effect on mechanical properties. Creosote compounds emitted during particleboard manufacture include acenapthene, naphthalene, phenanthrene, fluorene, and 1-methyl-naphthalene. Although the manufacture of particleboard from recycled creosote-treated wood is technically feasible, a safe and useful product with large potential market has not yet been identified.

Some issues for future research in this area include methods for accurate identification of preservative before re-use in a process and liability for unforeseen future health issues arising from manufacture and application.

Chemically Modified Wood Furnish

Protection against decay and insect attack may also be accomplished by manufacturing the wood composite from chemically modified wood furnish. One example of this chemically modified wood concept is outlined by Rowell *et al.* (*60*) who observed enhanced resistance against termite and fungal attack of aspen flakeboards made from acetylated flakes bonded with phenolic or isocyanate resins. The mechanism of biological resistance in acetylated wood is likely due to decreased moisture sorption and substrate blocking. Further, Youngquist *et al.* (*61*) reported that internal bond strength declined in flakeboards made from aspen flakes acetylated with acetic anhydride because of poor adhesive penetration into the flakes. It has, however, also been reported that particleboards made from acetylated poplar particles using thioacetic acid exhibited greater dimensional stability and decay resistance, but also significantly higher internal bond strength, maximum load and tensile strength compared to untreated boards (*62*). Hence, inclusion of an acetylation step in the processing of composites like OSB or particleboards may be an interesting alternative for obtaining better decay resistance. Isocyanates have also proven to be an effective modifying agent, which greatly increases the water repellency of wood, presumably due to the formation of chemical crosslinks between the resin and the wood constituents.

Chemical attack on wood constituents

Besides the fact that the inclusion of preservatives in wood composites can induce problems related to the adhesive and the resulting bonding properties, the added chemical may also degrade the wood substance itself, especially when the product is subjected to elevated temperatures and varying humidities. A well-known example of this is the rapid and severe failure of acidic fire-retardant-treated (FRT) plywood used as a roof sheathing material in the eastern US during the 1980's. In this case, the combination of acidic-fire retardant chemicals, high temperature, and high humidity that occurs in roof applications caused acid hydrolysis of wood components and subsequent rapid degradation and strength loss of the plywood (*63*). The hemicellulose constituents appear to be most affected by the FRT and exposure to high temperatures. Due to their lower hemicellulose content, softwoods generally exhibit higher resistance to chemical attack than hardwoods. It should be noted that this problem was not experienced by all FRT formulations. Improved formulations and standards have apparently corrected the problem.

Biodegradation Modes of Wood-Based Composites

Wood-based composite materials are increasingly used in exterior exposures, where conditions exist that are conducive to fungal, insect, and marine borer attack. Although the presence of resin in the wood composites may be expected to slow fungal attack, it has been reported over the past two decades that these composite products are experiencing failures due to fungal and insect attack (*14, 64–67*), particularly when they are exposed to high-moisture conditions. Only limited data are available on the effects of fungal and insect deterioration on wood based composites even though the increasing occurrence of fungal attack in wood composites has been reported (*14, 68, 69*).

Wood composite degradation can be analyzed in two categories:

1. Degradation of the "solid wood component" of the composite (e.g., veneer in plywood, wood particles in plywood, fibers in MDF, etc.)
2. Degradation of the adhesive

In general, the fungal flora of wood composites appears very similar to fungi isolated from solid wood products since the majority of the composites consist of "solid wood" with only a few percent resin by weight depending on the final product. Schmidt (*68*) reported the lack of detailed information regarding the fungal flora map (isolated species of decay, mold fungi and insects in different locations) of wood based composites throughout United States. Some decay and mold fungi species isolated from in-service wood based composite materials are shown in Table V.

Fungi and bacteria can degrade some natural and synthetic based adhesives used in wood composites. ASTM D-4300 (*70*) specifies how to determine the susceptibility of adhesive films to biodegradation and whether the adhesive will carry sufficient anti-fungal properties into the glueline to prevent growth of fungi frequently present on wood composite materials. It was also reported (*70*) that *Aureobasidium pullulans* and variety *malanigenum* are very common stain and soft rot fungi on wood products which deteriorate plastics and paint and might present a problem in adhesive degradation.

Chung *et al.* (*69*) investigated the effects of mold and stain fungi on the mechanical strength of fiberboard and concluded that fiberboard attacked by several mold fungi showed approximately 12-18% weight loss, resulting in about 50 percent loss of strength. They also mentioned that mold fungi consumed hemicellulose and α-cellulose, whereas lignin was hardly attacked.

Goodell *et al.* (*72*) reported that no significant weight loses occurred in wood-cement composites exposed to common white- and brown-rot fungi. However, a change in permeability, increase in moisture uptake, and some alteration of surface properties were observed. They also concluded that only wood in surface regions would be accessible to fungal attack.

Although there is information available on the chemical resistance of many wood composite adhesives, it is still very difficult to find any chemical exposure or long-term durability study focused on specific wood preservative chemicals. Information on the interaction of adhesives with the components of CCA [arsenic acid (As_2O_5), chromic acid (CrO_3), copper oxide (CuO)], or components of creosote [benzene, hexane, toluene, xylene, naphthalene] is needed. This type of specific information would be very useful in predicting the short and long-term chemical resistance of adhesives and gluelines when exposed to specific wood preservative chemicals (*73*).

Zabel and Morrell (*71*), Schmidt (*68*), and Smulski (*6*) have highlighted the importance of moisture entrapment for decay initiation in wood based composite materials. Particularly, misapplied coatings and over-layment, which are supposed to prevent moisture uptake, can result in the entrapment of moisture inside composites thereby promoting fungal activity. These problems are generally associated with faulty design and poor construction practices which need to be addressed.

Table V. Some examples of mold and decay fungi isolated from in service wood composite products:

Agent type	*Isolated fungi*	*Composite*	*Reference*
Decay fungi	*Phanerochaete chrysosporium and Schizophyllum commune*	Panel products	*71*
	Antrodia xantha, Gloeophyllum trabeum, Lentinus lepideus, Phanerochaete chrysosporium, Pleurotus ostreatus, S. commune	Wood based composites	*14*
	S. commune, Trametes hirsuta, Stereum spp., Pycnoporus cinnabarinus, P. chrysosporium	Aspen OSB	*68*
Mold fungi	*Alternaria sp., Aspergillus sp., Aureobasidium sp., Chaetomium sp., Diplodia sp., Epicoccum sp., Paecilomyces sp., Penicillum sp., Trichoderma sp.*	Panel products	*14*
	Alternaria sp., Aureobasidium sp., Penicillium sp., Paecilomyces sp., Trichoderma sp. Diplodia sp., Neurospora sp.	Aspen OSB	*68*

Acknowledgements

The authors would like to thank the following individuals for providing information on current wood protection practices being implemented commercially in the wood composites industry: Mike Hunsaker and Rob Brooks, Willamette Engineered Wood Products Industries, Inc.; A. William (Bill) Boehner and Paul Merrick, Trus Joist, A Weyerhaeuser Business; Louis Wagner, American Hardboard Association; Sheldon Q. Shi, APA-The Engineered Wood Association; and Gary Heroux, The Composite Panel Association. We also thank the reviewers and editors for valuable comments on the first draft of this chapter.

References

1. Forest Product Laboratory. *Wood Handbook–Wood as an engineering material.* Gen. Tech. Rep. FPL–GTR–113. U.S. Department of Agriculture, Forest Service, Forest Product laboratory: Madison, WI, 463 p, 1999.
2. *Protection of Wood-Based Composites*; Preston, A. F. Ed.; Forest Products Society: Madison, WI, 1993.
3. Laks, P. E. In *33rd International Particleboard/Composite Materials Symposium Proceedings*; Wolcott, M. P.; Tichy, R.; Bender, D. F. Eds; Washington State University: Pullman, WA, 1999, pp 151–158.
4. Winandy, J. E. *Forest Prod. J.* **2001,** *51*(2), 47–54.
5. Louisiana-PacificInner-Seal®Siding Litigation; http://www.lpsidingclaims.com/, 18 January 2002.
6. Smulski, S. *Forest Prod. J.* **1999,** *49*(5), 8–15.
7. Laks, P. E.; Richter, D.; Fears, R.; Ghaoui, L. In *Proceedings of American Wood Preservers Association*; AWPA: Granbury, TX, 1998; Vol. 94, p 121 (abstract only).
8. Felton, C. C.; De Groot, R. C. *Forest Products J.* **1996,** *46*(7/8), 37–46.
9. Marra, A. A. *Technology of Wood Bonding, Principles in Practice*; Van Nostrand Reinhold: New York, 454 p, 1992.
10. APA – The Engineered Wood Association. Data file Form No. Q220X, 5 p., Tacoma, WA, 1999.
11. American Wood Preservers' Association. 1999. Standard for preservative treatment by pressure processes of structural glued laminated members and laminations before gluing. AWPA C28-99, AWPA, Granbury, TX. pp 106–111.
12. ICBO AC78, ICBO Evaluation Service, Inc. Whittier, CA, 1999.

13. ICBO ER-4891, ICBO Evaluation Service, Inc. Whittier, CA, 2000
14. Eaton, R. A.; Hale M. D. C. *Wood: Decay, pests and protection*; Chapman & Hall: London, 1993, 546 pp.
15. ICBO ER-4890, ICBO Evaluation Service, Inc. Whittier, CA, 2000
16. ICBO AC62, ICBO Evaluation Service, Inc. Whittier, CA, 1998.
17. ICBO ER-5548, ICBO Evaluation Service, Inc. Whittier, CA, 2000.
18. Huber, H.A. *Forest Prod. J.* **1958** *8*(12), 357–360.
19. Padmanabhan, S.; Kamal, S. Z. M.; Ananthanarayanan, S. In *Protection of Wood-Based Composite Products*; Preston, A. F., Ed.; Forest Products Society: Madison, WI, 1993, pp 26–32.
20. ICBO AC143, ICBO Evaluation Service, Inc. Whittier, CA, 2000
21. ICBO. ER-5490, ICBO Evaluation Service, Inc. Whittier, CA, 2000
22. PolyClear 2000™, Permapost Products Co., Hillsboro, OR, http://www.permapost.com/cwp.htm, 18 January 2002.
23. ICBO ER-5180, ICBO Evaluation Service, Inc. Whittier, CA, 2000
24. Laks, P. E.; Palardy, R. D. In *Protection of Wood-based Composites*; Preston, A. F. Ed.; Forest Products Society: Madison, WI, 1993, pp. 12–17.
25. Quinn, P.; Pendlebury, A. J.; Conradie, W. E. 1993. In *Protection of Wood-based Composites*; Preston, A. F. Ed.; Forest Products Society: Madison, WI, 1993, pp 57–62.
26. Personal Communication, Paul Merrick, Trus Joist, A Weyerhaeuser Business.
27. Personal Communication, Louis Wagner, American Hardboard Association.
28. Annual Report (Phase II). Engineered Wood Composites for Navy Waterfronts. Prepared for Office of Naval Research under contract N00014-97-C-0393.
29. Laks, P. E.; Palardy, R. D. In *Proceedings of Wood Adhesives 1990*; Forest Products Research Society: Madison, WI, 1991, pp 150–154.
30. Laks, P. E.; Haataja, B. A.; Palardy, R. D.; Bianchini, R. J. *Forest Prod. J.* **1988,** *38*(11/12), 23–24.
31. Sean, T. G. Brunette and F. Côté. *Forest Prod. J.* **1999,** *49*(6), 47–51.
32. Lee, S.; Wu, Q.; Strickland, B. *Wood Fiber Sci.* **2001,** *33*(3), 425–436.
33. Murphy, R. J. In *Proc. Wood Preservation in the '90s and Beyond*; Proc. No. 7308; Forest product Society: Madison, WI, 1994, pp 89–94.
34. Bridaux, V.; Charrier, B.; Fauroux N.; Charrier, F.; Goncalez, J. *Holzforschung* **2001**, 55, 559–562.
35. Morris, P. I.; Cooper, P. *Forest Prod. J.* **1998,** *48*(1), 86–88.
36. Kimmel, J. D.; Janowiak J. J.; R. T. Baileys, P. Merrick. In *Proceedings of American Wood Preservers Association*; AWPA: Granbury, TX, 1995; Vol. 91, pp 100–107.

37. Merrick, G. P.; Baileys, R. T.; Watt, J. L. In *Proceedings of American Wood Preservers Association*; AWPA: Granbury, TX. 1997; Vol. 93, pp 405–413.
38. Manbeck, H. B.; Shaffer, K. R.; Janowiak, J. J.; Blankenhorn, P.R.; Labosky, P.; Baileys, R. T.; Webb, D. A. *Wood Fiber Sci.* **1995,** *27*(3), 239–249.
39. Vinden, P.; Fenton, P.; Nasheri, K. Inter. Res. Group on Wood Pres; Doc. No. IRG/WP/3329: Stockholm, Sweden, 1985.
40. Murphy, R. J.; Dickinson, D. J.; Turner, P.; Wickens, P. J.; Hashim, R. In *Protection of Wood-based Composites*; Preston, A. F. Ed.; Forest Products Society: Madison, WI, 1993, pp 49–56.
41. Hashim, R.; Murphy R. J.; Dickinson D. J.; Dinwoodie J. M. *Forest Prod. J.* **1997,** *47*(1), 61–66.
42. Hashim, R.; Murphy R. J.; Dickinson, D. J.; Dinwoodie, J. M. *Forest Prod. J.* **1994,** *44*(10), 73–77.
43. Acda, M. N.; Morrell, J. M.; Levien, K. L. *Wood Fiber Sci.* **1997**, *29*(2), 121–130.
44. Mitchoff, M. E.; Morrell J. J. *Forest Prod. J.* **1991,** *41*(9), 11–18.
45. Ruddick, N. R. J.; Louwers, R. N. In *Protection of Wood-based Composites*; Preston, A. F. Ed.; Forest Products Society: Madison, WI, 1993, pp 63–68.
46. Haataja, B. A.; Laks P. E. *Forest Prod. J.* **1995,** *45*(1), 68–70.
47. Hiziroglu, S.; Kamdem, D. P. *Forest Prod. J.* **1995,** *45*(11/12), 66–70.
48. Weyerhaeuser ChoiceDek® Classic, manufactured by AERT, http://www.choicedek.com/clasbul2.htm, 18 January 2002.
49. Vick, C. B.; Kuster, T. A. *Wood Fiber Sci.* **1992,** *24*(1), 36–46.
50. Vick, C. B.; Christiansen, A. W. *Wood Fiber Sci.* **1993,** *25*(1), 77–86.
51. Vick, C. B.; Geimer, R. L.; Wood, J. E. *Forest Prod. J.* **1996,** *46*(11/12), 89–91.
52. Munson, J. M.; Kamdem, D. P. *Forest Prod. J.* **1998,** *48*(3), 55–62.
53. Mengeloglu, F.; Gardner, D. J. *Forest Prod. J.* **2000,** 50(2), 41–45.
54. Zhang, H. J.; Gardner, D. J.; Wang, J. Z.; Shi, Q. *Forest Prod. J.* **1997,** *47*(10), 69–72.
55. Vick, C. B. *Forest Prod. J.* **1995,** *45*(3), 79–84.
56. Huang, C.; Cooper, P.A. *Forest Prod. J.* **2000,** *50*(6), 49–56.
57. Smith, R. L.; Shiau R.-J. *Forest Prod. J.* **1998,** *48*(2), 44–48.
58. Zhao, J. H.; D. J. Gardner. 2000. In *Wood Adhesives 2000 Extended Abstracts*; Forest Products Society, Madison, WI. pp 115–117.
59. Zhao, J. H.; D. J. Gardner. 2000. In *Wood Adhesives 2000 Extended Abstracts*; Forest Products Society, Madison, WI. pp. 118-120.

60. Rowell, R. M.; Esenther, G. R.; Youngquist, J. A.; Nicholas, D. D.; Nilsson, T.; Imamura, Y.; Kerner-Gang, W.; Trong, L.; Deon, G. In *Proc. of the IUFRO division 5 forest products subject group S.5.03: Wood protection*; May 16–17; Honey Harbour, ON, Canada, 1988, pp 239–266.
61. Youngquist, J. A.; Rowell, R. M.; and Krzysik, A. *Holz als Roh- und Werkstoff* **1986,** *44*(12), 453–457.
62. Kumar, S.; Morrell, J. J. In *Protection of Wood-Based Composite Products*; Preston, A. F., Ed.; IUFRO Symposium of Wood-Based Composite Products; Forest Products Society: Madison, WI, 1993, pp 33–37.
63. LeVan, S. L.; Winandy, J.E. *Wood Fiber Sci.* **1990,** *22*(1), 113–131.
64. Shupe, T. F.; Dunn, M. A. *Forest Prod. J.* **2000,** *50*(5), 10–18.
65. Johnson, B. R.; Gutzmer, D. I. Research paper FPL RP, U.S. Department of Agriculture, Forest Service, Forest Products Lab. (0258): Madison; WI, 1990, 28 pp.
66. Ziobro, R. J., R. T. Baileys. 1999. In *Proceedings of the American Wood Preservers' Association*; AWPA: Granbury, TX, 1999; Vol. 95, pp 158–162.
67. Zahora, A. R., A. Preston, K. Archer. 2000. In *Proceedings of the International Research Group on Wood Preservation,* 31st Annual meeting, Kona, Hawaii, 14–19 May 2000, pp. 1–9.
68. Schmidt, E. L. 1993. In *Protection of Wood-Based Composite Products*; Preston, A. F., Ed.; Forest Products Society: Madison, WI, 1993, pp 23–25.
69. Chung, W.-Y.; Wi, S.-G.; Bae, H.-J.; Park, B.-D. *J. Wood Sci.* **1999**, *45*, 64–68.
70. American Society of Testing Materials. ASTM D-4300-97, West Conshohocken, PA, 1997.
71. Zabel, R. A.; Morrell, J. J. *Wood Microbiology: Decay and its Prevention;* Academic Press Inc.: London, 476 p, 1992.
72. Goodell, B.; Daniel, G.; Liu, J.; Mott, L.; Frank, R. *Forest Prod. J.* **1997,** *47*(11/12), 75–80.
73. Tascioglu, C.; Goodell, B.; Lopez-Anido R. Submitted to Forest Prod. J. 2001.

Chapter 26

A Brief Overview of Non-Arsenical Wood Preservative Systems

Tor P. Schultz and Darrel D. Nicholas

Forest Products Laboratory, Mississippi State University, Mississippi State, MS 39762

The principal U.S. wood preservative is currently chromated copper arsenate (CCA). However, the three CCA suppliers in the U.S. and the U.S. Environmental Protection Agency (EPA) have agreed to restrict CCA-treated wood to industrial use by 2004. This chapter briefly reviews non-arsenical wood preservative and/or treatment systems which are commercial in the U.S. and/or other countries or may be in the process of being evaluated as viable candidates.

INTRODUCTION

The major U.S. wood preservative is water-borne chromated copper arsenate (CCA), with oil-borne pentachlorophenol (penta) and creosote used to a lesser extent. CCA, along with small amounts of ammoniacal copper zinc arsenate (ACZA) used in western North America, are currently used to treat about 80% of all preserved wood products in the U.S. (*1*) and at this time are essentially the only systems used to protect lumber for residential applications. CCA is an inexpensive and highly effective preservative and several EPA and other studies have found that CCA-treated wood poses negligible risk when used in residential construction, gardens or for items such as playground equipment. However, recent public perceptions in the U.S. on possible arsenic exposure led to a rapid agreement to restrict CCA to only industrial applications by 2004 in the U.S. Since the largest market for treated wood is residential, this will reduce production of CCA-treated wood products by about 70%. Canada will also shortly restrict CCA applications. Most European countries have already limited CCA use with further restrictions being considered, and Japan had earlier quickly changed to preservatives that do not contain arsenic or chromium.

Three second generation (non-arsenical) water-based copper:organic biocide mixtures are currently commercially available in the U.S. and/or other parts of the world as CCA alternatives for ground-contact and above-ground residential use. These are the alkaline *copper quat* (*ACQ*), amine *copper azole* (*CA*), and *copper bis-(N-cyclohexyldiazeniumdioxy)* (*Cu-HDO*) systems discussed below. Based on data from many years of lab and field tests, together with commercial experience in the U.S., Europe, Japan and Australia for *ACQ* (and in Europe for *CA* and *Cu-HDO*), these systems appear suitable alternatives to CCA for intended applications. In addition, all of these systems are free of chromium, another metal which the EPA desires to reduce usage of.

While copper(II) is not as acutely toxic or mutagenic as are arsenic or chromium, it may be subjected to potential environmental concerns. Many of the new copper:organic systems apparently leach more copper than CCA-treated wood; this may be due to the absence of chromium which, in the CCA system, partly oxidizes the lignocellulosic material to provide relatively strong binding sites for copper. (Formulation research may diminish this problem in the future, however.) Also, the future disposal of metal-treated wood products may be relatively expensive and onerous, whether it be CCA or the newer second generation copper:organic systems. [We consider the disposal of CCA-treated wood, rather than arsenic exposure, to be the major problem.] While research into, and mitigation of, the potential effects of copper are in progress, it is possible that copper-based preservatives systems may face future restrictions in the U.S. This is already the case with several European countries which, for their own particular reasons, will soon require non-metallic, totally organic wood preservative systems.

The objective of this chapter is to provide a brief overview of non-arsenical biocides and/or processes with the potential to protect wood and which may fill the void left by CCA as its use is restricted.

BIOCIDES

As discussed in the chapters by Preston and Leightley, the market value for wood preservatives is relatively small and has a high liability potential.

Consequently, chemical companies have little incentive to develop new biocides solely to protect wood. Instead, many of the organic biocides discussed below were first developed for agrochemical markets and subsequently examined as wood preservatives. Furthermore, treated wood can be attacked by a wide variety of fungi and insects, and a preservative system must protect wood for a lengthy period during which the biocide level in the wood product can be reduced by various leaching and/or degradation mechanisms. Thus, preservative systems must be thoroughly tested by laboratory studies followed by lengthy outdoor exposure of wood treated with the system. While many biocides initially show promise, extended outdoor exposure studies may indicate shortcomings. Even after years of commercial use it is possible that an unforseen problem may arise with a particular system at a particular location and/or application.

Organic biocides are usually considerably more expensive than CCA and most are not water soluble. Thus, to reduce costs and environmental concerns future biocide systems will likely use the minimal biocide retention necessary. In addition, new totally-organic systems will likely require the development of economical water-based emulsion formulations which are stable in a commercial operation. Finally, many of the new organic biocides are highly effective against many of the wood-destroying organisms but weak or inactive against other agents. Thus, future wood preservative systems will likely combine two or more biocides and/or insecticides. Additional requirements for any new biocide and wood preservative system are listed in the chapter by Leightley.

The objective of this chapter is to briefly overview potential biocides and, thus, the many references which provide efficacy, toxicity, formulation and depletion data are not cited. The compounds or systems are listed alphabetically. Only those biocides which are already commercial or, in our opinion, appear to be viable pending further studies, are discussed. Additional references are available (*2-5*).

Amine oxides (Fig. 1) have similar structures and properties as the *quaternary ammonium compounds* (*quats*) discussed below. They have been used as surfactants for some time but have only recently been examined as potential wood preservatives. *Amine oxides* have three alkyl groups (versus the four present in the *quats* discussed below), with typically two of the alkyls being methyls and the third a C_{12} - C_{16} alkyl with a coordinate covalent bond linking the nitrogen to oxygen. *Amine oxides* are water soluble but complex to acidic sites in wood to become fixed and thus should have good stability and leach resistant properties in wood. Another advantage is that *amine oxides* are surfactants and could possibly be used to develop oil-in-water emulsions for formulating organic biocides. These compounds will have a relatively low cost and, provided biological activity can be demonstrated, may be potential candidates for combining with other organic and/or copper biocides. No commercial preservative product is currently available.

The *azoles*, or more properly triazoles, include *cyproconazole* (*[1H-1,2,4-triazole-1-ethanol, α-(4-chlorophenyl)-α-(1-cyclopropylethyl)]*), *propiconazole* (*(2RS, 4RS)-2-(2,4-dichlorophenyl)-2-[1-1H-(1,24-triazole)methyl]-4-propyl-1,3-diaxolane*) and *tebuconazole* (*(3RS)-5-(4-chlorphenyl)-2,2-dimethylethyl-3-(1H-[1,2,4-triazole]methyl)-3-pentanol*). They are highly active against wood decaying fungi, readily soluble in hydrocarbon solvents and exhibit good stability and leach resistance in wood. Although *azoles* have a high cost per unit weight, their high activity makes them comparatively cost effective. Disadvantages include minimal activity against sapstain and molds and no insect/termite activity and, thus, combinations with other fungicides and/or termiticides are necessary. These advantages and disadvantages suggest that *azoles* would be good candidates for combining with other biocides, and this is indeed the case. *Copper azoles* (*CA*), and other commercial preservatives in Europe which consist of an *azole* plus other biocides, are discussed below.

Borates (*Borax, boric acid, disodium octaborate tetrahydrate [DOT], sodium borate*) are inorganic biocides containing boron as the active component, often as a mixture of borax and boric acid. *Borates* have extremely low mammalian toxicity and a broad range of activity against both wood decay fungi and insects. *Borates* are also inexpensive and readily soluble in water. However, due to borates' water solubility the use of borate-treated wood is limited to applications with minimal or no leaching exposure. As a sole preservative it is currently used commercially in Hawaii and many other parts of the world. *Borates* are also a component in two of the non-arsenical water-based copper:organic systems used in Europe, but the boron apparently quickly leaches out in most applications. In addition, chromated copper borate (CCB) and other similar systems are used in some European countries, and research in the U.S. has examined the combination of boron and copper. *Borates* are also used as a diffusible biocide for the remedial treatment of millwork and related applications in many countries. Due to the many advantages of *borates*, several groups have examined ways to reduce leaching. It may also be possible to use a *borate*/water repellent combination for applications with limited leaching potential.

Busan 30 (*2-(thiocyanomethylthio) benzothiazole, TCMTB*) is an organic biocide with a broad range of activity against both wood decay fungi and insects, is readily soluble in hydrocarbon solvents and exhibits good leach resistance in wood. It may undergo slow microbial degradation, however. *TCBTB* is currently only used for sapstain and mold control in the U.S. and Europe.

Chlorothalonil (*2,4,5,6-tetrachloroisophthalonitrile*) is an organic biocide with very low mammalian toxicity, broad activity against wood decay fungi and insects, relatively low cost, and good stability and leach resistance in wood. Based on these favorable properties a major research effort in the 1990's examined *chlorothalonil*, either alone or in combination with the insecticide chlorpyrifos, as a replacement for penta. This work resulted in encouraging efficacy data. However, the poor solubility of *chlorothalonil* in most organic solvents made formulation difficult and interest has waned. At the present time *chlorothalonil* is only used commercially for sapstain and mold control in the U.S.

Copper azole (*CA*), without or with added *boron* (*CBA*), consisting of the combination of *copper(II)*, *boron*, and *tebuconazole* (or *propiconazole* and *tebuconazole*), is one of the major non-arsenical water-based wood preservatives for above- and ground-contact use in European and Asian countries. *CBA* is listed in the American Wood-Preservers' (AWPA) Standards (*6*) as *CBA-type A*, with a copper:boric acid:tebuconazole composition of 49:49:2. A modified formulation of *CBA* which does not contain *boron* has been just introduced in the U.S. *CA*, along with *ACQ* discussed below, will likely be one of the principal CCA replacements in North America. Formulation of *CAs*, and *ACQ* and *Cu-HDO*, often includes ethanolamine to minimize metal corrosion and improve biocide distribution but which increases the cost and the potential for mold on wet lumber.

The organometallic *copper bis-(N-cyclohexyldiazeniumdioxy)* (*Cu-HDO*) (Fig. 1) exhibits good stability and leach resistant properties in wood (but the borate component quickly leaches). A water-based *Cu-HDO* product has just been proposed to the AWPA for above-ground use in the U.S. and is formulated with an organic amine with 93.6% of the copper as copper(II) carbonate and the remaining 6.4% copper as *Cu-HDO*, and a CuO:boric acid:HDO ratio of 4.38:1.75:1. A similar product is one of the major above- and ground-contact wood preservatives currently used in many European countries.

Copper bis(dimethyldithiocarbamate) (*CDDC*) is formulated with copper(II), ethanolamine and sodium dimethyldithiocarbamate (SDDC). Since copper reacts rapidly with SDDC to form an insoluble complex, a two-step treating process is required with this preservative system. This results in a very stable, non-leachable preservative with good activity against most wood-destroying fungi (but not copper-tolerant fungi) and insects in both above- and ground-contact applications. However, *CDDC* requires a dual treatment which adds to the cost. A treating facility was constructed in the U.S. but later converted to another use and *CDDC*-treated wood is no longer available.

Copper citrate (*ammoniacal copper citrate, CC*) is formed by the combination of copper(II) and citric acid. It is effective against most wood destroying fungi and insects but is weak against copper-tolerant fungi and may leach some copper. Thus, *CC* may be best suited for above-ground applications. It is currently listed in the AWPA Standards (*6*) but only very small amounts are available.

Copper quats (*alkaline copper quat*, *ACQ*, *amine copper quat*, *ammoniacal copper quat*) are the combination of copper(II) and *quaternary ammonium compounds* (*quats*) discussed below, with a CuO:*quat* ratio of 2:1. The products can be formulated in aqueous solutions using either aqueous ammonia (e.g. *ACQ-type B*), or an organic amine (e.g. *ACQ-type D*) which is relatively expensive and thus increases the cost (*6*). [Chemical Specialities, Inc. has just won the 2002 (U.S.) Presidential Green Chemistry Challenge Award based on their *ACQ* work.] Aqueous ammonia formulations enhance preservative penetration of refractory species such as Douglas-fir, while amine formulations provide color uniformity with pines. *ACQ* has been commercially available in the U.S. (as types B, C and D, depending on the formulation and type of *quat*) and Australia for about 10 years, and even longer in Europe and Japan. As with *CA*, *ACQ* is expected to shortly be one of the major U.S. preservative systems.

(Bis)-copper-8-quinolinolate (*oxine copper, copper-8, Cu-8*) is an organometalic compound with very low acute mammalian toxicity, excellent stability and leach resistance, and broad activity against both decay fungi and insects. *Cu-8* has been used for over 30 years in the U.S. It is very insoluble in water and most organic solvents and is thus difficult to formulate. An oil soluble form is available using the relatively expensive nickel-2-ethylhexoate. A water soluble form made with dodecylbenzene sulfonic acid is available but the solution is highly corrosive to steel and requires special handling. *Cu-8* is currently the only biocide listed in the AWPA Standards (*6*) for treating wood that will come in contact with foodstuffs. A small volume of *Cu-8* is used in the U.S. for above-ground applications and for sapstain and mold control, and minor amounts are sold to homeowners as a brush-on preservative. Some research on adding additional copper(II) to form the mono form of *Cu-8* has recently been conducted.

Copper naphthenate is an organometallic biocide made by combining copper(II) with naphthenatic acid mixtures. *Copper naphthenate* is relatively low cost and has been used for about 50 years in various biocide applications, including as a wood preservative during the WWII era. It has low mammalian toxicity, exhibits broad activity against decay fungi and insects, is readily soluble in hydrocarbon solvents and has good stability and leach resistance. Research examined *copper naphthenate* as an alternative to penta and creosote in the 1990's and found that it was a good preservative system. This led to commercial treatment of southern pine and Douglas-fir utility poles at some treating plants. Surprisingly, a small fraction of the pine poles experienced severe decay - some as early as two years after installation. Although the exact cause for the failure remains unknown, one conclusion was that *copper naphthenate* - and the other non-arsenical preservatives - may be more sensitive to handling in the treating plant than the CCA, creosote and penta systems. The problem seems to have been corrected and *copper naphthenate* is still used to treat some utility poles. Another commercial product used in several countries is the combination of *copper naphthenate*, *borate*, water and a thickening agent to form a paste which is applied as a remedial ground treatment to utility poles followed by an outer wrap of tarpaper or plastic. Small amounts of *copper naphthenate* are also sold over-the-counter to homeowners. *Copper naphthenate* imparts a green color to wood;

for applications where this color is objectionable the colorless but slightly less effective *zinc naphthenate* can be used. A water-based system is available in the U.S. for non-pressure applications and which may be available in the future for pressure treating

Dichlofluanid (*1,1-dichloro-N-[(dimethylamino)sulfonyl]–1-fluoro-N-phenyl-methanesulfenamide, DCFN*) is a fungicide which is commercially available in Europe for control of stains and molds. *DCFN* is used in paints and wood stains in Europe. The authors do not have information on its efficacy and stability in wood or commercial usage in the U.S.

The isothiazolone *4,5-dichloro-2-n-octyl-4-isothiazolin-3-one* (*Kathon 930*) is a biocide with moderately low mammalian toxicity, broad activity against wood decay fungi and termites, is soluble in hydrocarbon solvents and exhibits excellent stability and leach resistance in wood. Extensive research has shown that *Kathon 930* effectively protects wood in ground contact and it is a potential penta alternative. It is currently not used as a commercial wood preservative.

Imidacloprid (*1-[(6-chloro-3-pyridinyl)methyl]-N-nitro-2-imidazolidinimine*) is an insecticide available in Europe. In termite tests in Gainesville, FL, and Hilo, HI, *imidacloprid* had greater efficacy than chlorpyriphos.

3-Iodo-2-propynylbutyl carbamate (*IPBC, Polyphase*) is an organic biocide with low mammalian toxicity, is readily soluble in hydrocarbon solvents and has a broad range of activity against wood decay fungi, but has no insect/termite activity and may be slowly degraded in wood. It is currently used in above-ground applications for millwork and similar applications and in combination with *DDAC* for sapstain/mold control (*NP-1*). *IPBC* has been examined as a component in other wood preservative systems but interest has recently declined in the U.S. In Europe many combinations of *IPBC* and *propiconazole*, or *IPBC*, *propiconazole* and *tebuconazole*, both solvent- or water-borne, are used in above-ground applications. Also, *IPBC* is the biocide in many of the over-the-counter wood preservatives sold as brush-on systems in the U.S.

Polymeric xylenol tetrasulfide (*PXTS*) is an oligomeric material consisting of a mixture of alkylphenol isomers linked by chains of 2-10 sulfurs (Fig. 1) with a degree of polymerization (of phenolic units) of less than 22. This biocide has many of the same characteristics as creosote (dark color, fairly viscous so treating solutions need to be heated, and low cost), and essentially the same efficacy as creosote in protecting wood in marine or ground-contact exposures. One advantage of *PXTS* is that, being a oligomer, *PXTS*-treated wood will have minimal leaching which will reduce the loading needed to ensure long term protection. Thus, *PXTS*-treated wood may be suited for environmentally-sensitive locations such as aquatic pilings. *PXTS* also exhibits low mammalian toxicity on tests run to date. If test results continue to be favorable and a suitable formulation developed *PXTS* may be a viable replacement for creosote or penta in certain industrial applications.

Several *quaternary ammonium compounds* (*quats*) have been extensively examined as potential wood preservatives, including *didecyldimethylammonium chloride* (*DDAC, Bardac 22*) and the *alkyldimethylbenzyl ammonium chlorides* (*alkyl benzyldimethylammonium chlorides*, *benzalkonium chlorides*, *ABACs*, *ADBACs*). The latter are usually sold as a mixture of compounds with C_{12} - C_{18} alkyl groups. The *quats* have very low mammalian toxicity, are inexpensive, have broad activity range against wood decay fungi and insects, are readily soluble in both water and hydrocarbon solvents, and exhibit excellent stability and leach resistance due to chemical fixation reactions with wood. However, their efficacy is only moderate and when used alone to protect wood may be ineffective. Due to their surfactant properties and low cost *quats* are often combined with other biocides. For example, copper and *quats* are the active ingredients in *ACQ types B, C* and *D*, and *DDAC* plus *IPBC* (*NP-1*) is a commercial sapstain and mold agent. Q*uats* will undoubtably continue to be considered in the development of future wood preservative systems.

Several *synthetic pyrethroids* (*Permethrin, Cypermethrin, Cyfluthrin, and Deltamethrin*) are available. These compounds have low mammalian toxicity, exhibit good efficacy against insects (but are not fungicidal) and are soluble in many hydrocarbon solvents. In the U.S. research on the combination of a *synthetic pyrethroid* and fungicide has been conducted but no commercial applications exist at this time. In Europe several combinations of a *synthetic pyrethroid* and other biocide(s) are used, including *benzalkonium chloride*, *permethrin*, and *tebuconazole*, and *cypermethrin* and *tebuconazole*.

Tributyltin oxide (*TBTO*) is an organometallic biocide which is fairly nontoxic to mammals and exhibits good activity against fungi and insects, is soluble in most hydrocarbon solvents and has good leach resistance. It is used commercially as an above-ground treatment for millwork and related applications in the U.S. and other countries. However, it has been shown to undergo slow dealkylation to form dibutyl and monobutyl compounds with greatly reduced fungicidal properties and, consequently, *TBTO* should not be used to treat wood which will be exposed to high decay hazard conditions.

Zinc borate is an inorganic biocide with very low mammalian toxicity and cost, broad activity against fungi and insects and much greater leaching resistance than the *borates*. Thus, *zinc borate* is commercially used in the U.S. as an in-process biocide for many wood composites including plastic wood and particle-board (see chapter by Gardner et al.).

To summarize, for the residential U.S. market *ACQ* (*B, C* and *D* formulations) and *CA* [without *boron*] will likely be the principal water-borne CCA substitutes for treating lumber and some wood composites for both ground-contact and above-ground uses in the near future. The above systems are about three times more expensive than CCA and are more corrosive to metals, requiring some capital expenditures by treating plants which convert from CCA to *ACQ* or *CA*. These systems generally appear to be about as effective as CCA at protecting wood in ground-contact and above-ground applications in North America, based on extensive field trials. *CC* is listed for both ground-contact and above-ground but may be best suited for above-ground use and will likely have limited availablility. *Cu-HDO* has just been submitted for standardization [use] for above-ground applications in the U.S. *Zinc borate* is currently used as an in-process preservative for some wood composites, and *borates* could be more widely employed to treat lumber for non-leaching residential applications. Water-based *copper naphthenate* might be available in the future and could conceivably be used in certain residential applications. The presence of copper(II) could potentially impact the use and disposal of wood treated with copper:organic systems in the future. Consequently, totally organic systems, or *borates* for non-leaching applications, may be required in the future for certain applications and/or regions of the U.S.; this trend is already apparent in parts of Europe. While some organic biocides and/or biocide combinations have potential, at the present time no totally-organic water-based system suitable for treating ground-contact wood for the U.S. residential markets has been proposed. Future totally-organic systems will likely consist of the combination of two or more biocides and possibly additional non-biocidal additives and/or water repellents which are discussed below. The development and subsequent testing of effective totally-organic emulsion-borne preservative systems suitable for ground-contact use in the residential market will be arduous and require considerable time and resources, particularly in North America where alternative biocides have not been extensively used and where severe decay hazard locations or Formosan termites can exist. These systems will likely be very expensive relative to CCA. The performance of all preservatives, including CCA, varies in different parts of the world and a particular preservative system may not be suitable in certain applications or locations. Finally, as mentioned earlier even after years of commercial use it is possible that an unforseen problem may arise with a particular system.

For industrial markets CCA will still - on paper - be available in the near future. However, the current negative public perceptions of CCA appear to be also affecting the industrial market. Several alternatives are available, including the first generation penta and creosote systems, the second generation water-borne copper-based *ACQ* and *CA* systems, and the oil-borne *copper naphthenate* system which is already commercially used to a limited extent. Several other totally organic oil-borne systems could be relatively quickly developed based on biocides such as *chlorothalonil*, *PXTS*, *Kathon 930* or an *azole/insecticide* combination, but would require a lengthy test period to ensure their effectiveness. CCA's high efficacy under extremely high hazard conditions suggests that CCA may continue to be available in the North America market, albeit at greatly reduced levels.

As mentioned earlier, for economic and environmental reasons biocide retentions of the newer systems will be reduced to as low a level as possible. However, the use of very low biocide retention levels is questionable because of the inherent variability in biocide retentions among individual boards in a given treating charge. Consequently, when a load of lumber is commercially treated the biocide retention will vary among the individual pieces and, thus, some fraction of the lumber or poles can have a relatively low biocide retention. The industry and appropriate regulatory and professional agencies will need to carefully ensure that any future proposed system has sufficient biocide level(s) for satisfactory performance; if any of the newer systems perform less than satisfactory the resulting negative public perceptions of wood-based construction materials will likely mean an even greater loss of market share to non-woody materials such as steel studs.

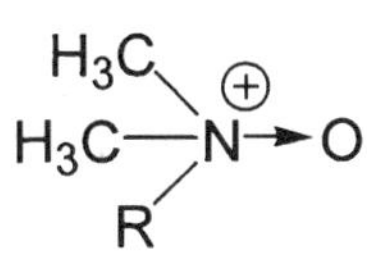

R = C_{12} - C_{16} Alkyl

Amine Oxides

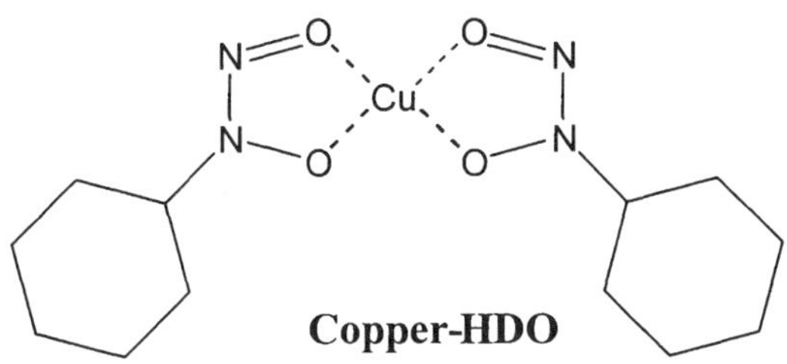

Copper-HDO

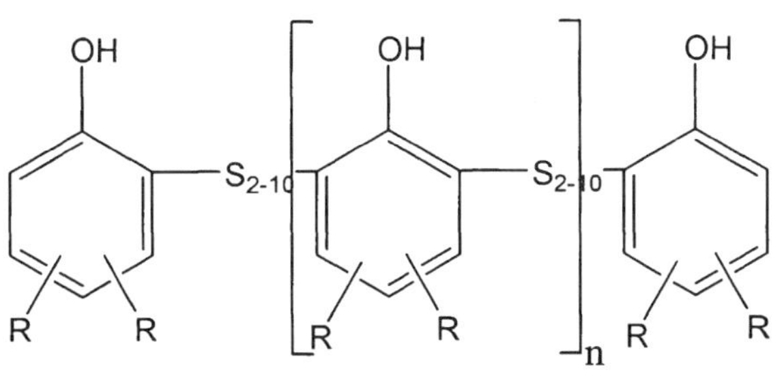

R = H or Alkyls, n < 20

Polymeric Xylenol Tetrasulfide (PXTS)

Figure 1. Selected compounds with the potential as wood preservatives

NATURALLY DURABLE WOOD

The heartwood of some woods are naturally resistant to fungal and/or insect degradation. Many environmentally-conscious homeowners have considered building outdoor decks or purchasing outdoor furniture or playground equipment made using durable woods. While the use of naturally durable wood is feasible in certain applications, this approach has some limitations. Two of the major drawbacks are that the durable woods most commonly used in North America, western red cedar, redwood and cypress, are not highly durable and thus may have a relatively short service life in high hazard areas such as the southeastern U.S. (In Europe, larch and pine heartwood have gained some markets as "naturally durable woods".) Secondly, the annual growth of these woods is nowhere near the volume of pressure-treated wood annually produced in the North America, and public policy may restrict harvesting of native durable species on public lands. Alternatively, tropical wood species with decay resistance properties have been imported into temperate countries in lieu of preservative protected native lumber. It can be argued that this simply exports the problem of wood protection and sustainable wood use to other countries that do not limit the harvesting of durable wood species. Other concerns include the possible presence of the non-durable sapwood in decks or playground equipment; the variable durability in heartwood, even from lumber cut from the same tree; and the misconception that naturally-durable woods are benign to humans. [Some of the general U.S. public apparently perceives the biocides used to treat wood as inherently dangerous, yet they have no concerns over using some of the same biocides discussed above in personal care products such as shampoos and also consider naturally-durable wood to be benign. For example, creosote is a distillate of coal tar and composed of polyaromatic hydrocarbons and phenolics. Some of these compounds are reported to cause cancer and, consequently, several groups have requested that the EPA prohibit use of creosote. However, coal tar is a component in several medicated shampoos where use exposes individuals to relatively high levels of these same chemicals. Other biocides in shampoos include *IPBC* and *quats*, discussed above. On the toxicity of natural wood; quite a few people have died from using branches from the oleander bush to cook hotdogs. In contrast, the health hazards from CCA wood are minimal and only in extreme cases where misuse of treated wood has occurred, such as burning CCA wood indoors, have humans or animals suffered ill effects. The dust from western red cedar is considered very harmful and commonly leads to respiratory problems and nasal cancer due to the presence of toxic extractives (which impart durability to this wood), but this wood can be safely used provided minimal safety precautions are taken.]

NON-BIOCIDAL ADDITIVES

In recent years our understanding of the basic wood decay process has been enhanced considerably. As a result it is now known that wood decay fungi utilize both enzymatic and non-enzymatic systems to degrade wood, with metals - either free or as organometallics - as co reactants in many mechanisms. Many of the non-enzymatic reactions utilize free radical reactions to degrade wood. The above knowledge, and the fact the extractives in durable heartwoods are well known to have excellent antioxidant and metal chelating properties, suggested to us that antioxidants and/or metal chelators might be useful in wood preservative formulations. Subsequent laboratory experiments showed that the use of antioxidants or chelators alone were not effective as wood preservatives. However, further studies showed that when these compounds were combined with organic biocides they enhanced the activity several fold (see chapter by Green and Schultz for data). Thus, it appears that the efficacy of relatively expensive organic preservatives can be significantly enhanced by adding these relatively inexpensive and often benign compounds. Extensive field trials are now underway to evaluate the potential of this concept and the preliminary results are promising.

WATER REPELLENTS

In addition to protecting wood from biodegradation, a good wood preservative system should enhance the weathering characteristics of wood. Other than UV degradation, the swelling and shrinking of wood resulting from water sorption and desorption is the major cause of wood degradation that occurs when wood is subjected to exterior exposure. As a consequence, the addition of water repellents to wood preservative systems is highly desirable. In addition to improving the weathering characteristics of treated wood, durable water repellents also provide some degree of protection to wood against biodegradation by reducing the moisture content. Furthermore, water repellents have the potential to enhance biocidal activity by reducing leaching and lowering the moisture levels in wood exposed to rain. Because of these attributes, and the low cost of wax-based water repellents, it is anticipated that future wood preservative systems will incorporate durable water repellent components. Indeed, lumber treated with several formulations of a wax-based water-repellant and preservative combination is commercially available in North America, and the use of linseed oil-treated wood (with no co-preservative) is being studied in Europe.

THERMAL TREATMENTS

In the last decade a considerable amount of research in Europe has been devoted to improving the durability of wood by heat treatments. As a result of this effort, a number of processes have been developed and commercialized. A common element of these treatments is subjecting the wood to temperatures close to or above 180°C for several hours in a low oxygen atmosphere. The oxygen levels are minimized by immersing the wood in a nitrogen atmosphere, or water or oil, during the heat treatment. These treatments cause some chemical degradation of the wood and, consequently, a reduction in mechanical properties – particularly impact bending strength which is generally reduced 50% or greater. The process also increases the permeability of the wood. On the positive side, the treated wood is considerably less hygroscopic with reductions in EMC values of up to 50%. The wood also exhibits significantly better dimensional stability, exemplified by reductions in shrinking and swelling in the range of 50-90%. The chemical modification also improves the decay resistance of the wood.

Although the durability of heat treated wood is not equivalent to that of wood treated with preservatives, it does appear to perform satisfactory in some above-ground applications. Furthermore, the improved dimensional stability improves the weathering characteristics which is highly desirable for some applications. However, the reduced wood toughness limits uses to non-structural applications.

REFERENCES

1. Micklewright, J.T. *Wood Preservation Statistics 1997.* Am. Wood Pres. Assoc.: Granbury, TX 1999.
2. Schultz, T.P.; Nicholas, D.D. *Wood Preservation in the '90s and Beyond.* Forest Products Soc.: Madison, WI **1995**; p. 169-173.
3. Nicholas, D.D. *Wood and Cellulosic Chemistry*, 2nd Edit. Hon, D.N.S., Shiraiski, N., eds. Marcel Dekker, New York **2001**; chp. 21.
4. Suttie, E.D.; Bravery, A.F.; Dearling, T.B. *Int. Research Group/Wood Preservation Document* **2002**, IRG/WP 02-30289.
5. Edlund, M-E.; Jermer, J. *Int. Research Group/Wood Preservation Document* **2002** IRG/WP 02-30297.
6. Anonymous. *American Wood-Preservers' Association Standards 2001.* Am. Wood Pres. Assoc.: Granbury, TX 2001.

Indexes

Author Index

Subject Index

B

C

D

E

F

G

I

M

N

O

P

Q

R

S

X

Z